π

to a million places

Dr. Jason A. Davies, Ph.D.
Published by J. A. Davies Study Guides

Pi to a million places
Compiled by Dr. Jason A. Davies, Ph.D.
Published by J. A. Davies Study Guides

ISBN-13: 978-1470109615
ISBN-10: 1470109611

Disclaimer: Every effort has been made to ensure the accuracy of the information contained within this publication, however if you choose to make use of the information contained within this publication, you do so at your own risk.

$$\pi = mc^2$$

where **m** is the area of a circle of radius **1/c**

(geek humor)

π

---

to a million places

3.

```
1415926535897932384626433832795028841971693993751058209749445923078164062862089986280348253421170679
8214808651328230664709384460955058223172535940812848111745028410270193852110555964462294895493038196
4428810975665933446128475648233786783165271201909145648566923460348610454326648213393607260249141273
7245870066063155881748815209209628292540917153643678925903600113305305488204665213841469519415116094
3305727036575959195309218611738193261179310511854807446237996274956735188575272489122793818301194912
9833673362440656643086021394946395224737190702179860943702770539217176293176752384674818467669405132
0005681271452635608277857713427577896091736371787214684409012249534301465495853710507922796892589235
4201995611212902196086403444181598136297747713099605187072113499999983729780499510597317328160963185 9
5024459455346908302642522308253344685035261931188171010003137838752886587533208381420617177669147303
5982534904287554687311595628638823537875937519577818577805321712268066130019278766111959092164201989
3809525720106548586327886593615338182796823030195203530185296899577362259941389124972177528347913151
5574857242454150695950829533116861727855889075098381754637464939319255060400927701671139009848824012
8583616035637076601047101819429555961989467678374494482553797747268471040475346462080466842590694912
9331367702898915210475216205696602405803815019351125338243003558764024749647326391419927260426992279
6782354781636009341721641219924586315030286182974555706749838505494588586926995690927210750930295 5
3211653449872027559602364806654991198818347977535663698074265425278625518184175746728909777727938000
8164706001614524919217321721477235014144197356854816136115735255213347574184946843852332390739414333
4547762416862518983569485562099219222184272550254256887671790494601653466804988627232791786085784383
8279679766814541009538837863609506800642251252051173929848960841284886269456042419652850222106611863
0674427862203919494504712371378696095636437191728746776465757396241389086583264599958133904780275900
9465764078951269468398352595709825822620522489407726719478268482601476990902640136394437455305068203
4962524517493996514314298091906592509372216964615157098583874105978859597729754989301617539392846813 82
6686386894277415599185592524595395943104997252468084598727364469584865383673622266099124608051243 88
4390451244136549762780797715691435997700129616089441694868555848406353422072225828488648158456028506
```

0168427394522674676788952521385225499546667278239864565961163548862305774564980355936345681743241125
1507606947945109659609402522288797108931456691368672287489405601015033086179280986092087476091782493858
9009714909675985261365549781893129784821682998948722658804857564014270477555132379641451523746234364
5428584447952658678210511413547357395231134271661021359695362314429524849371871101457654035902799344
0374200731057853906219838744780847848968332144571386875194350643021845319104848100537061468067491927
8191197939952061419663428754440643745123718192179998391015919561814675142691239748940907186494231961
5679452080951465502252316038819301420937621378555966389377870830390697920773467221825625996615014215
0306803844773454920260541466592520149744058073251866600213243408819071048633173464965145390579626856
1005508106658796998163574736384052571459102897064140110971206280439039759515677157700420337869936007
2305587631763594218731251471205329281918261861258673251259198414848829164470609575270695722091756716
7229109816909152801735067127485832228718352093539657251210835791513698820914442100675103346711031412
6711136990865851639831501970165151168517143765761835155650884909989859982387345528331635507647918535
8932261854896321329330898570642046752590709154814654985946163718027098199430992448895757128289052923
2332609729971208443357326548938239119325974636673058360414281388303203824903758985243744170291327656
1809371744403070746921120191302033038019762110110044929321516084244485963766983895228684783123552658
2131449576857262433444189303968642622543879678028073189154411010446823252716201052652272111660396
6657309254711055785376346682065310989652691862056476931257058635662018581007293606598764861179105
3348850346113657686753249441668039626579787718556084552965412665408530614344431858676975145661406800
7002378776591344017127494704205622305389945613140711270004078547332699390814546646458807972708266830
6343285878569830523580893306575740679545716377525420211495576158140025012622859413021647155097925923
0990796547376125517656751357517829664547791745011299614809304659441132962107340437518957335961458901
9389713111790429782856475032031986915142870808599048010941214722131794764777262241425485454035321571
8530614228813758504306332175182979866223717215916077166925474873896654949450114654062843366393079003
9769265672146385306736096571209180763832716641627488880078692560290228472104031721186082041900042296
6171196377921337575114959501566049631862947265473642523081770367515906735023507283540567040386743513
6222247715891504953098444893330963408780769325993978054193414473774418426312986080998886874132604721
5695162396586457302163159819319516173538129741677294786724229246546800980676928238208699640048242435
4037014163149658979409243237896907069779422362508221688957383798623001593776471651228935780615881617
5578297352334460428151262720373431465319777741603199066554187639792933441952154134189948544473456738
3162499341913181480927777103863877343177207451445277709212019051660962804909263601975988281611332
3166636528619326686336062735676303544776280350450777235547105859548702790814356240145171806246436267
9456127531813407833033625423278394497538243720583531147711992606381334677687969597030983391307710987
0408591337464144282277263465947074458784778720192771528073176790770715721344473060570073349243693113
8350493163128404251219256517980694113528013147013047816437885185290928545201165839341956621349143415
9562586586557055269049652098580338507224264829397258473163057775606088764462442468579260395352773
4803048029005876075825104747091643961362676044925627420420832085661190625454337213153595845068772460
2901618766795240616342522577195429161629919306455377991403734043287526288896399587947572917464263745
2540790914513571113694109119393251910760208252026187593518877058429725916778131496990090192116971737
2784768472686084900337702424291651300500516832336435038951702989392233451722013812806965011784408745
1960121228599371623130171144484640903890644954440061986907548516026327505298349187407866808818338510
2283345085048608250393021332197155184306354500706484556527939757175461395398468339363830437
4611996653858153842056853386218672523340283087112328278921250771262946322956398989893582116745627010
2183564622013496715188190973038119800497340723961036854066431939509790190699639552453005450580685501
9567302292191393185680344903982059551002263535361920419947455385938102343955449597783779023742161
2711172364343543947822181852862408514006660443235888569867054315470696574745855033323334210730154594
0516553790686627333799585115625784322988273723198987571415957811196358330059408730681216028764962867
4460477464915995054973742526901049037781986835938146574126804925648798556145372234787330390468838384
3634655379498641927056387293174872332083760112302991136793862708943879936201629515413371424892830722
0126901475466847653576164773794675200490751552781965336213239210669160136358155907422020203187277605
2772190055614842555187925303435139844253223415762336106425063904970086562710953591946589751413104
2276930624743536325691607815478181152843667957061108615331504452127473924544945423682886061340841486
3776700961207151249140430272538607648236314143462351897576645216413767969031495019108575984423919862
9164219399490723623466844117394032659184044378051333894525742399508296591228508555821572503107125

126683024029295252201187267675622041542051618416348475651699981161410100299607838690929160302884002 6
910414079288621507842451670908700069928212066041837180653556725253256753286129104248776182582976515 7
959847035622262934860034158722980534989650226291748788202734209222453398562647669149055628425039127
577102840279980663658254889264880254566101729670266440765590429099456815065265305371829412703369337 8
517860904070866711496558343434769338578171113864558736781230145876871266034891390956200993936103102 91
616152881384379099042317473363948045759314931405297634757481193567091101377517210080315590248530906 6
920376719220332290943346768514221447737939375170344366199104033751117354719185504644902636551281622 8
824462575916333039107225383742182140883508657391771509682887478265699599574490661758344137522397096 8
340800535598491754173818839994467862655165827658483588453142775687900290951702835297163445621296
404353211760066510124120065975585127617858382920419748442360800719304576189323492292279651098751872 12
726750798125547095890455635792122103334669749923563025494780249011419521238281530911407907386025152 2
742995818072471625916685451331239480494707911915326734302824418604142636395480004480026704962482017
928964766975831832713142517029692348896276684403232609275249603579964692565049349183609003238092934 5
958897069536534940603402166544375589004563288225054525564056448246515187547119621844396582537543885
690941130315095261793780029742076651479369896959469955657612186561967337862362561252163208628
692221032748892186543648022967807057656151446320469270682120738837781423356282360896320806822246801
224826117718589638140918390367367222088832151375560037279839400415297002878307667094447456013455641 7
254370906979396122571429894671543578468788614445812345935719849225693698115950591500610165525637567 8
566722796619885782794848558534397518744545512965634434803966420557982936804352202770984294232533022 5
763418070394769941597915945300697521482933665556615678736400536665641654732170439035213295435291694 1
459904160875320186837937023488868947191510716378529023463540077365949563051007421087142613497459561 5
138498713757047101787957310422969066670214498637464595280824369445789772330048764765241339075928434 0
196340391147320233807150952220106825634274716460243354400515212669324934196739770415956837535551667 3
027390074972973635496453328868944061196496162773449518273695588220573551766515898551909866065393549
481068873206859907540792342402300925900701731960362254756478940647548346647760411463233906551343306 8
449539790709030234604614709616968868850140834700454607429586991382966824681857103188790652870366508 3
243197440477185567893482308943106828702722809736240893996270607472645539925399442480811373694338872 94
063079261595995462624629707062594845569034711972996409089418059534393251236235500134949004364278527 1
383159125689892951964722875739469142725343669415236100453730488198551706594121735246258954873016760
029886592578662836124966552353382942878545234048308330701653722856355591525347844598183134112900199 2
059813522051173365856407826489427644113763938669248031183644536985891175442647399882284621844900877 7
697763127957226726555625962825427653183001340709223343657791601280931794017185985999338492354954000 5
709955856113498025249906698423301735035804408116855265317099570899427328709258487894436460050410892
266917835258707859512983441729535195378855345737426085902908176515578039059464087350612322611200937 3
108048548526357228257682034160504846627754500312620080589782548534694146977516493270950493463938
243222718851597405470214828971117779237612257887347188196825462981268685817050740272550263329044976
278944236216741191862694396506715157795867564823993917604260176338704549901761436412046921823707648
878341968968611815581587360629386058101712158552726648358400283404046547588040513080163363887421637140
643549556186896411228214075330265510042410489678352858829024367090488718190909494533144218287661810
310073547705498159680772009474696134360928614849417850171807793068108546900094458995279424398139213 5
055864221964834915126390128038320010977386806628779239718634452443734464463452700735921003154150
893679300816998053652027600727749674584002836240534603726341655425902760183484030681138185510597970 5
664007509426087885735796037324514414676703688098806097164258497595138069309449401515422221943291302 1
739125383559150310033303251117491569691745027149433151558854039221640972291011290355218157628232831 8
234254832611191280092825256190205263016391147724733148573910777587442538761174657867116941477642144 1
111263583553871361011023267987756410246824032264843464176639806375651349204530224081972785647198 3
963087815432211669122464159117767322532643356861461865452226812688726844596844241610785401676814208 0
885028005414361314623082102594173756238994207571362751674573189189456283525704413354375857534269869 9

4725470316566139919996826282472706413362221789239031760854289437339356188916512504244040089527198378
7386480584726895462438823437517885201439560057104811949884239060613695734231559079670346149143447886
3604103182350736502778590897578272731305048893989009923913503373250855982655867089242612429473670193
9077271307068691709264625484232407485503660801360466895118400936686095463250021458529309500009071510
5823626729326453738210493872499669933942468551648326113414611068026744663733437534076429402668297386
5220935701626384648528514903629320199199688285171839536691345222444708045923966028171565515656661113
7806101371500448991721002220133501310601639154158957803711779277522597874289191791552241718958536168
0594741234193398420218745649256443462392531953135103311476394911995072858430658361935369329699289837
9149419394060857248639688369032655643664264425676079147108699843157337496488352927693282207629472823
8153740996154559879825989109371712621828302584811238901196822142945766758071865380650648702613389282
2994972574530332838963818439447707794022843598834100358385423897354243956475556840952248445541392394
1000162076936368467764130178196593799715574685419463348937484391297423914336593604100352343777065888
6778113949861647874714079326385873862473288964564359877466763847946650407411182565837887845485814896
2961273998413442726086061872455452360643153710112746809778704464094758280348769758948328241239292960
5829486191966709189690983320121031843034012849511620353428014412761728583024355983003204202451207728
7253558119584014918096925339507577840006746552603144616705082768277222353419110263416315714740612385
0425845988419907611287258059113935689601431668283176323567325417073428017332230462987992804908514094
7903688786878949305469557030726190095020764334933591060245450864536289354568629585313153371838682656
1786227363716975774183023986006591481616404944965011732131389574706208847480236537103115089842799275
4426853277974311395143574172219759799359685252285745263796289612691572357986205734083757668373884266
4059909935050008133754324546359675048442352848747014434545419576258473564216198134073468541117668311
8654489377697956651727966232671481033864391375186594673002443450054499539974237232871249483470604406
3471606325830649829795510109541836230303094530973358344628394763047756450150085075789495489313993944
8992161255255977014368589435858775263796255970816776438001254365023714128346792610199558522471172201
7772370041780841942394872540680155603599839054898572354674564239058580216719031395262944554391131663
1345308939062046784387785054239390524731362012947691874975191011472315289326772533918146607300089027
7689631148109022097245207591672970078505807171863810549679731001678708506942070922329080703832634534
5203802786099055690013413718236837099194951648960075504934126787643674638490206396401976668559233565
4639138363185745698147196210841080961884605456039038455343729141446513474940784884423772175154334260
3066988317683310011331086904219390310801437843341513709243530136776310849135161564226984750743032971
6746964066653152703532546711266752246055119958183196376017991192035795820057956053023462677594
3936307463056901080114942714100939136913810725813781357894005595500183542511841721360557275221035268
0373572652792241737360575112788721819084490061780138897107708229310027976659358387589093956881485602
6322439372656247277603789081445883785019750843779362407828505270487581647032580129087839523245323789
6029841669225489649715606981192186584926770403956481278102179913217416305810554598801300484562997651
1212415363745150056350701278159267142413421033015661653560247338078430286552572227530499988370153487
9300806260180962381516136690334111138653851091936739383522934588832255088706450753947395204396807906
7086806445096986548801682874437861264538158342807530618454859037982179945996811544197425363443
9025100158882721647450068207041937615845471231834600726293395505482395571372568402322682130124767945
2264482091023564775272308208106351889152692889108455711266039650343978962782500161101532351605
5904211844949907789992007329476905868577878720982901352956613978884860509780859570177312981553
6814671769597609942100361835591387778176984587581044662839988060061622984861693533738657877359833616
1338413385368421197893890018529569196780455448285848370117096721255333875862158231013310387766827211
5726949518179589754693992642197915233857662316762754757035469914892904130186386119439196283887
67774322427680913236544948536676800000106526248547305865159899914017076938548318875014293890890568
5453076511680333732226517566220752695179144225280816517166776672793035485154204023817460892328391703
27542575086765511785939500279338959205766827896776445318404041855401043513483895312013263783692
8271937831265496174599705674507183320650345566440344904536275600112501843356073612227659492783937064
7842645676338818807565612168960504161139039063960162021536849410926053876887148379895599991109164
64644119185682770045742434340216722764455893301277815868695250694934610175685060167143354315814801
0545880656455013320375864548584032402987170934809105562116715684847780394475697980426318099175
09873998766973237695737015808068229045992123661689025962730430679316531149401764737693875314093

321614280214976339918983548487562529875242387307755955595546519639440182184099841248982623673771467
260616336432964063357281070788758164043814850188411431885988276944901193212968271588841338694346828
900666408063140777577257056307294004929403024204984165654797367054855804458657202227637840466823379
282710578431975334179501134727362577408021347682604502285157979579764746702284049995616015691089038
824502679265942055503958792298185264800706837650418365620945554346135134152570065974881916341359556
196496540321872716026485930490937874858906612725079482827693895352175362185070962977851461884327192
322381015874445052866523802253284389137527384589238442535472653098171578447834215822327020690287232
330053862163479885094695472004795231120150432932266282727632177908840087861480221475376578105819702
263097174950721272484794781695729614236585957820908307332335603484653187302930266596450137183754288
755797144992465403868179921389346924474198509733462679332107268687076802623991936196504409954216762
840914469856925715074315707938053239252394775574419918458261562518192155233709607483329234921034514
264374498055961033079941453477845746999921285999993996122816152193148887693880222810830019860165494
654261696858678837260958774567618250725992950893180521872924610867639958916145855058397274209809097
817293239301067663868240401113040247007350857282264271349463685318154696904669686939254725194139929
146524238577625500474852954768147954670070503479995888676950161249722820403039546327883069597624936
151010243655335223069061294938859901573466102371223547891129254769617600504797492806072126803922691
027772261054449221576504508120677173571202718024298106203776578837166909109418074487814040907551780
203856539099104775941413215432844062503018027571696508209642734814695726397884256008453121406593580
904127113592004197598513625479616063228873618136737324450607924411763997597461938358457491598809766
447093006546342423460634237476060431701260052055928493659414340814685298150539471789004518357551
541252235905906872648786357525419112888773717663748602766063496035367947026923229718683277173932361
200777452212624751869833495151019864269887847171939664976907082521742336566272592844062403021411371
922785269984698847702323823840055655178890876613601304770984386116870523105531491625172837327286760
072481729876376981633541507460883866364069347043720668865127568826614973078865701568501691864748854
167915459650723428773069985371390430026653078398776385323538182155355973253306860430106757608389086
049841888595138091030423595782495143985901131858358406674237029714978508414585308578133915627076
563907639473114554958322669457024941398316343323789759556808563629725386791327505554252449194358912
840504522653812179131914513500993846311770417971512283785460116035955402864405902496466930707769055
481028850208085800878115773817191741776017330738554758006560143377432990127286772530431825197579167
929699650414607066457125888346979796429316229655201687973000356463045793088403274807718115553309098
702550520768040630346086581653948769519600440848206596737947316808641564565053004988161649057838115
454850526600698230931577765003780700466126470602145750579320960204782561524714591896522360839664562
051955105223572397395128818164059785914279148165426328920042816091369377737222999833270820829699557
772737566715527113922588055201898762011416800546878550633471603734291703907986396522961312801782
679717289822936070288060908776866059325274637840539769184808204102194471971386925608416245112398062
131845412447820501107987607171556831540788545439041210873032402010685341947230476666721749869868547
678120512473679247919315085644477537985379973223445612278536846647513336573692387201464723679427
870042503255589926884349592876124007558756946413705625140011797133166207153715436006876477318675587
487839890810742953094106059694431584775397009439883949144323536685392099468796450665339857388878661
762944341410498889931600512076781035886116602029619363968213496075011164983278653531614516845769
871090029997698412632665023477167286573785790857466460772283415403114415294188047825438761770790430
015669867679576009909969366075294965152736349811896413043311467712338817406037317439705406703109
676574869535878967003192586625941051053358438465602339179674926784476370847497833655579007384191473
198862713525954625181604342253729962863267496824058060296421146386436864224724887283434170441573482
818333016405669596868665161143628426414974533349999480002606998758881593507357815195889900539051
085351035726137364034367534714104836017546488300407846416745216737190483109676711344349481926268111
739948250607394950735031690197318521195526356323984590998224986240670310768318446607291248747540316
796994113973877658998685541703188477539700943988349114922352093822787888098863359116081923
535557046463491132085918979613279131975649097600013996234445535014346426860464495862476909434704829
294140411146540923988344435159133201077394411184670648981066347241048239358274019449356651086
312567852977697734684303061462418035852933159734583038455410337010916767763742762102137013548544092
307190114731848574923331816720721372793556795284439254815609137281284063330393773562420016045664557414
588166052166608738748074243391212955877763906969037078828527753894052460758496231574369171131761347

388271941686066257210368513215664780014767523103935786068961112599602818393095487090590738613519145 9
181951029732787557104972901148717189718004696169777001791391961379141716270701895846921434369676292 7
459109940060084983568425201915593703701011049747339493877885989417433031785348707603221982970579751 1
914405109942358830345463534923498268836240433272674155403016195056806541809394099820206099941402168 9
090070821330723089662119775530665918814119157783627292746156185710372172471009521423696483086410259 2
887457999322374955191221951903424452307535133806856807354446995127203174487195403976107308060206099 2
580760202927314552520780799141842906388443734996814582733720726639176702011830046819000241308350088 4
658415214899127610651374153943565721139032857491876909441370209051703148777346165287984823533829726 0
136110984514841823808120540969125274580881099486972216128524897425555516076371675054896173016809613 8
038119143611439921063800508321409876045993093248510251682944672606661381517457125597549535802399831 4
698220361338082849935670557552471290274539776210493182014658008021566536067765508783804304134310591
804606800834591136640834887408005741272586704792583191274157390809143831384564241509408491339180968
402511639919368532255573389669537490266209232613188558915808324555719484538756287861288590041060060 7
374650140262782402734696252821717494158233174923968353013617865367370064216677813773995100658952887 7
427662636841830680190804609849090469763667335662328249513235278880615776828781595886691802389403330764
419124034120223163685778603572769415417788264352381319050280870185750470463129333537572853866058889 0
458311145077394293520199432197117164223500564404297989208159430716701985746927384865383343614579463 4
175922573898588001698014757420542995801242958105453108310462972829375841611625325625165724980784920
998979906200359365099347215829651741357984910471116607915874369865412223483418872292944633517865385
673196255985202607294767407261676714557364981210567771689348491766077170527718760110999081441130586 45
577910526843048114402619384023242709392498029335507318458903553971330884461741079591625117148648744
686112476054286734367090466784686702740918810142497111496578177242793470702166882956108777944050484 3
752844337510882826477197854000650970403302183106255614733211777117441335028160884035178145254196432030 9
576018694649088681545285621346988355444560249556668436602922195124830910605377201980218310103270417 8
386654471812603971906884623708575180800353270471856594994761242481109992886791589690495639476246084 2
406593094862150769031498702067353384834955083636065718677106080980426924713241000946401437360326564 5
184566792456669551001502298330798496079949882497061723674493612262229617908143114146609412341593593 0
958540791390872083227335495720807571651718765994498569379562387555161757543809178052802964200447215
396280746360211329425591600257073562812638733106005891065245708024474937543184149401482119996276453 1
068006631183823761639663180931444671298615527598201451410275600689297502463040173514891945763607893 5
285550531733141645705049964438093630843874484783961684051845273884032452024705685164657164771393 2
377551729479512613239822960239454857975458651745878771331813875295980941217422730035229650808917705
068259248822322154938048371454781647213976820963320508305647920482085920475499857320388876391601995 2
409189389455767687497308569559580106595265030362661597506622250840674288982659075106375635699682151
094966974458054728869363102036782325018232370845979011154847208761821247781326633041207621658731297 0
811230758159821248639807212407868878114501655825136178903070860870198758898074566439551574153631931
919810705753366337380832721527988493503974800158905194208797113080512339332219093046589051594208348 4
140187106035460379464337900589095772118080446574396280618671786101715674096766208029576657705129120 9
907944304632829472061595104309022214393718495606340561893425130572682914657832933405246350289291754
708725648426003496296116541382300773133272983050567240141851520418907011542885799208121984493 15
699905918201181973350012618772803681248199587707020753240636125931343859554254778196114293516361223
496661522614735399674051584998603552953329245723888101362023476246690558164389678630976273655047243
486430712184943734853006063876445662773656271562137974146998113287461411771455244470899714452
288566292442402301847912054784985745216346964489738920624019435183100882834802492490854030778638751 65
911302873958787098100772718271874529013972836614842142817105531796543076504534324600535636147261818 09
699769334826407743519992868632383508875668359509726557481543194019557685043724800102041374983187225
967738715495839971844490727914196584593008394263702087563539821696205532480321226749891140267852859 9
673405242031917978999057188219493913207534317079800237365909853755202389116434671855829065537118979
526262344924833924963424497146568465912489185566295893299090352923333643574352030707701010843882 90
759834217018554228386161721041760301164591870539367447472059985025582891833692922337323999480437108
419659473162654825748099482509991833069676569367159689364493348864744213580840700660883597235039523
401795825570360163690098671132109798897070517287558551912699306730992507040702456850778679069476
612629080225163313639952117098452809263037592242674257559982892783704744452189363203489415521044597

261883800300677617931381399162058062701651024458869247649246891924612125310275731390840470007143613
62316992371698481325542009145304103713545329662063921054798243921251725401323149027405858920632175 8
94943454890684639931375709103463327141531622328055229729795380188016285907357295541627886764982741 86
16421878988574107164906919851162815285486794173638906653885764229158342500673612453849160674137340 1
7357277995634104332688356950781493137800736235418007061918026732855119194267609122103598746924117283
749312616339500123959924050854375698507957046226646190001035004901830415354584283376437811198855 6
318777792537201166718539541835984483052037628194407615941068207169703022851522505731260930468984234
33152732131361216582808075212631547730604423774753505952287174402666389148817173086436111389069420 27
908814311944879941715404210312908470940802540232924945493878640230512927119097513536000921971105
41209668311516328705423028470073120658032626417116165957613272351566662536672718998534199895236884 8
30999302757419916463841427077988708874229277053891227172486322028898425125287217826030500994510824 78
35729056919885554678860794628053712270424665431921452817607414824038278358297193010178883456741678 11
398954750448339314689630763396657226270437932167454218245570625247972199786685427989779923395790575
81890622525473582205236424850783407110144980478726691990186438822932305382318559732869780922253529 59
1017341407334884761005564018242392192695062083183814546983923646413639891012102177095976704908305081
85470419466437131229969235889538493013635657618610606222870559942337163102127845744646398973818856 67
46260879482018647487672722722206267646533809980196688368099415907577685263986514625333631245053640261
05696055131838131742611844201890885319635698696279503673842431301133175330532980201668881748134298 8
68158557781034323173306478498321062971842518438553447262012823457071698530518326179641178579608888 1
50329602290705614476220915094739035946649169162353968092019457817589108893199211226007392814916948161
52738427362429809823406320024402449589445612916704950823581248739179964864113348030247577752197089 32
77226234948601504665268143987705161531702669692970492831628550421289814670619533197026950721437823 04
768752802873541261663917254925170010714180854800636923259462019002278087409859771921805158532147 39
2653251559035410209284665925299914353791825314542905984158176305892796690989699111643811878094353 71
52133226144362531449012745477269573939348154691631162492887357471882407150399500944673195431619385 54
85207665738825139639163576723151005560372633948672082078086537349424401157996675073607111159353139 5
9197120948964171553024531364770942094635696982226637752099945168450643623824211853534887989395673187
80660610788544000550827657030558744854180577889171920788142335113866292966717964346876007704799953 78
8338787034871802184243734211227394025571769081967030920182401884270570460926225641783752652633583242 4
061253311529423457965569502506810018310900411245379015332966156970522379210325706937051090830789479
99900499935322153622748476603613676979778567386584670936679588583788795625946464891376652195882869
33801836011632857855855819555604215625088365020332200245137621582046181067051953306530600605010548 8
71672453779428313887163139559690583208341689847606560711834713621812324622725884199028614287284956
87963932546428534307530110528571382964370999035694888528519040295607346131138263878897551788562424 9
98748316382804046848618938189590542039889872650697620201995548412650005394428203930127481638158530 39
64399254702016727593285743666616441109625663373054092195196751483287348089574777527834422109107311 1
35182804603634719185565572957144747682552857863349342885311874944000322969069775831590385803935535
21358860079600342097547392296733106493956018122378128545843176055617338611267347807458506760630482 2
9409653041118306671081890303110887172816751957967534718853722930961614320400638132246584111115775835 8
58113501856904781536898137718472814751998350504781297718599084707621974605887423256995828892550419
37958260616211842368768511418316068315867994601652057740529423053041201780313357263267054790338401257 30
59123396018801378254219270947673371919872878532421248921183470876629667207272325650565129333126
05950577727542471214683128329820723617505746738701282095755443059683955556868611883971355220844528
52640081252027665557677495969626612604565245684086139238265768583384698499778726706555191854468698 46
94784957346226062942196245570853172277652309895545019303773216664918257815467729200521266714346320 9
6378918523232150189761260343736840671941930377468809992968775824410478781232662531818459604538533438
3911449677531286426092521153767325886672260404252349108702695809964759580579466397341906401003636190
40420331135793365424263035614570090112448008900208014780566303710154122328914657229314506770716706 43
5568274377439657890679726874384730763464516775621030986040927170909512808630902973850445271828927496
89212106670081648583395537735919136950151620189088874842107980768991148046692706509407620465027725 2
86507289053285485641453003000569378547786109696920253886503457718317663686859223681448475276498468
82194973972970737187188400414323127636504814531122850990020742409255859252926103021067368154347015 2
52348786351643976235860419194129697690405264832347009911154242601273438022089331096686367898694977 99

400126016422760926082349304118064382913834735467972539926233879158299848645927173405922562074910530 8
53153718291168163721939518870095778818158685046450769934394098743351443162630317247747486897918209 2
394808331439708406730840795893581089665647758599055637695252326536144247802308268118310377358870892 4
061303133647737101162821461466167940409051861526036090925219472188909181073358719641421444786548995 28
58234394705007983038853886083103571936000277119455802191194289992272235345877075662469261776631788551
4435021828702668561066500353105021631820601760921798468493686316129372795187307897263735371715025637
873357977180818487845886650433582437700414770414934927438457587107159731559439426412570270965125108
115548247939403597681188117282472158250109496096625393395380922195591918188552678062149923172763163 2
183398969380756168559117529984501320671293924041445938623988093812404521914848316462101473891825101 0
909677386906640415897361047663450006807710565671848628149637111883219244566394581449148616550045676
982690308911185687986929470513524816091743243015383684707292898288460222373014526556798986277679680
914697983782687643115988321090437156112997665215396354644208691975673700057387649784376862876817924 9
74694384274652563163230055513041742273416464551278127847577724575203863437542828256714128858345443 5
13256205446424101103795546419058116862305964476958705407214198521210673433241075676757581456990693 0
46047522770167005684543969234041711089888993416305585515788735343081552081177207188037910404698306957
86854739376564336319797868036718730796939242363214484503547763156702553900654231179201534649779290 66
241508328858395290542637687668968805033317227800185885069736232403894700471897619347344308437443759 9
2503417880797223585913424581314404984770173236169471967551535331977549971627856631190469126091825912 4
989036765417697990362375528652635733763526969344354400473067198868901968147428767790866979688522501
6369498567302175231325292653758964151714795595387842784998664563028788319620998304945198743963690706
827626574851043911223261879405994155406320713198985707677611053236062986748037791537675115830432084 9
872092028092975264981256916342500052290887264692528466610466539217148208013050229805263783642695973 3
707053922789153510568883938113249757071331029504430346715989448786847116438328050692507766274500122 0
035262037094660234146489983902525888301486781621967751945816771876275720050543979441245990077115205
15461993050983869825428464072555409274031325716326407929341833421470904125425353523248021932277075355
5467958716383587501815933871742360615511710131235256334858203651461418700492057043720182617331947157
0086757853933607862273955818579758725874414025420771054753612940474601000940954446459662881486915903 89
9071865980563617137692227290764197755177720104276496949611056220592502420217704269622154958726453989
227697660310524980855759471631075870133208861463266412591148633881220284440694169488261529577625325 0
198703598706743804698219420563812558334364219492332759372212890564209430823525440841108645456940496
9271494003319782861318186188811118408257865928757426384450059944229568586460481033015388911499486935
43603022181093446470646400022362550573612946262960961987605642599639461386923308371962659547392346241
345977957485246478379807956931986508159776753505539189911513352522987361127791827485420086895396583 5
9421963331502869561192012298889887006079992795411188269023078913107603617634779489432032102773359416
908650071932804017163840644987870171537567811853213284082135110754952829497493621460821558320568723 21
8557406516109627487435098092230211609982633033915469494644491004508572580925089745074896760324090768 98
36529406579201983152654106581368237919840906457124689484702093577611931399802468134052003947819498 66
20262400890215016616381353838151503773502296607462795291038406868556907015751662412928722448827194293
3100485482445458071889763300323252582158120328276920028147624318286221710543528983482082734516801 8
61317195933247110746622285087106661177034653528395776259977446721857158161264111432717943478859090 892
8084866949141390977167369002777585026866465405659503948678411107901161040085727445629384254941672596
054871172359464291058509995021495879311219613590815882620682332156153086833730838173279328196983 87
50870834838804638847844188412697454370937329836240287519792080232187874488287284372737801782 7
0080587824107493575148897891173974612932035108143270325140903048746226294234432757126008664250833 31
876886507564292716055252895449215376517514921963671810494353178583834538652556566406572513635750643 5
32365089367904317025978781771903148679638408288102094614900797151377170990619549696400708676721530
048672631475510537231757114323217411441168062286420638890621019235522354671166213749969326932173704 31
059872250394565749246169782609702533594750209138667377289443869640002811034402608471289900074680776
48440887113413525033678773167977093727768634421173226463784769787514433209534000165069213 0
546476890985050203015044880834261845208730530973189492914625322933612431514306578264070283894098416
029503092418971209716016492656134134334222988279099217860426798124572853458013382609958771781131021 6
73402565627440072968340661984806766158050216918337236803990279316064204368120799003162644491461902 19
4582296909921227885539487835383056468648816555622943156731282743908264506116289428035016613366978240

5177015521962652272545585073864058529983037918035043287670380925216790757120406123759632768567484507
9151147313440001832570344920909712435809447900462494313455028900680648704293534037436032625820535790
1183956490893543451013429696175452495739606214902887289327925206965353863964432523883275224996059869
7475988232991626354597332444516375533437749292890058117578635555562693742691094711700216541171821975
0519831787137106051063795558588905568852887989084750915764639074693619881507814685262133252473837651
1929901561091897779220087057933964638274906806987691681974923656242260871541761004306089043779766785
1966189140141449252704808819714988015420577870065215940092897776013307568479669929554335636139847738 06
0394368895887646054983871478968482805384701730871087111776115966350503997934384693391197898871091565 41709
1330826076474063057114110988393880954814378284745288383680794188843426622207043872288741394780101 77
2139228191199236540551639589347426395382482960903690028835932774585506080131798840716244656399794827
5783650195514221551339281978226984278268342838183350750926261054872570092407004548848569295044811073 80
9965474815689139353809434745569721289198271770207666136024895814681191336141212587838955773571949863
1721084439890142394849665925173138817160266326193106536655350414730708044149391693632623737677709585
0313255990095762731957308648024601370121232702053374260705314244820816813030639737873664248367253 9837
4876909806021827857862165127385635132901489035098327061725893257536399397905572917516009761545 90447
7169226580631511102803843601737474215247608515209901615858231257159073342173657626714239047827958728
1505095633092802668458937649649770232973641319060982740633531089792464242134583740901169391964250459
1288134034988106354088875968200544083643865166178805576089568967275315380819420773325979172784376256
6118431989102500749182908647514979400316070384554946538594602745244746681231468794344161099333890899
2638411847425257044572517459325738989565185716575961481266020310797628254165590506042479114016957900
3383565748692528007430256234194982846791447632277400552946090394017753633565547193100017543004750 47
1914489984104001586794617924161001645471655133707407395026044276953855383439755054887109978520540117
5169747581344926079433689543783221172450687344231989878844128542064742809735625807066983106979935260
6933921356858813912148073547284632277849080870024677763036055512323866562951788537196730346347012229
3958160679250915321748903084088651606111901194843412350124646928028805969613428351188471544977127847
3361766285062169778717743824362565711779450064477718370221999106695021656757644044997940765037999954
8450027106659878136038023141268369057831904607927652972776940436130230517870805465115424693952651271
0105292707030667302444712597393995501462840467643136339978259184541176413327906460636584152927 01903
0276017339474866960348694976541752429306040727005090395031485229213925755948450788679779252539 31765
1564161971684435243697944473559642606333910551268260615957262170366985064732812667245219890605498802
8078288142979633669674412480598219214633956574572210229867759974673812606936706913408155941201611596
0190237735525556300606247983261249881288192937343476862689219239777833910733106588256813777172328315
3290825250927330478507249771394483389252520811756084529686439085862154101760001179857293813998258 3
1929367910039184409928657560599358910002969864460974714718470101531283762631146774209145574041811590
8800064943237855839308530828305476076799524357391631221886057549673832243195650655460852881201902363
6447127037486344217272578795034284863129449163184753143504139209610087601539703366179371539630
3719925365047090858251393686346386336804289176710760211115982875539940120076013947033661793715 39630
6139863655492213741597905119083588290097656647300733879314678913181465109316761575821351424860442292
4453041131606527009743300884990034745054184067734260358340960860553374627609356568531097609942383
4738222208729246449768456057956251676557408841032173134562773585605235823638953203853402484227337163
9123973215995440828421666636023296545469470357718487344203422770665383738750616921276801576618109 5420
0977083636043611105924091178895403382012426523948926864398089261146354145715351943428507213534530183
1587562827573389826889825355779929572765452293915674775666760510878876845534936306682780506462 281359
8885879259940946446041705204470046315137975431737187756039815962647501410906658866162180038266989961
9655805872086397211769952194667898570117983324406018115756580742841829106151939176300591943144346 051
5404771057005433900018245311773371895585760360718286050635647997900413976180895536366960316219311325
0223851791672055186056926351803625121457592623836934822266589557699466049193811248660090997981 28573
4940066155521961122072030922776462009993152442735894887105766238946938894464950939603304543408421024
6240104872332875008174917987554387938738143989423801176270083719605309438394006375611645856094312951
7597713935396074332172489221267045808183331764165818269526410587289244774003594700092686626596 51422 0
6300785920024882918608397437325538490983964326147000532423540647042089499210250404726781059083 6440074
6638002087012666420945718170294675227854007450855237772089058168391844659282941701828823301497155423
5235917748186285929676050482038643431087795628929254056389466219482687110428281638939757117577 86915

43016505860296521745958198887868040811032843279867198621306205559855266036405046282152306154594474
48990883908199973874745296981077620148713400012253552224669540931521311533791579802697955571050850747
38747507580687653764457825244326380461430428892359348529610582693821034980004052484070844035611678I7
1705128133788057056434506161193304244407982603779511985486945591520519600930412710072778493015550388
953603382619293437970818743209499141595933963681106275572952780042548630600545238391510689989135788?2
00194117863556821491185282078521301255185184937115034221595422445119002073935396274002081104655302O7
93286725474054365271759589350071633607632161472585140764205302004534018357233829266191530835409512O2
26329160544261236191970516138393573266937601569144299449437448568097756963031295887191611292946818?8
49363386473927476012269641588489009657170861605981472044674286642087653347998582220906198021732116I4
23041947775499073873856794118982466091309169177227420723336763503267834058630193019324299639720444?1
79288122854478211953530898910125342975524727635730226281382091807439748671453590778633530160821559?1
13141442050914472935350222308171936635093468658586316485557586244781862010871188976065296989926932?8
17870557643514338206014107732926106343152533718224338526353202177354407152818981376987551575745693?7
27150488469793619500477720970561793913828989845327426227288647108883270173723258818244658436246580S9
25603381052156062061557132991560848920643403033952622634514542836789828807425142256745180618414956?4
68611163540497189768215422772247947403357152743681940989205011365340012384671429655186734415374161S0
4256325671343024765152252912618035780169240326699541746087592409205100704469340936591017813485783569444076
04702325407555577647284507518268904182939661133101601311190773986324627782190236506603740416067249?2
4901374332172464540974129955705291424382080760983648234659738866913499197840131080155813439791948528
30436739012482082444814120954437738983200598649091595053228579145768849625786658859991798675205545S
80990045564611787552493701245532171701942828846174027366499784755082942282020232901221630102309772151
56944642790980219082668986883426307160920791408519769523555348865774342527753119724743087304361951I3
9611908003025587838764420608504473063129927788894272918972716989057592524467966018970748296094919064
87646937027507738664323919190422542902353189237729316673608699622803255718530891928440380507103006?4
77684786324319100022392978525537237556621364474009676053943983823576460699246526008909062410590421S4
53927904411529580435334500256244101006359530039598864466169595623518780606885137234627079973272331?3
4693971456285542615467650632465676620279245208581347717608521691340946520307673391841147504140168924
12131982688156866456144538028753933116023229255598644910429953564009956784953409351152664540244187759
49316930560448686420862757201172319526405023099774567647838488973464317215980626787671838005247 69688
40849891850614900343240347674268624595239589035858213500645099817824463608731775437885967767291 9526
11121385919472545140030118050343787527766440276218941017576872680428176623860680477885242887430 2591
4524707395054652513533945959878961977891104189029294381856720507096460626354173294464957661265195349
57018600154126239622864138977967333290705673769621564981845068422636903678495559700260798679962610 19
03933126376855696876702929537116925637116268706840872893922571451248113137867664902425161099027747 10
90335933309304948380597856628844787441469841499067123764789582263294904679812089984857163571087831 19
18486302545016209298058292083348136384054217200561219893536693713367333924644161252231969434712064 17
37549121635700857369439730597970097176266666422667431117762174040688681310351899112271339724036880 09
96862922546450063852886203938005047782769128356033725482557939129852515068299691077542576474883254
14121328006267170940090982235296579579978030182824284902214704811112401860761341515038756983091865?2
78065889668236252339784527263453042041880250844236319038331838455052236799235775292910692504326144?9
50109861088899914658551881873582528164302520939285258077969737620845637482114433988162710031703151?3
44023095263519295868406872002372372358558061201000213740851154484912685841268659379714913382057849800698
25519574020181810564129725083607035685105533178784082900004155251186577945396331753853209214972052?6
07831260281961164858098684587525129997404092979768176639914653386108937587952214971731728131517932?0
4431121815871023518740757222100123768739268371234237032342167051708018562372526732547033324875754482
96757345001932190219919960797983733836732425761039389853492787774739805080800115547640610535222023
2540944367718794565434040673589649101761077594836454082348613025471847648518957583667439979150851285
80206078205554462991723202082229148869530997299742974711553718589242384938558528204704308104882624648
78805330427146301194158989632879267832732245610385219701113046658710050008328517731177648973523092?6
6123458887310288351562644602367199664455472760831011878838915114934093934475007302585581475619088139
8752357812331342279866503522723367131725801544849703600795698276263923441071465849578024141408
15840522953693749971066559489445924628661996355635065262340533943914211127181069105229002465742360411
300936918892558657846684612156795542566054160050712664176605687427420032957716064344860620123982169

827172319782681662824993871499544913730205184366907672357740005393266262276032365975171892590180110429038427418550789488743883270306328327996300720069801224436511639408692220745320244624121155804354542064215121585056896157356414313068883443185280853975927734433655384188340303517822946253702015782157373265523185763554098540332326382319219892171177449469403678296185920803403867575834111518824177439145077366384071880489358256868542011645031357633355509440319236720348651010561049872726472131986543435450409131859513145181276437310438972507004981987052176272494065214619959232142314439776546708351714749367986186552791715824080651063799500184295938799158350171580759883784962257398512129810326379376218322456594236685376799113140108043139732335449090824910499143325843298821033984698141715756010829700658306521134770680368060953229719909944512090872757762253510409023928877942463048328031913271049547859918019696783532146444118926063152661816744319355081708187547705080265402529410921826485821385752668815558411319856002221351588872103656960875150631875330029421186822218937755460272272912905042922597877106678738400006167721546384412923711935218284998243509208918016855272981564218581911974909857305703326676464607287574305653726027689823732597450844796495456480307715981539558277791393736017717422996027353102768719449444491793978514463159731443535185049141394155732938204854212350817391254974981930871439661513294204591938010623142177419918406018034794988769105155790555480695538785400664533759818628464199052204528033062636956264909108276271159038569950512465299960628554438383303276385998007079292284665950355121124528408751622906026201185777531374794936205549640107300134885315073548735390560290893352640071327473262196031177343394367433857591245081493357369116645412877811745402305475066713651825828489809951213919399563324133656677709800308191027204099714868741813466700609405102146269028044915964654533010775469541308871416531254481306119240782118869005602778182423502269618934435254763357353648561936325441775661398170393063287216690572225974520919291726219984440964615826945638023950283712168644656178523556516412771282691868615572716201474934052276945957121983149433816221140069363074304441732847861017777438379770372317952554341072234455125555899986461838767649039724616795901810003509892864120419516355110876320426712979826529425882951141275841262732790798075597518515768412647422094797218433093529726652100156625145529947451276315509176367302594621329301904028379542463232585503010969706922720227074863419005430265068121414213507154175057508639907673946335146209082888934938376439392569060040673114220933121959362029829723511632593867722414779116295727807523950506251581603133359382311500518626890530658368129988108663263271980611271548858798093487912913707498230575929091862939195014721197586067270092547718025750337307993971343595326461952699965963856549175904583335857991020127132045839032008538788816336376851820837278851311752277696097879621423721625452145912818317982160441113116714069148271709810154577819392023115638719508050246797257924976057726259133285597263712112019057207714091486450740949267180358151575715140503976109638467555692989703835473141002238025834687673501297754132795320609711545064842121859364909979177668747744818828706323155158650328981642282882327468661065927321979071623846421534898524762167890502609980452664839295423572873439776804957740914495383915755654854590589764951985138010079580107837599457752991967005476022525520344539887125387801719607181640781248478472579124078254434316168234523957068951427226975043187363326301110305342335821609333191218806608268341428910415173247216510553558499932245487307788229905252342384613515209769384610425828479634753418375620030149157032796853018686315724884015266309835689593636346574353217834931998255421173084677452970858395076164582296303244243282377374505170285606980678895217681981567107816334052667595394249262807569683261074953233905362230908078014554919837355377748702290390181429373115293346444681512129450975965343062842153914457271186149000176505581770953024688752632501197052094761594167687277844720019278913725184162285778379228443908430118112149636642465903363419454065718354477191244662125939265662030688852005559912123536371822692253178145879259375044144893398160865795539255481793756681046474614105142498870252139936870509372305447734112641354892806841059107716677821238332810262185587751312721179344448201440425745083063944738363793906283008973306241380614589414227694747931665717623182472168350678076487573424904155762821758397297513447899069658953254894033561563161674032647246921250575911625152965456854463349811431767025729566184477548746937846423373723898919206620485118943788682248072793520220517965453437572741639107919772952950812942922205347717304184477915673991738418311710362524395716152714669005817000026330104526435478659032907332054683388720787354447626472952976901709120078741837367350877133769776834963444252419949951388315074877537433849458259765560996555954318040920178497184684549703762120885243770138537576814243126434423982152941645378000492502762765150789085071265997036708726692764308377229685985169122305037462744310852934305273078865283977335246017463527703205938179125396915621063637625882937571373840754406469647831007045806134467312715911946

08435935825987782835266531151065041623295329047721740835593497237585521380483050900096466760883015 4
06128243087406455944318534137552201663058121110334531207450868243394321590435944303124312274713858 42
03039010607094031523555617276799416002039397509989762933532585557562480899669182986422267750236019 32
57974726742578211119734709402357457222271212526852384295874273501563660093188045493338989744157149 054
41825597380080871565281430102670460284316819230392535297795765862414392701549740879273131051636119 137
57700892956482332364829826302460797587576774537716010249080464230183652416175665560016085912153455 62
67602192689982855377872583145144082654834844094784631787773747945358016996077940556870119232860804
11309046293508718271259346687127666948738998245985277864995691654640294589350649643358098247659651 65
14209009867552038083092032304873427034682887516040715466538346196112230137594515792526967436425319 273
90036038608236450760286988274976187235754767628899507521148048525279508450339558570838130476937881 32112
36742813194879502280663201700224603319896719706491637411758548518784840120548446725885140156272501 9
82171906696081262778548596481836962141072171421498636191877475409650308957099470934337856981674465 8
28267911940611956037845397855839240761276344105766751024307559814552786167815949657062559755074306 52
10853015979080733437360794328667578905334836695548680391343372015649883422089339997164147974693869 6
90548008919306713805717150585730714881564992071408675825960287605645978242377024246980532805663278 70
41926768467116266879746348695046450974042019373945325296420647831361206202636498199999498405 14
38682852589563422643287076632993048917234007254717641886853513723326678792173834754148002280339299 7
35793615241275582956927683723123479898944627433045456679006203242051636928258844308543830720149562 1
06460533238537203143242112607424485845094580494081820927639140008540422023556260218564348994145439 95
04109805918179488826280520664410863190016885681551692294862030107388971810077092905904807490924271 41
01893354281842999598816966099383696164438152887721408526808875748829325873580990567075581701794916 19
06114001908553744882726200936685604475596557476485674008177381703307380305476973609786543859382187 22
05839023444435088674998665060404658743460053318274362961778625180818931443632512051070946908135864 40
51922951293245007883339878842933934243512634336520438581291353927308652909778330067126179813031 67
94385535726296998740359570458452230856390098913179475948752126397078375944861139451960286751210561 63
89760088800927461158608002078033415914517970730368351969777660763737853330102041201120469886092093 39
08536577322239241244905153278009550558664594776344822609860748132973026309750288121035177231244650 953
49653693090018637640940943498373132513218620802148099226850294845466181471557444709669530177690443
42720318927706047177845279391604722815343798035396798614243709566832214914654380145938292773939603 2
75404800955223181666738035718393275707714204672383862461780397629237713120958078936384141479298205 880
65522129262093623930637313496640186619510811583471173312025805866727639997263579078063818813069156 36
62741254312595899361196476261014056350339952314032311381965623632719896183725484533370206256346422 3
95276694356837676136871196292181875457608161705303159072882870071231366308722754918661395773730546 0
65997437810987649802414011242142773668027513909593134041558262667895108467761186659576601659981780 8
94149875549762843878561002637965431783136344635383814161151902064991335487331311150227006819301352 9
59597164019719605362503355847998096348871803911161281359596856547886832585643789617315976200241962 15
52896297904819822199462269487137462444729093456470028537694958859591606789282491054412515996300781 36
83674902093749157329627700286568293444313423473512392982591667393066832292325794707209634807372692 141
28874666045146148785034614282765991608090398652575717263081833494441820193533385071292345774375579 3
44062178711330063106003324053991693682603746176638565758877580201229366353270267100681261825172916 60
82025418928859352444910701382062115538277935652969145765020486432828655579347072096348073726921 186
89546732276775133569019015372366903686538916129168888787640752549349249733427181117889275993159671 93
54758988097924525262363659036320070854440784544797348291802082044926670634420437555325050527522837 7
88870408043353192340768563010934777212563908864041310738178533383160381352808281190408325644018 42
05374679299262203769871801806112262440900924264198582086175117711378905160914038157500336642415609 52
16328197122335023167422600567941281406217219641842705784328598028233505982820819666624903585778994
03331522748177695284368163008853176969478369058067106482808359804669884109813515865490693331952239 4
36328792399053481098783027450017206543369906611778455436468772363184446476806914282800455107468664 53
92805399409108754939160095731619715033166968309929466349142798788422572206971488755806374803086299
51184731871247772919100702275888934869394562895151802965372150409603107761289831263589964893410247 036
03664505868728589051406841238124247386385427908282733827973332688550493587430316027474906312957234 97
42611221517417153133618622410913869500688358989623492763173164783400774608866555987333821138299287 7
69114594921841920877160060684728746736818861675072210172611038306717878566948129487850489430630861 69

9487987031605158841082823512741535385133658953329486294944950618685147791058046960390693726626703865
1290520113781085861618888694795760741358553458515176805197333443349523012039577073962377131603024288
7200537320998253008977618973129817881944671731160647231476248457551928732782825127182446807824215216
4695678192940982389262849437602488522790036202193866964822156280936053731780408637272684266964219299
4681921490870170753336109479138180406328738759384826955358077395761447997270003472880182785281389 50
3217986345216111066608839314053226944905455527867894417579202440021450780192099804461382547805858048
4424164047750315360549065914300781583724301213751156228401583864427089071828481675752712384 67824595
3433444962201009607105137060846180118754312072549133499424761711563332140893460915656155060031738421
8701570226103101916603887046614388977363187809407115275281746895764015810470169652475577409164 4568
6777171585005832699434016772021567677240681283665652641229824394651331973591997094032759385026 695574
7023181320324371642058614103360652453693916005064495306016126782846849443739716671766123104 897503188
573216555498834212180284691252908610148552781527762562375045637576949773433684601560772703550 9629049
3924870884062810679436224187047470083688426710225583024035998416459511224852726336326451140173952480
8619463584078375355688562224711552094722536709260679735100056554938122457548372854571179739361575
6167641692895805257297522338558611388322171107362265816218842443178857488798109026653793426664216990
9140563634322493013348679881548866286650523469972355747384248305904236771432787923164224038777643301
9260019228477831383763253612102533693581262408686669973827597736568222790712583247888642369 34639616
4363308730139814211430306008730666164803678984091335926293402304324974926887831643602681011309570716
1419128306865773235326396536773903176613613159655535849993968060651559219367599777179330197446881483
7110320650369319289452140265091546518430993655349333718342529843367991593941746622390038952767381333
0617747629757494386871697845376721949350659087571191772087547710718993796089477451265475750187 1194870
7387367858902006173733210736932201632062843206367110209695058576117396163232621177089454262146098584
1023781321581772760222273813349541048100307327510779994899197963883530734444347532975914263768 40544
2264784216063122769646967156473999043715903323906560726644116438605404838847161912109008701019130726
0710441141432419767968285478855247794764818029597360494937004795960402927462992035720997619501403483
1538094771460105633449988208221205872815107291829712119178764248803546723169165418522567292 3442918
7128163232596965413548589577133208339911288775917226115273379010341362085614577992398778325083 550730
1998184590295983559892605532996737704917224549353296833000223015171226575787524058832249085821280
8974790932610076257877042856500699617621217684547899644070506624171021332748679623743022915535820078
0141165348065472488230615000339206898379476625503654982805329662862117930628430170492402301985719978
9488368971830438051821744191476604297524372516843354112170386313794114220952958857980601529387527537
9903093887168357209576071522190027937929278630363726876582268124199338480816602160372215471014300737
7537792699069587121289288019052031601285861825494413353820784883465311632650400764242839087012101 5194
2319616522684220037112304643067344206474771802135307012409886035339915266792387110170622186588357
8121093517977560442563464999787251125440854522748109148743072598696020402759411789425812818821 5995
2359658979181144077653354321757595255536151885607203193465072968079907939637149617743121 19402
0212975731251652537680173591015573381537720019524445436200718484756634154074423286210609976132434875
4884743453966598133871746609302053507027195298394327142537115576660002578442303107342955153394506048
6222764966687624079324353192992639253731076892135352572231008898193391686682789482811704726245 01948
4097009757609209837240900747179733407881418251958425980962417476101382526439551352593118850456362641
8830033853965243597416931322894719878308427600401368074703904097238473945834896186539790594118 59931
0356168436869219485382055780395773881360679549900853123259442529724486666766834641402189915944 565309
4234406506678519484177667794704720419588220432953803263105374948831221803912796784461001397267538921
9511917183658766252808369005324900459741094706877291232821430463533728351995364827432583311914445901
7809607782883583730111857543659958982724531925310588115026307542571493943024453933187017992360816611
3054262539958338979429716020703387678150330102801200959972522222808014235710947603519255444349299867
6781789104555906301595380976187592035893734197896235893112598390259983102671933041892151096891562 2506
9659119828323455503059081730735195503721665870288053992138576037035377105178021280129566841984140362
8727256232144287543022100094727210734741349755141907370433182766261772759968888260272252471 336833534
5281669277959132886138176634985772893690096574950243625907724124930843008717556926257586506
9912016659622436080242870024547362036394841255954881727272473653467783647201918303998717627037515724
6499222894679323226936191776416146187956135566995677830682903163589699430763335083349079062410 02025
0613405734430069574547468217569044165154063658468046369262127421107539904218871612761778701425886482

5775223889184599523376292377915585744549477361295525952226578636462118377598473700347971408206994145
5807190802135907322692331008317595106590191212947954086036407573587502058902087045796700070552625058
1142066390745921527330940682364944159089100922029668052332526619891131184201629163076894084722356436
6808182168657219688268358402785500782804043453710183651096951782335743030504852653738073531074185917
7056103973950626403554422751561011072617793706347238049900666922161971194259120445084641746383589938 2
3994651739550900085947999013602667426149429006646711506717542217703877450767356374215478290591101261
9157555870238957001405117822646989944917908301795475876760168094100135837613573591356924455647764464
1786671153919513576961048649224900834467154863830544779143300976804868783481846727337584368927243104
4740680768522786251856501920882638132336231487333671474652045087662761495038994950480956046098960432
9123358348859990294526400284994280878624039811814884767301216754161106629995536681931232874257020 63
7383520200868636913117334697317412191536332467435256308713473027921749562270146873258678917345583799
6435135880095935087755635624881049385299900767513551352779241242927748856588856651324730251471021057
5352516511814850092750476845518252096331899068527614435138213662152368890578786699432288816028377482
0355060160298940091197138501798716836337441392759736440170070147637066557035043381211135764150184518
2141361982349515960106475271257593518530443214879643214590712680576243842864712219618709178560783936144 51
1383335649103256405733898667178123972237519316430617013859539474367843392670986712452211189690840236
3274114966012434830989299417380305884171666130730400067588380432111555379440605497721705942821514861
6567277124090338772274562909711013488518437411869565544974573684521806698291104505800429988795389902
7804383596282409421860556287788428802127553884803728640019441614257499904272009595204654170598104989
9675045119364711727722024361026140797508096869751766002371472884168110203102346856117264476612347 76
2785219024120256994353471622666089367521983311181351114650385489502512065577263614547360442685949807
4396932331297127377157347099713952291182653485155587137336629120242714302503763269501350911612952993
7858646813072264860082070881333538193703682598867893321238327053297625857382790097826460545598551318
3668884462826513379849166783940976135376625179825824966345877195012438404035914084920973375464247448
8176184070023569580177410177696925077818493386672557898564589851056891960924398841569280696983352 240
2256345704973122452693544193837004843183357196516626721575524193401933099018319309196582920969662476
6768365964701959575473934551433741370876151732367720422738567427917069820454995309591887243493952409
4441678998846319845504852393662972079777452814399418256789457795712552426826089940863317371538896262
8896294021121088844273765686245276121303710173008513571540453304150795944777614359743780374243664 69
7324713841049212431413890357909241603640631403814983148190525172093710396402688909498325722979546404
2701757722904173234796073618787889913318305843069394825961318713816423467218730845133877219086975104
9428437693250249816566738162606159417682525099937416728839517440669325495340310145222531618900923 53
7648637848288134420987004809622717122640748957193900291857330746010436072919094576799461492920042798
1687729426487729952858434647775386906950148984133924540394144680263625402118614317031251117577642829
9146445334089209769616990983726523617874560589470496817013697490952307208268288789073019001825342 58
0534342170592871393173799314241085264739094828459641809361413847583113613057610846236683723769591349
2615824516221552134879244145041756848064120636520170386330129532777690912318648020067556905682295016
3549319923059142463962170253297475731140942201801993680350264956369558664259067626856873721103391 567
9383989576556519317788300024161353956243777784080174881937309502069990089089932808839743036773659552
4891300156633294079071396154645340887915103006513219344866732482759079468078798194250195826223203 95
1312520141099605312606965554042486705499867869230217469890095478507256729787947698888310934874644264
0071818316033165551153427615562240547447337804924621495213325852769884733626918264917433898782478 927
8468918828054660998230368993978341374758702580571634941356843392939606819206177333179173820856243 6433
6335598634944968907810640196740744365836670715869245521182997893804077137501290858646578905771426 8335
8276897855471687184427726120509266486102051535642840632684818072879407171279668200606072755955590 404
0233178749447346454760628189541512139162918444297651066947969354016866010055196076873353965116149 30
9375709685545593815137895690392510149532656281470119983269922000663928753747131352364215892651262040
7288771657383840322196460541053444364216656224456504299901026568692727914275293117208279393775132610
6052881235373451068372939893580871243869385934389175713376300720319760816604464683937725806909237297
5234867029169104263692620901996052041210240776481903160140858635584276095370865581642739953493465463
1450404019952853725200495780524656251154109252437991326627136090994029022620062836752132305065 18393
4057450112099341464918433236465693717259144893241590062420206128857329261335968087265000456282 84557
57459659212053034131011182750130696150983551563200431078460190656549380654252522916199181995 96027523

27702249855738824899882707465936355768582560518068964285376850772012220347920993936179268206590142165
56159253067379445689490708532635681968318617722682499114726157320358076462981162440133167378927886892
29032593349861797021994981925739617673075834417098559222170171825712777534491508205278430904619460835
21740200583867284970941102326695392144546106621500641067474020700918991195137646690448126725369153716229079138540393756007783515337416774794210038400230895185099454877903934612222086506016050035177626483161111533255877050735412792499909859373473787081194253055121436979749914951860535920403830235716352727630874693219622190064260886183676103346002255477477813641012691906569686495012688376296907233961276287223041141813610060264044030035996988919945827397624114613744804059697062576764723766065541618574694522722923822875186799156983390747671146103022776606020061246876477728819096791613354019881402757992174167678799231603963569492851513633647219540611171767387372555572852294005418617851765023075446938693078734991103521825329297260445532107978877114499898709115112372506042387537348412570860640690520584521227545338480082053024504565176695185769132000428167580549248117805198326460324457928297301291053183856368212062155312886685649565126138922613670640939533345705269869596923503530942245438652786776730275404027022463844835532399147513634410440500923303612714960813554905315390210022995957565837058381261965683144286057956696621547216956208700137201784210061300488007048333251327931122325071486302069512453950037357233468070946564830892098015348787056334910923660575540508641115214414814346304372732710450277686619531078583233348578402971609252153260925589326556006712143594642550659967717703884453961816328796144608177892217183690088012677820743010642252463448074543004764429885553409062185153654355474125476152769772667769772777058315801412185688011705028365275543214803488004442979998062157904564161952721278450928489806426497420790579129069217807298769477975112447305991406050629946894280931034216416629935614828130998870745292716048433630818404126469637925843094185442216359084576146078558562473814931427078266215185541603870206876980461747740080832434366538235455510944949843109349475994467267366535251766270677219418319197719637831026993367508376005716345464367177672338758864340564487156696432101428259564534984138841289042068204700765159691684303899934836679354254921032811336318472259230555438305820694167562999201337317548912203723034907268106853445403599356182357631283776764063101312533521214199461186935083317658520471123643312267651299643171325217513553261867681942338790365546890800182713528358488844411176123410117991870923650718485785622102110400977699445312179502247957806590653296596940383987369907240797679040826794007618729547835963492793904576973661643405359792219285870574957481696694062334272619733518136626063735982575532496509982065072362236088236059283418558480269584137725589708388994291054980033111388460340193916612218669605849157148573356828614950001900759112521880039641976216355937574371801148055944229873041819680808564726571316152162920044988031540210553059707666636274932830891688093235929008178741198573831719261672883491840242972120043496552694272640255964146352591934840067586769035038232057293413298159353304444649682944136732344215838076169483121933311981906109614295220153617029857510559432641468505452684975764807800922133581137819774927176854507553832876887447459159373116247060109124460982942484128752022446259447763874949199784044682957360968534549843266536862844489365704111817793806441616531223600214918768769467398407517176307516849856359201486892943105940202457969622924566644881967576294349535326382171613395757790766374569570259738800438415805894336137106551859987600754924187211714889295221737721146081154344982665479872580056674724051122007383459271575727715218589946948117940644466399432370044291140747218180224825837736017346685300744985564715420036123593397312914458591522887408719508708632228264282288463184371726190330577714761565144382230679184738603914768310814135827575583643597721650028277803713422869688787347950960311088991961433866640684506974207877002805093672033387232629637856038653216432348815557557018469089074647879122436375556668678067610544955017260779114293083128576125448919444494732448190937953690080820638463167822508093531810406570254327604385703505922818919878065865412184299217273720951032422510797180778330426090867942734489573555925272380551144043800123904168771644518022694916816419274011064516224311017000566911217331894234005479596846698042980173625704067332821299621536848814041021944634246462207455756439604529853130714090846084996537678037932018991408658146621753193376659701143306086250098295669176388460567629729314649711493704624463593698924386187298507775928792722068554807210497817653286210187476768728860859626112660504289296966535652516688885572112276802772473708917389639977225756489053340103885593112567999151658902501648696142720700591605616615970245198905183296927893553030393468121976158218398048396056252309146263690039848912438618728950777592879272206855480721049787417653286210187476768728856048491395603494803767270363169210073508340738652616845074824964485974281349364803724261167042668708319250409976153190768557703274217850100064419841242073960400139603601583810565928413684574119102736420

274163723488214524101347716529603128408658419787951116511529827814620379139855006399960326591248525 3
0849369031313010079997719136223086601109992914287124938854161203802041134018888721969347790449752745
4288072803509305828754420755134816660927879335665212556201399882496284787262144323628536765025914 50
4683776352825876521391564809721419296755493843755826002531685363567313792624758780494459441834291727
5698837622626184636542743497662411138451305481449836311789784489732076719508784158618879692955819 73
3250699951402601511675529750575437810242238957925786562128432731202200716730574069286869363930186765
9582513264991459502609170693475194089753576404168309119884645247361895605647942635807056256328117892
6966302647953595109712765913623318086692153578860781275991053717140220450618607537486630635059148391
6467656723205714516886170790984695932236724946737583099607042589220481550799132752088583781117685214
2693347869218952406226579210436203488529262679840139532164587911515790504605797108389833718640380244
1751134722647254701079479399695355466919726763255229914654933499663234185951453609803440922122067 1
2567698723427940708857070474293173329188523896721971353924492426178641188637790962814486917869468177
5917171506691114800207594320120619696377951032270890295660855622254526026104607361313688690092817210
6819861855378098201847115416363032626569928342415502360097804641710852553761272890533504550613568414
3775854429677977014660294387687225115363801191758154028120818255606485410787933598921064427244898618
9616294134180012951306836386092941000831366733721530083526962357371753307386533382048421903081864491
8409372394403340524409554558016406460761581010301767488475017661908692946098769201691202181688291 04
0870709560951470416921147027411339005225334083481287035303102391969997859741390859360543359969707 5604
4601342424536824960987725813110247327985620721265724990034682938868723048955622532044636026398542252
5841646432427161141981780248259556354490721922658386366266375083594431487763515614571074552801615967
7048442714194435183275698407552677926411261765250615965235457187956673170913319358761628255920783080
1852068901515047133403861003100559148178521103847545429333891884441205179439699701941126951195265649
1959418997541839323464742429070271887522353439367363366320030723274703740712398256202466265197409019
9762452056198557625760008708173083288344838183107005451449354588542267857855191537229237955494333410
1744201696000909664156127322977702212179518683763590822551288164700219923488640439591530184640047143
2118636062252701154112228380277853891109849020134274101412155976996543887719748537643115822983853312
3071751132961904559007938064276695819014842627991221792974834890186847167650382732855205908298452 9
8062592503521284519259279865935061329619467962523739725655841578537445675589980324054921869628884903
3256085145534439166022625775512916200772796835937530454181080729285891989715381797343496187 23
2927614747850192611450413274873242970583408471112333746274617274626582415324271059322506255302314738
7592517247873228814914559156050363345754242337791603749525024930223514819613811625639114156103268449
5807250827343176594405400926976526394435798634797097431244982719331138638731596363612186234972614055
6079920628316999420072054811525353393946076850019909886553861433495781650089961649079678142901148 387
6456821749140756237676184537751440314754112067601607264605568592577993220703373333989163695043466906
9482843662998003741452762771654762382554617088318981088688684785370553548046935095588180253605297 4079
3538676511195079373282083146268960071075175520614433784114549950136432446328193346389050936545714506
9008644834401804283633905135781572739733453728426337217406577577107983051755572103679597690189954
9413019599957301790124019390868135636813177594438763207986880037160730322054742357226689680188
21234243918858416897227765219403249322731479366923400484897605903795809460419427961378253378122 3
9476461478329269765451622902817011004378460387565441517394339600489153188175766505009516974024156447
7129363661425394936888423051740012992055685428985389794266995677702708914651373689220610441548166215
6804219838476730871787590279209175900695273456682026513373111518000181434120962601658629821076663523
3617740078377834237091526440630540718078433580617226911055500204151316967304684921335683726540037 5
0982908936461204789111475303704989395283345782408281738644132271000296831194020332345642082647327623
3830294639378998375836554559919340866235090967961134004867027123176526663710778725111860354037554487
4186935197336566217723592293967764632515620234875701137957120962377234313702120310049651521119760131
7641940820343734851285260291333491512508311980285017785571072537314913921570910513096505988599993156
0863655477403551898166733553588004821466509974143376118277772335191074121757284159258087259131507 4606
0256349037772633739144613770380213183474470311130326702969171335047701632100616622783002726928335459
1179141944780874825336071440329625228577500980859960904093631263562132816207145340610422411208301000
8587264252112262480142647519426184325853386753874054743491072710049754281159466017361225904405189 91
6002298278017960351940800465135347526987776095278399843608069089891978396935532179980139135442552717 9
1022539701081063214304851137829149851138196914304349750018998068164441212327332830719282436240673319

6554692677851193152775113446468905504248113361434984604849051258345683266441528489713972376040328212
6602535166939140820499473204860216277597917712347510975024030789357599377150950217516935558270725339
1189233407022383207758580213717477837877839101523413209848942345961369234049799827930414446316270721
4796117456975719681239291913740982925805561955207434243295928989808052923336641541925636738068949420 1
4712413405250722040617943552525552250087487900865683145428351677505422948032274830440564385815919526
6675828292970522612762871104013480178722480178968405240792436058272467443076721645270313451354167 64
9668901274786801010295133862698649748212118629040337691568576240692963724930972016287072001898354 23
6903641492702369162938743724803298550451120891928798298744678641291594175316756025334533106267452545
0711418148323988060729714023472552071349079839898235526872395090936566787899238371257897624875599044
3228895388377317348941122757071410959790047919301046740750411435381782464630795989555638991884773781
3413470702467473621120489862269918885174562517325193413520381115863350123981149079167366284475675141
6105041097350585276204448919097890198431548528053398577784431393388399431044446566924455088594631408
1751220331390681596592510546858013133831521764182104334297888261196304431113887962587460902261309 00
8499754303957712432306169062629194039214397402708947776637024881554993224588259790206312574369109463
9325280624164247686849545532493801763937161563684785982371590238542126584061536722860713170267474013
1145261063765383390315921943469817605358380310612887852051546933639241088467632009567089718367490578
1630851581381619668822220475043759061433804072585386208365517699842677452319582418268369827010 62374
1493836349662935157685406139734274647089968561817010605511048809715548591186171896680259735417 0542398
5135560018723035070609946424217149393196046527424050882225579381519135438571253258540493946010865
7937980586201433660788252197178090258173708709164604527279771535099103407364250203863867182205 228796
9445838765294795104866071739022932745542678566977686593992341683412227466301506215532050265 534146099
5249356050854921756549134830958906536175693817637473644183378974229700703545206663170929607 591989627
7324230902523974438610142630986877339138825186843165010279649114977375828889134503411488659 486702154
9210108432080783428089417298008983297536940644969903125399863919581601468995220880662285408 41486427
4786281975546629278814621607171381880180840572084715628986836919333186427845453795671927239 7972364
6516675920110579956639625985355127635587681402134098290162968734298507924718460568748283313 812591619
6247615690287590107273310329914062386460833873926302391590003557609032477281338887339178 09696
6601469615031754226751125993315529674213336300222964906480934582008181061802100227664580 400278213 36
7585730190113717546727630590443531313190360924809724642792845554991349000518029570708291 90525567818
8991389962513866231938005361143264294610248954072404857123256628889317221164329478161905 5486805494
3441034090680716088028227959686950133643814268252170472870863010137301155236861416908375 67574637239
7631857570381094433905645644685241830281481079983769185121272019350440418046047216269 39445788377 0901
0597469321972055811407877598977207200968938224903032368305186284168948968997813751 793762321511125
2349734305240622105244234353732905655163406669506165892878218707756794176080712973 78133518711793 1650
0331555238224877306534441794534153952024244970341012087407218810938826816751204229 9404948179449472 7
3289477011157413944122845552182842492224065875268917227278060711675404697003793 96187877969488255
5614674384392570115829546661358678671897661297311267200072971153613027503556 16781776544228744211472 9
8816148027052438068176535732755786025058470840132088379328160687690813004924 9147368251703538221961 90
3901499952349538710599735114347829233949918793660869230137559636853237380 67035911442432685615121 0940
4259582639301678017128669293283231057658851714020211196957064799814031505 63304514156441462316376 3809
9044028162569175764891425697141635984393174338420733792378123369304301289 2626375382667795034169334 3236075
0024817574180875038847509493945489620974048544263563716499594992098088429 47903636662975260032438563 5
2945844728944547166209297495496616877414120882130477022816116456044007236 35158114972973921896673738 2
6472047226422212420165601502849713063237958143025160136948255617047809357 90889657134926158161346901 8
0696508955631012121849180584792272069187169631633004485802010286065785859 126997463766171446393415956
9539554203314628026518951167938074573315759846086173702687867602943677780 50024467339133243166988035 4
0732323882818475010516141331189537036488442269027047805240749063492028254 75505400345716018407257453 69
3814553117535421072655783561549987444748042732345788006187314934156604635 29797794550753593047956872 0
9316724536547208381685855606043801977030764246083489876101345709394877002 94617579206195254925577510 9
0385251714885252656710453498134198033906415298763434695420256080277614421 91431899213930883454313176 96
8510184010384447234894886952091943531906506555354617335814045544837884752 52625394966586999205841765
2780125341033896469818642430034146791380619028059607854888010789705516946 21522877309010446746249797 9
9926271209516847795684825833410226647721084336243759374161053673404195473 89641978954253350363018614

0095153476696147625565518738232924685473569358028960115367917873035531593783630822486151777705415775765617593585120166929431111388635821596676188303261041646517148469793854226216871614001223782137797413126897726671299202592201740877007695628437393220108815935628628192856357189338495885060385315817976067947984087836097596014973342057270460352179060564760328556927627349518220323614411258418242624771201203577638889597431823282787131460805353357449429762179678903456816988955351850447832561638007495169908624710001974880920500952194363237871976487033922381154036347544862648459561597551937654101150140670012269274743938885899438597302454148010612359080362745852884935632515853843832424932526660875889083187007091002373771065769850564339288543376583425967506537150053335144899082938877373520514593304962653141514138612443793588507094468804548697535817021290849078734780681436632332281941582734567135644317153796781805819585246484008403290998194378171817730231700398973305049538735611626102399943325978012689343260558471027876490107092344388463401173555686590358524491937018104162620850429925869743581709813389404593447193749387762423240985283276226660494238512970945324558625210360082928664972417491914198896612955807677097959479530601311915901177394310420904907942444886851308684449370590902600612064942574471035354765785924270813041061854621988183009063458818703875585627491158737542106466795134648758677154383801852134828191581246259933516019893559516796893285220582479942103451271587716334522299541883968044883552975336128683722593539007920166694133909116875880398882886921600237325736158820716351627133281051818760210485218067552664867390890090719513805862673512431225691635790227732870541084203784152568328871804698795251307326634027851905941733892035854039567703561132935448258562828761061069822972142096199350933131217118789107876687204454887608941017479864713788246215395593333327556200943958043453791978228059039595992743643913778664940964048777841748336432684024628923406260081908081804390914563519368560630450891422896452199877988493474777291327972660276584016678901364905087411421268619698620441269652829810870454798615595453380212011556469799767857389201862435993267776894540605082818838227909833627167124490026761178498264437703300208184459000971723520433199470824209877151444975101705564302954282181967000920251561584417420593365814813490269311151709387226002645863056135260579256092733226557943543080833726283791830216591836181554217157448465778420134329982594566884558266171979012180849480332448787258183774805522268151011371745368417870280274452442905745182346749195641885512444213377835214238659799259882032870851093383868299065719946149062902574276860388505110326385445404191849588665854504057132362968106914681484786965916686184287567984600418687622980555629630459532279230516167215919686758495236352989357885077460815373214546424987923105116763577494946229525694976603594739624309953433104049942096778838270027144784940690370732491064444151696053256565058677875741747211082743577431519460675798356362914332639781221894628744779811980722564671466405485013100965678631488009030374933887536418316513498254669467331611812336485439764932502617954935720430540218297487125110740401161140589991109306249231281311634054926257135672181862932786138833718028535056503593195274140086951092616754147679266803210923746708721360627833292238641361959412133927803611827632410600474097111104814000362334271451448333464167546635469973149475664342365949349684588455152415075637660508663282742479413606287604126439138285194564026431532258586240431418386695906332450630003922213192647625962691510044576953014440546180378575030368621246227863975274666878012100339298487337501447560032210062235802934377495503203701273846816306102657030087227546296679688089058712767636106622572235222973920644300935243272281008599730951325286306011054979156447918400461804674620892892568091293059296064235702106152464620503248966593987324933967376952023991760898474571843531936646529125848064480196520162838795189499336759241485626136995945307287254532463291529110128763770605570609531377527751867923292134955245133089867969161529073841302167573238637575802008035365575728002754490327953079900799442541108725693188014667935595834676432868876966610097395749967836593397846346959948950610490383647409504695226063858046758073069912290474089879166872117147527644711604401952718169502882973353371485309289370463844208932997711258568408440683993404568902678751600877546126798801546585652206121095349079670736553970257619943137663996060606110640695933082817187642604357342536175694378484849525010826648839515970049059838081210522111091943323951136051446459834210799058082093716464523127704023160072138543723461267260997

870385657091998507595634613248460188409850194287687902268734556500519121546544063829253851276317663
922050938345204300773017029940362615434001322763910912988327863920412300445551684054889809080779174
609243933491264116424009388074635660726233669584276458369826873481588196105857183576746200965052606
929263548291499045768307210893245857073701660717398194485028842603963660746031184786225831056580870
703055675958613417007454029656876347741764310517510367328692455585820823720386017817394051751304379
486882232004437804103170921034261674998000073016094814586374488778522273076330495383944345382770608
760763542098445008306247630253572781032783461766970544287155315340016497076657195985041748199087220
908756860377835919947193433527729472855379257876848323011018593658007172911869676176550537750302930
383070644891281141202550615089641100762382457448865518258100581403453201247547232290875475070785765
732542844459353044992070014538748948226556442223696365544194225441338212225477497535494624827680533
369832841561386923634433585538684711114304982483989918031654586382893537991305352228334301379533729
401625762322808113849949187614414132293376710656349252881452823950620902235787668465011666009738275
660405446941653422239052108314585847035529352219928272760574821266065291385530345549744551470344939
868634294596584310241907859236802245607639367841662705185551787029040735573046206396924533077957822
594971042018804300018388142900813703945050734278701312446686000922778518110409115117293748736278874
907465285565434748886831064110051023020875107768918781525622735251550379532444857787277617001964853
035551676552091193393437628662846198440262952521836785223674751088097815070989784130862458815226609
355140187449583692691779904712072649490573726428600526114035812310760066995185361248627467563758962
299116496066876508261734178484789337290563739007878617925351440621045362506404637288156982323175005
962610809219552111508593029556549675388626129723399146283584760486276270273097392020014322487075823
735491524608560821032888297418390647886992327369136004883743661522351705843770554521081551336126214
911815615301758882573594892507108792621286413924433093837973338678061317952373152667738208580247014
335270092438032669517421195076708843263464427125589077468635821626166042741315170212458586055062331
149316464691394656249747174195835421860774871105733845843368993964591374060338215935224359475162623
188685307822821763983237306180204246560477527943104796189724299533029792497481684052893791044947004
908649918727234541350810019838818646736093925719305119686456018557824502182310658894379865224320506
737996619695547244058592241795300682045179537004347245176289356677050849021310773662575169733552746
302943031203596260953423574397249659211010657817826108745318748031874308235736991951563409571627009
924449297419054898515196586647041482251063353679497152510229341882585117371994499115097583746131010
550506419772153192935487537119163026030328586585284801935092258757755974252765840117213423236480840
021143536367542046375182552524944329657043861387865901965738802868401894087672816714137033661732650
205786539157807030887142615190750014925761129276751930967284539711602136063030905422439663206743235
279788933232440577919927848463333977737655901870574806828678347965624146102899508487399692970750432
753029972872297327934442988646412725348160603779707298299173029296308695801996312413304939350493325
123550710544611825911411164545347103298810478440677801380771314654000993863064812666143308582068113
583831916954555825942689576984142889374346708410794631893253910696395578070602124597489829356461356
788983472419979478564362042094613412387613198865352358312996862269849680408436655606876954501274486
314050547353517468730098063227804689122468214608067276277084024022661554850240089528916571176174390
033758487784291128962324705919187469104200584832614067733751027195653994697162517248312230633919328
079838007484857265161234349332733566644733585564302532088392434828787608861649432893939916639921048
078477770480457284914563033532650700295890626591549850940797276756712979501009822947622896189159144
152003228387877348513097908101912926722710377889805396415636236416915498576840839846886168437540706
121039062501281076637990479008796477869697341704752534421563903872012388063236880370179493089549
007763315230635483742568166533616066419800301882871237674818983302468363714883092592833759022789425
806008728603885916884973069394802051122176635913825152427867009440694235512020156837777885182467002
651708500929623747726813694284350062938814429987905301056217375491826799732177353039368928062100253
962688074980926434580116557158867004435039765053234782873273688408635400027406767838219635222653929
093980736739136408289887220177767471681181958561337215831190546829360832369761134502817578302093484
982925000895682630271263295866292147631422333517930933879513570953463771836840924444220963193312956
203055755173400679737406141621079236334238056468500920371671526425563718538895714164197723874226105
666739699717316816941543509528319355641770566862221521799115135563970714331289365755384464832620120
424338016558626985610224606460693307938478588143674070005997697036490192733288261353293631124036506
986521606389872502672380874033967443978302582969842568967418643361349794752455262912465228424192430

338810358005378702399954217211368655027534136221169314069466951318692810257479598560514500502171591 3
3177516099578655519818861932112821107094422872404424811534060558959583558152320121846058205635926993
0347885113206862662758877144603599665610843072569650056306448918759946659677284717153957361210818084
1547273142661748933134174632662354222072600146012701206934639520564445543291662960602783089068118790
0908152950636267820756143888157813511346953663038784120923469428687308393204323338727754968052103028
2154432472338884521534372725012858974769146080831440412586818154004918777228786980185345453700652665
5649170915429522756709222217474209122067206566229898060328916720687436549482461086973672255474048128 89
2424718543236057534116728507575520571311566979545848873987422281358879858407813135060548290551482785 2
9489112190538319562422871948475940785939804790109419407067176443903273071213588738504999363883820550
1683402777496070278047882048819122206368886368110435695293006521955282615269912716372738884189932871 0
56346468822739828876319864570983630891778648708667618548568047676255267541474285102814580740315299 21
978145577568436811101853174981670164266478840902626828244482580275320945499150451851771654631180490
4567985713257528117913656278158111288816562285876030875974638494352756766121689592614850307853620 45
27450775295063101248034180458405943292607985443562009370809182152392037179067812199228049606973823 87
4331262673030679594396095495718957721791559730058869364684557667609245090608820221223571925453671519
18348725874239194108904441159599327600445065562064611646556654875942473692523369559930303550958176 26
176231849561906494839673002037763874369343999829430209414073618947932692762445186560239559053705128 9
781634554233201149759948962784243274837880327014816769526211894750064051497555986502930048676050 2801
049153788541390942453169171998762894127722112946456829486028149318156024967788794981377721622939 5437
811004448060797672429276249510784153446429150842764520002042769470698041775832200970202916573472515
8290463091035903784297757265172087724474095226716630600546971638794317119687348468873818665675127 929
857501636411314627530499019135646823804329970695770150789337728658035712790913767420805655493624646
412600243796845437773902647251281941632007684873625176406596754069362175887930785591647877272747 3927
2002910342949562447661308200729250734529170764226621047637863169954237455117456522022783240096803
52466766319086101120674585628731741351116229207886513294124481547162818207987716834634132236223411 77
88231027659825109358892359162055108763298087993165172528938001237817434896832151590562493347370208 63
2232100111863739577056747386710217321237522432524162635803437625360680866916357159455152781780392 1774
322823436633772811186390511893075901666650742952758384008544635419317190531363659724905158409106582 2
018147347990223590671381469051160519223012694823161134179944714833040862484426913950236717341242512
386402665725813094396762193965540738652422989789782198637918299709557924747320303239116410445906907
977862315158349593035305923789817515891457650408025109479123421758482841881950138546165680301755035 5
800549448948848713516053755934023457489793596186244233832140603009593710558845705251570426628460035440
2823678768550982678161765520375795655481677896038924498355608791541177749423573400764161093294003899
982199267257086957320668749742248020233075251876502559684207606932299885875798988964607443817881700
815489522651672283440452772191069914157646394852311267947308658031950764551976754628957428881796812 09
002638714525785831527761510908863170424369568056787301523542780479341426649522383370711751126537550 3
94237209878466804913947344653071407962259728713050307725871487557050258257346686666138023514260561161
974055434365486908054448792959702875903582540978268359866446586045694241390729095266249932902973 44
056816068380572662605727088440703471496060064561454070734432782514087474275506722304845357006092214
3900029929816082117170479176145051910081326703752149307405678533111060583529127810073917499491978451
12915913681107394055175208019630530935307402485095537725003670546651623304304250874423242604632115 07
899733692998540704156261041976700202415094892411856092409637604429612002364590706449770627207919019
235964807048923636979860198283087284228564752353162882791324295524814447505521909672046080689545181 7
122049303218537406272474215197403057690436026863607807920047762324295518294735220272443763390277 2139
208776706571624163975178585925442692342835327432885633685078965196207251941655606187037055021846284 5
4342578503800900553745182929584404649198386857934839611512971605816657450967036774958366666931218 87
6367964494361713041603724305065848513174926405585519401800518090847521186822461697614924323831948643
441590855801103707031120150224341607315792952875293683582039700338911211417106852193665897894595031 54
389589011530382714300192958907414994359289408309707707638269391544840370450368618966975811201852319 2318
686599680385838123703291562075788359487809416882055316051281901526475928075749581545642213414593781 6
7056992868299895611982533837157880480478704584175394665497690173220310890070303362911767308448450372
1456696444014695451738574341578101586187838392785526093991305702555755590609470514980934877733200727
9757303824598946680968080822221348485873822999281794090825665209581655472475244566743697594474686376 33

2428904269776106791933910983300422310293728298798903209391092682836306173610173878123679898645149311
7024371282858826304862988844922074156406071470591374055246657569718702173552872454394277148091793644
3765063786186132434863579741125852086345992780368879249835436329845768765016506511534500869572123950
7544785683173631557153527046524235259737513408825461609661440746675514226836031959801072152463551069
1718713357316854856312808578344356236709596509499469688206611851180860342028213318012494109915026014
3545001743273079362511307029825049941799428445114647932915459955590958780762163666859179106543596606
5253525320273650725989121255686842802077246487722010996631829559552903393312284364864475973560859840
7609472983895424339326231532399189818522436218083129543365340748288634656185048106322888050567378445
6200094146560349928087940511531005758712955257196411150685034077371060438037125957559698594936205847
7512026354947347534748189262254190352671614429284899857536740692165271630086060654373736823556588626
4863436891532180955722044567771373683104580755845296128328326063196297285279666743629748008213186279
2186904428434263073576070399969430789508147269730253817375694922751795354326156912040594832286094999
2364412287881226419148504856328072066418557059520375303322916894489427578306090910852410601400683274
2055839697738231507349961087587637042555649640868550719422563449667324306562592504745817627332818160
1701969816654242637876360145303594653845032547667499973734083566513818602515652028363738917101654541
4882674448009105704186162568758491712088614135727961109088292970229692128180978798951391504270936786
4449831964201345668339087759430064424856230121246145116979219396344095080832292812942704365991464827
4998437594211302041829730841717881309037955854560324717081919530277146579455547554475428443440813938
8908609776017857389307518616906505018077165001840744325854024184369205181753930398444374604547815105
6625176412759821181939661101850562880559425660600323121618099462212930100247091334715068226843045 8
6803009042428616820255621409460879000651910994955708158165058298833407394660844575657806366902728434
6201858732825292479650528668410850353851983752363745912562277954902905579070302839501048548359298434
2814487304358047053315081510503001521428117175393649133166172621235405527863308002083177055630294963
5942016543330940941771963262341193871051615701017980535516793708602913667569860971241203685838129576
9530779814136570017476135696986146068491439699573837631695824602513342108072621713601943018087320988
8551415024163818325975259593165531865833117126857941527206612218422661411825154657484878312610347834
5467492583087299854474212064509523324505083749314961665552517971680209917200264093749219075699368963
3028131964720896358177173555848592706524504862516419540550801343510323389813378302497701822275490638
1499964723334079613041469739476372650869273347108415685608430921316240434629863920841660055904598506
4912435052647660676003444416181864036700837741141019320588935559865867007786367189694408962232 13740
3411359719913135946553085446692367652589012108413774324821918127478478922872648929700323718 7345615
7981599834839100412601050746964599430331978810634913923812490503061433407918328004063907098672596197
0983112659601474737253352052685371177421465540058739241280726173649051987133680677239525707813606866832
6139501432950947485159472466752720168431658660880751276858475554118438116901162200555211348448896066
8259227431319007963011587084670117654935393046563356225311244727796669005831190616101972663073970542
5314398184573794494867801346812178759390769996020298396567728784690573640156401504769644899394754 14
7460833991869688927115694234549265124664550779255402810503762203596753055860185649205606287909076945
3339208800894947782889485112215474323019138324556299388102061449026687601020775321091568497783074 0859
6498579671526170100394754945399176987913235465501064073558169994097562481499674432784292027626441897
9391815839456270817330158210022551965989876937616401986120746675504886111085572676450705262244613022
2335852072273626485057289238815884938754535229184380884061757286220950122506515863104258884134
3554319737298562177530720226294755248304444534043488887858117034134534252235143140787797284676018 15
8322079774518092934219318958124828326589500407048552600998937839003419141630446391638805496587865 0
1375046341695655156618298878630705842306967660254053024811471007899784211830489010464056896539702885
5955309255586360521589573751140895649058441567749371058596480143158746144912505492531911646538215851

9737009328019453032057262845265804604633781663142993307664664653076059054896288872418971606022588261
75775399220551315093772006248630855628204935757527244995567089221634233983602565328731029194007041176
9192208500151167356701019589710017970195781208929109694177543699043682025630240548226254019056965077
1058157424072149633956036527028333440730575007367456226058464988611510168961218111905847171446106871
976101745658737379674069713742328753839030317200200207205928488785123911746471673743737923283881966
2016876221913462338937625995270256721386221124589802121305014072889043003225355040958668187241393699
3819306914874471718664618311194260316166407037731648700186479960024304400324224180940227853330901150
9880870678268835317200767522553138008818780431690190072804831799287414125476123089606833095828377667
688287576886830929760010119745338983319525886196301329170943858166153741717944963191771543125069598
534812856846193776698942774591709188025200127499055594072896965947933316722436215678967769670803522
903901848573080627567086765862710476940920356593025352743419865927002227049233186829991560936413757
0049885373045963961527346293969749517480626964517930187199867885375814159757993148066085572325683743
0528276417567005028804048942989958094810353483393414492788592526219241554723199714338508663732092663
272824351493364070458968385234562474436117525967676597222439206357507471552918102762614012992480
4228839902978799254185174991296302839907296355885798905933177959087690739056460256235335672215522594
6883829845288292296627513716242217295467867071584092418408414755758253938524096330205134970474069539
9567897981727860920462286839735779815111868152659884606949758965481314651150392626377749513761557248
1951161198772503445647107385134359273555387124623755981938132142384415819290700463897716838872079163
617414324970791096581627464297170728717251427459073525886970953462682016908535610894489840710058192030
2176945120771774588795519510473384184739980796306767885845167575729904306971542642383498009870869933
670912108394453506245922432312348278549660374657188014892937945147870540607924575900601296221239287
2001721558866634573497140953372115165598575794172441989092616701610161155783431502546032878119842402
748460851072240667677876085524761777383308950261006438835055020545623436167859451941795669874968515
2448838475136181806671083161655642093692705206118985172926171417144346555087063060635510129494003097
5916779915842604919712095432270267843265429657240327208871432199964553132025871096771651258496696255
2698607311763718207498827399770601991362093083230736838206455732563765982912578131492224220427971241
4416299512659456397927593803838047826231604243253991328511230322470375619423217330478540785762440132
91717992979240783390715757981426816864655384296484739920588863165593941986789696284044734496802407709
2831376408103352255242717404107673565424441004483347440101726441052954787296345898640501203608024451
190350994974493973617181575277093780209236668135841636268319263406714182797421342546225054156000509
5967404561684045177174795279035325493258912048338574659009678173041600052108893461076875400424197780
3082885181200173369559127137714195011361304409753279190504891583246399143483531648681548579178632935
1239255525102111827885736960604276931301469661433449624232031143824837056335327938588952676720766889712
744358156320881066501495681435587965769098577659027687074536297636497555344961730807816098710324801
3795136170367763457594975686208013996374551762425147780628722265971455482906769295713643572152674468
9878894188207512922257565091435528288746141950978624275278815715664007637210378031940430958442725492
6998716923433189002214150311399876526068876156674021019720171960239086108297492763956954115303227546
0173870795625993579785302443476716399591462317931239989986928437975702492369551587297683854005227651
495614447105971962889888157109415171701518114736435940051162462021311748007919837497001004713634
3252328157891135545045337190527506822915618500332846956792622620819044247334036250389279207158596003
9363153368842724375366799698647934741133198328619441460653922784099903143840354565047056789552024827
1760118743356346902435030856313095590552503904927316133117349225846446090245350791901844112993216997
70451832853586480428556822208737213161649058630325636891308410376021567992702000532235543980465311933
97754590440450785680213984650096934295473102692499475864660580901669984160084646087293943808274308285
8174796941728729940391101319267557389798409136425347969494348037770334643495847686298259010347072786612
186230019866079877826842459338356389195702068535216032116352300649874460020017041305698536515466875
2023859375183280372851143274811699683692849220447380570633496618711240947835915869626858643589141359
8542535776887749327436345147544886408688180303696524317556883002058607732569597160864854158344684324
8996307701137134467515693024488548207712413355773230694945806726784523594363150787272815790157307003
3178796854442679525719023623274614262868732738009497741122856237663214904653294072026197539071740422
25953924288816455979657003095714138910693684503626823105398674375324005270153347458933256795149418545
3780882706345729596216908538353537038141811557381637820903256151986974535764641212549807600515614170
729804699481359348315056811664279321933532798227147157673401860887215189799669350252700757556099719882

863064285448128275139280694702750148163289727314347348528529504604883271673978981563678804780443602 1
090073207273697493446304997314425715604331336903876181009488731207134827108158898574832658542075100 7
795311832686170803707093592761493678253085834048235100363216637895742620255035011686154340737950451 6
482896755698358935522020173679548075781909502697981271148703431190363112246128295303820512870430929 4
719745946908210256347889954317715243796962112812245034260663992688521330791963702777804488579205730 4
699080092344018663811325209712309647605998994792575985100817303960682221997532730160658262852758257 6
695078547260349382981335825281786706085126560022688717811253597829337347791412736284188656175920832 8
794474109697038798547369840254580632948350223593935435874802239897609162962501104739311694491006669 0
723063469313016971182063253526924404384009372428442820970936485690946892008737175325255703054353928 8
727812301139808093867015474885803445631871319602678548793893316205007675264112044390237583342724298 6
996547863685341028488573702547255023656634186809190383886707879072084036194021646701215348379781518 3
282647257862881520710108149955898033811896156944175676134071704653851217090212377788433364965187211 9
905407581877394397528364143953044245913903178813004188791887114553148267469987055587931040240388884 0
838506873416250716572741851349520849636709555425043948394804597915622828248378793415272036226336956 1
180555637107681488889361927574265993582355943153088793305276755874751236550659884833490561517722031 9
868024351719937686100361102312568364256079597410574153628297180046497485737183786390370390153973749 1
116546854997164539451541767161167171454017651905650525206622778831290457196932059902413753959838619 8
260320549583950167555250964413711822256149601400302035407899209698677507867200038074267970530307167 9
932296015648622808518403352350170608589512912223246117830253163628943946073652771336511631646446199 0
990212249224123153061376785863736315526002503488487813233001910893996167027314149992651194574263 6
676196500243473717272902846220979839487106598227000995491887769618850543265321180221944428222842515 2
556141187434018041946141394514712872527592391255964437356833972896331267678234910356332961294719101 5
157143115795490933903236141191865475237624721531102079391584874220582274734320173558507712243796985 7
796549158062795027409771688611480761631516185530685669245717176922044366843312739893379411162972245 1
699985468562215702417594711769952916550211685500108985761934639455908826270775311465775223884634351 9
376539734984802454976076024403080844890106838786972612370978357245166801171485983679405529046198262 1
656691720274262854823933960018254599409254308169691032978411234022885600190549342750223185294712829 6
609693976813734197704278121300147328677605719405969799275512461718434954967112872481183465420642 3
187145518241528676305675131162677177350617511245463387994265291270105789956718057214365579183506917 7
793070407573290439749499582241062381051491765023850418273009662017175094059080540957283755406335152 2
199658207573513157592361539863949521115586400098097552610538382568992721584785041746065161511337 8
883360976012114848700556016581249247068256844272045472896309420306650445298646223594226008554991589 1
499530604984280345794927570094979594506023787750194706246323949549578230822830668408188025210766390 7
423097372091628533717680621644693543231791785530583317142084798863034084657264269395570026857605753 9
347888587094600582723230519108117514234912687336585960799891732928915896001815091816337400806034752 0
000515117510290122992487096154592802620607616982721810291673155489294237408519674330791660784990557 8
210193571366243599088361385980851615641747694605478554008109535067080308969763041529464682332105328 7
237438944115685176271711636309401479909649945635459295013073900362682100732637008235615069126964318 3
517162543903046989893142615442635951136346605737865495124457475262167895470362890483048499680403772 2
513431937373441236618586944588064018584073147633792940386340435919419872355526301565460805186876068 0
431608451284591604244132698791253856029915996727876619519505317648831346932573668946443825581391084 8
620966374267457983130122343872583124422033094571457541470479293875858238997738515213523723895596643 1
122356432626286011147489086817159281066872708400820337718692153523526926347226809082598988400262081 5
217828261122931311820866007099686036540981832680755824776760950410975861436243552161945353029200 25
466736799648504337133349520820150711992589266389795769858707901856123791578843744690378710590011 2
550210038845311923652965599461900474846620642347942329670060529003709175578188708193522146871427235 2
776325598980869487211138459800141238421638278244127365424467488333816797162011288619141540193671290 9
478990264664431560983729615019686242282506723061667209435465714251493086424887785986827595887490605 0
772602509518295367651811823686169447243607837642947624692263194989219646440683169287661615060508138 4
631941511620257790786307180123115945586036896562526554223346234454507394788690268159497513116885143694 1
521021688319044616862976333252298638518188500492869357276476682385556446365544964006317648285575785866 0
610228551564859908820958689444362546986795238226861159699100563660829267915337538160661122478695313 2
615853187176388598937792918890299879387978100036973078489592706254104848593158543233956831042390299 07

0263443797875691855434089764407601308444819786250794764408301349424358342818859152592934714363175333
7495897010728735012707889804816350456766676932075530518404324461007403216764718360837084750651269307
0766084982529900031785030585368213951273503863824605642510337775580986446433980171862081426630741725 9
2226000531091342681074670129014301654101064933212283790827515001003530015654597508323772965439697382
0477416265710657408216499606262274961879533479070659889748717795643340648417456457479069251701494998
1009535341354890875483632757952240720698629102467170357925144176670388660990698572626058124082533622
5218992000418975745765315123000064445710539170177168863548333305192158205594611735771632113223393196 5
3203861990051161781713340010705766526899197081692022194647043237953564118660639205586090344570641517
9778214505472227885298721019785884607004742002846887379584422894997433562671877991721137916164492 54
1329715652879529532639759538535920950138633380507561369530899547584883024261962758985941513780515805
0257675404017857958524488311721050892770892272734319738238846873071682302487886885855101080735227814
0537140652075810727084816726397709873145516264691142328610303693298433030032367616271426406758780673
1883971515002798163374779078775038307986759404591073921034587404219617034925808189907205961291586420
2028857340091149552388651079113714953346397639881839488045300750747403722809368205354304949519483328
3347007516197900868728543994629815756058916376247230691628711111376760864803237524596694930411753946 13
6464337804671165055504670671836221285795040867615630427626711429999113487698447050370637900181096888
6297217579517324338027806174704963020424929166191718862433555992820932439194457118863215563201616542
4705537593869662465633412154101403228690090301591328858088312412428828763738724283803859071029274863
3351503090445328052597795658920554562434297982794134891756382400771612173324736428540160610044337641
4572207859217155914010378320201321338330963807789040957238105588293927963743816606868351950592770195
1536160172215890428785678482068291944169871819286273080704441630396254713053284388372730384393758826
1221162583602728961624559041896770247453827583966522993712351630489833012421417455788591594256059792
4277218199085562798486056174536844789237969079755945551546468531630244623256740348958454625674 4582
0204245739199425309426422450420268090381501526836024125598075975236481628093048912746151196231546 1140
0822056396780658535407668688227542650381225999162076017089556747444652423445201766165032594566591296 6
7863426213799192229614586714224824928860476803210864779941004100600339069727523736254602774296007347
8803835668752200348245769490845686269605771570191917489226063520812973879744383548328613693956245039
2976805783223402171676555917766840375723484409461762931288492689936871389838822271060279037990019045
5833600797392774109266557392331470259092338906543884223513241153880185592349576139930223919645504503
6935292701156630515335191864186482344249991927202729534595990630487236080415957600296681211168317236
6038110542803591445720248256456105714055462420821343520948108417158289572445072063546816002305120140
8480543587422526171017681853883557558717415424775449772221214192613155252691091755633319323222432185254
2218272914915981058368970250352281300214119248601424806807975369964777193949068046835528083473276103
0604940973309169031678309793463661183278453186871646268073883365670456601042376850580139507443647963
9222841126979451347730049249878649636367940992912327125289775191817542796280608493237552081536111132
4033971316550439188796019838213858500077324246177884918758145964264233788979333081948816004011312652
5635693244693984006368903152547229239941447437706993189247936317800831026114195 48543
6051577871600495578865657970665885510428824663630572077789022667770425126815719795332251076389036819
7628440286102588053923393294746720240885412764923864476021611626208242129916603622991849237822363009
8347811952291382184732634228575912097980547585592591837983368017874112426447460022562414980691400 74
0979721023278539575615128345806165411117926710427990579394497134946328950456512868847841871758020504
5832838748531373691135102550620102775345809439105001021833973245650472889476879298925945019875076712
2363791875864427622114966061151280487096488630562284408393694438721692120849200851558381251071074795
7208093746942459731172811721051928903896370394235776862127668210931827636649984042124938144097959863 1
1422543648396549998347908430702176438555435125743682822815303222238083476795111355701480631820045322
0723794891863572149160264252699399467101533381426847706275852035240992079720869914 53730
1095516415033176282001969164115460268207236692552751418429969920538534330730680573723805041671972 21
1273740507892726634063885068673445856077326664838457802771891147580132310551987813652185190714 6068
1389868867103147598264611293795439526682772599448359025974458786876849646268348443441413591771 4587
7660880778435718393293719373932364083563375766884682111179935055410208556188490102016005056395416 87
4510822060355541081766646052412496622442280454524321603203601946413560979200195902404979292367329892
4553990101980112140290868699920575891777188074146122205024728585715367530747814389730571787268366360
1576136100772286319638852646235125538077319459563567965382364999265518043307963596211067455282 521429

02629498265675533527310046878865731047246649332656792733134512295505918623293739332608607745135077 53
09015744438294873397796053228493583013618379586264803212973684748175164769136621103603695091066650 5
17171150827820093278835872259839404630637631811808904423626219988123682680785795262197216687201745 5
17472627818032683058548803970977047934831035439855907843552776676033139884605271503138856332467688 92
71045985519328951319167823857735772658100479825639355193520055204080028705967824973937478860528356 493
59149783803779649600052124458347790017560424658666519980770288394385163809550430492196032443609034 00
85174660429627430976838715194598264473594023424821104475729111777958773134155360952759570898612586 77
14562523994500759380206093550248920084767332293085742222550206455690239126543663578524272429056053 20
57540308210145123820921746697579765347571725014658374788480805377351504222240429576036137543248619 96
55891939220504699982106293160967565179075132296077785755331026585842576086686764535520927748275567 54
51771699508789411805936305249944967012375980065534998739666395394471150596981015127193331184076792 3
27185395398097640485278467438723164329100290654953086128333026640075801296184992070220022559721569 57
58837616878436434679275586357397225356488413306011928957464280935785808113233143311528748217976603 97
12579528900364071989233281316116404169377366280132597382223742681891764895964227033803905929596496 9
64821331144731667650419678110849096646942571706945700787126401448652242846948897617256746535220506 1
62107300101926248314682120355169950152200731638400413203033324231216708268546893175843663043078435 07
85928104478492663952652398718644173380085681692321347429754583269402161253332837900960648627785494 12
66795136740458774169455961407626566250299006922672678760365871379327960418488393933934692635434154 80
95183623323317522937035210291464133127520371176654872063473892329378510729029514462927415467619479
42747166916030497829288961474587026499797079206387240825023006425544995904011974108535167844409018 80
64629374835443961440035352331030404117845722890295818058103212374382589870274737040106837771592512 6
45357065083009214792583498924751274536220061058545759973693135297078143742841340551954446721489415 05
74528391716037154530825255584342025125424166244575245629445791076971715214709518505500355054390631 6
88258105785074635656204791466768055698434552023887437983561462197088087021064210572937785147135883 73
77388240765280451914271374881105597447183100939375197659802100241012511230813682603384744910877161 32
28576602639388492849598823656572024263572024635720266494491262914191713064628059566982549360326132 0
19252804346170439028926027993140436137026582012131285148815857311178210413103357288871817295262712 0
00814750640268304641898876974787917317370381399918882424169942121527760415859567119094180737347933 10
99709283155468165639527101046113762540664495861838546889628998967783295501114314995936803982223037 13
63295742321735744647342109741497143641994731958840052638726959231836423254918455955045343778467094 70
45095942012021142208641912790493599452137392487110743231495113804293793655436372172634819075711353 12
70930795272952211247953149896990808946657476955651243605611420086639905609900038030250612423607753 02
93413472890501316772809713162683495963409292243031195084878867103533520023712730202916592975252657 03
92104214963495238570856057234346215769569851340683045483315459075364711469968242091023214311717692 27
73853477041779407644100130100487699267072113205231853822744487024332710398781147912754608083611568 77
92151311310450083663631007517511025900280864277150209627136623974010752884454683316182115027892643 07
29763557610551124620332480053105995111505431484829553433985305742724517378865271930007323217362375 8
73273148909109455374027048118555719905168393874535206797085921189640785489504109405699659887159886 33
62077955045219321563361246853031747054439402941829263552401554523160986825531389701880153970459625 01
69179664812501555932311482673005633835797260328601778474149600456972578349563205873287301245145557 634
52302986481495441009078835329801207012654109525184606662017674204525736799469077190845378748206080 290
48251670176619820730618331239219353569004070521549893903446593880904750772416954365185807506649045 94
43188862978723571601302248135220460100963521459820863974927551284769435499620039164489791974379020 95
71888632002475020791023790730729637463263366745942755637845536913673455240148971259094803685662823 21
00500394007310663207525728314711519263328928520696723934717509829526021254947643301953574383509258 28
31113391115390633766173373072363027988986998579945016512369067548837988929400605162826140048150469 48
28140330839164342486509396354589091328059511163345503656348245191505831794980831827341795050772717
33594966337188214919283787116463903566925779942457394355473044935559396848032790208614196815082606 48
10924688543383329866390745478052636291615627980313878282707451630327863907566533621719750632224064 75
94597535966732006038982629300007612514947980089561124525695598275854857690124636865949422427271771
51849641751071598416357207241224371968067203927064789427894217128426413342711831847944133460647243 14
11501550985511712414668243312352062840657226926060904747919644729752832274596981963277872816259540 120

20538073295825004974459308097824095299129654233184987798800771681631986086512088315867256506594414061
84468374963189291374599342160348482288315828973094216147368925585169927155311558888760072170341024445
87440208443428273004673097955556668115013003388895838023146431382900260076322850347583078087889551803
13981020762788985174353478225120846759497430024437895842895680752663203627696299460180834941994912 70
65591308400058626563996391104068510412820071532462564263714563557576945284927112635577196325065896 54
55364821254592633552572925952814993415878776515692231191510233734407169916564763982000896984629843 99
77593853981121332181032819896994579261764935829748373387752352859464035138238230626945363458100319 36
72502069828078204334411752831573143426398964163471270530347756991558003118159180911378802688385475 769
72923398882860323029977043066628869553012102727057633959897689410249968479498168420119925613480756 44
04065594623837087236888125489491487948734808614168105521140018455170084444842948477550732736642827 222
06336582401745498808291301883914015680905000089454653733000327477290917507461785951579953202237285
23592040074251522563861667562031883981176186119602216284743190797025036745928280467817853664739356 00
35403827828184576694782337457113822121932616729501042706940952026502805228985909350023944908745626 20
53452217311940957783019536051850385496140621825306182530516287337062111989390244889753863581809944 918
15784878336528865436542248302027892417049689651104172759475017812267858143917486494243573009091712 64
87716059592097445811462955422310022008512052258976477811482703942677666427827462593951174380719861 87
22265586504030028469146927864680031836034638172640570270742262034297187555809938687124046562233389 14
64658305543013155095285109726300508051882652726853353729373385691826937171677303161186474948104242 15
12791591014606569795333133774095936749326441463702424725453933501300928336485407069840343991212454 24
92755802997988240920664464042585966200888741916498773027540372920421581093781471313622628866669454 74
21244955284909149219337193623402943371255755699886529662364503535192026777637942482082860568936231 52
15231788501452131321491469868548359447046865850109813142058926764161151621094053567807368100897342 458
72932705210853572676380564228840929665884477795279546710735193295474713015079220840328232204428944 67
82183965471109021173407251397247573570085553127432199967512595825680632358808838843662032622661914 14
93474043649800024739833209241183866742960029460701418388178110714282439657796388439864782313715424 9
89472583041145149526872423618996763058816820846327437441210390552765218710735564525713360114558045 58
56845586504328599176765196193271143498665407745145004730727117147957122757201812886446440777517460
32824231733853376532989810442322404677246320475683756101981314200580876897513405948054826877288477 6293
84645496040270370508539419092769937066804551719416040376351180185513657545109524703460226002074142 8
23849481782254906365992084749037583205744677959106755660640775009347129817005818769408027992690469
49872117634151914882518670439557310017937100046657292183728487979715692278888397041982545657064289
08985827958625659901375968750078569853420944399597152366767355991155709006141301885395600693305082 61
15788315979018829128777653969640675392080848528229047556190518043755987513405948054826877288477 6293
74687098568014236137076370467423618029218679592476977765292629290417983927505343294338447653339 85012
28283627985150263745427966717714841975733906572871543054321575235449320534653754238204844850884634 59
08533866772925385204449844413686375189411768486261360368197837635133932540806852269214743073291 34467
62529322640845330845493864715156181394136343503648177947550976339255988278690369632386330342579445 292
29237752032874489020040532668139354752855017464531717214599508145561364692526650227115337381817597 85
57950419880754858113362891549009039080607754157573613737559880187573075362487370012912238261134381 03
92343723135368988915337494937863249849417642814170452840829693991724323286772564150483765773114493 35
21553852300178110827616363037090205259503779092534110470570046565251977925679331410886632640592623 17
88931260315285758716424211903337987257758742901290175936269722343148935725724188379418627686456 57
86869202760143980501638714352047767388090057892836338177973884573441001499664332358222579253571 1059
48560789182401521998285226946509587631492471279520164467647402704689545435103069826179991402234 07285
48915468068420957432075066211544876266446579863446388023258636088691875944227152142965066416138 4963
81502797217307126592057826600278471814003420926569307030904457024594675649018527813931481315092 036
41049845969060225314474822945707025270436340614114355128412226936650125425237207439401827725330 0432
91521517059974545931259468212143510622763303318504339488951276720637291512493681935703191046935 72905
27628876878250048505480059732307532652277925524199131596179115220694196854791873415669978109670 25629
93993208164501771417349056433986521998639055709352119852439067986150214883692843873982018760222 85471
23039494596615725875096503200712476657593813721248011341535506167547203695791055974610671125417 11745
36954301471914199373197227971690211613572625243116472289366644142621243854981362369496357128211 60368
5441607108231775107801298304253814190892249208595364610821395648113205316073707772070605993498150342

```
4064077512331512158999246297497845474385785595227089267102479199199645043040166005621762962340149282
1816115205046438140512010176327979026932712227012592708163045794086959388503088585777767698805771202
7746185837281858599701177211160371098273932414719793766386484316000841579272530611640850151500165203 0
0200142743376390418788622635274702258984849469077694747613276391052599405660382382371636943555470658
1748273071824741827263627240462399440284444736424586444751046902997652674973443569857085390578191599
5859960967506128309101947488656507512613971363429766415834913042083009508511004140745574437849278 97
6072610576974181963369679075518838322017344376439805368296268732851893953081597213840998753657746635
4932531139362559789543000911914267407538592546901579734191837104016999179009456783596285732244714 79
0732045696471978631549086284123332517481278482880984876102210097427834751646279055393851966889569 61
0876062872957459088920170238672074010602453894151954739328142466223126892362650272056402643021776903
1895555206112711463146717038915773390065452869232720808111578757374991035324446693616535175221246886
6080593973805468948675560258870687103081189892202421749529345821953530099156135536073159095673469906
9924874268001953821752462105349862701061321590757260240804300827868356293198384271052198354727511764
2330279958926872730531183558056875276124091974244476335680956874844410454670283523651415276562700804
3630974774537670980820873498038498259924881067029775494995352282995165465985068742832817628520857196 139
3797828505779014996232139220462341524168238038894466242673730018965433764765036341251828509512088864
8562947143987795665592807491648962562185926751464943766872976956204472866005326043254455055024566143
1957804705747148460072968182372287977560496089158178686729363237902415792047283646970210313097518009
7841598550007055364938753212574961674875872583259259576150743391862284379883013460445408808178096 85
4911945411934702689650599198604099765321119658106296655005501161836517062029288087760919846167316442
6864197089230648463056754573872024760165257760852977210933584453871074027292591915246267625381797
8693064215340131633701135735635111098141821129662210736726269615672674830775248874448416766573702400
4850839370255838591012266948358093154547916601645691486305239359779324467255886717141604853503871149
0317607553732194472830582219155807880752453696932744601747360524205864696869757706121867761972058749
1045165142715495423853920232526975123495465463090613294600566507283098728033873735155375223563183570
2537006494092638080317374634854036114660004846876242310894723791650074517970524862846727663375517303
6873683856440370498066179092008317107882104981833155261485053735407503510822393924744563010969204227
8844737169688950911185736926890336659718522537770329622016708106551812675800940852515068477579219138
9321380928696119531220905038018107658748836831788278142527862618796676068219770390932600672961512755
7125278643706989835444409613917379035454851804039733137480525387910955583040048153480453918785403824
3236907304310274062641777762657301034703384021129669084818046162496487394734584412155302581522214994
5822249941941954725641031750211442208652302802213424093193932727678195990608112598623967333945898961
9071679777780259511631477576264028588262514815821643994413506196081175890461951158539082613354960388
0323713522245169681180597512189590028591737572256609522495280407827130270045377437267855325204850397463
5573946460980485658930184822341614986583150346608218622360580194811455490351547426626606129502687840
9754779810472682395693147248760982803450811893834040961534314863011248676465315478758454946522227531
8773560890835043837081120882441759938586466301597048117253004020305813409944755157637054103501416
6861912485252694933482978510811147232987404539612754022221009584405087230662326888849704223456700 01
1949751859796494099148971385362279458874076099043285422812773058183040249451087063369869468674008948
1097539710090849476830410711529550638887652495994260773886347394552511448972036104793757254472
3966023547748127494160698351013147640236419491461059805563757044651556671236525682870157445284762 02
0781753972337164096986264920557668761564457744644694924577346729725557053882859078923175970676863982
4966294555601938731527103627201242931201764252246448031819544683337639946131383614457041608883422253
7155878358070161156027177541424723331527813566940098980044458238998420064074895892389238927522891473
2945531240424775520838052379510123938435858775254202172068286659998570098429393846007329623842629
0797218233372747669464015269204881430422739438838699801236503400880952451272600136152570415774 9789
5464274592866962164154275190720789657656762040870872910259298877128340580613171820687950962735523 08
0228036658853093027046194006144644918627856642449420816201203832761116962244213863973117511301189185
3169915158165025834281284874149275360507355014927516496556894986881445782807241540090116176936589862
8113745927903225784890933976881608670857002995345721579420980997220532145751427154112209398869874562
8011653320792545519698519103842815726835120109236795242906867995456830838859301306672185211353641724
4228370492060364815444971779988618739061970126506684370640425124459951909006226082179845415139874086
1561892465930844027470147101672547160166860173976919976620111199893015535406281778132823867987398831
```

854809365141752690405027399232695322939310360456984250594710877602232101677467927935625307683377220
692980995213327549341076406829369625653809798299221502007619065671332333071917531109537696743144582 7
047452191856565617305618532160042594645538561688375993453276738278878122231537281113417355451707355 3
208276044077452544230785453748112596654635574596043270368542157386962244479609259367500830989140006 8
538363588177874864271068825787874079928341825197714084223048949791551797676782746847540849289938647 63
498391753924459329312913808073876500505220066666272734384454049896801183432553499976250119217678755 8
098067233241678261782570891163017980881955837910754011805096216010930804225701805492976467841153876 9
143070882475312172313794037236592877104345544469626659999262339332986411371001268040811602769694022 87
13650729810644525201655173386046865040621292457892714722742676386142682367640856411947662651437101 3
938556806427007782969596804860775179492212156291738671635464988985383575153249743153835413991322136 505
155138410903090275543323644120225300770428211147141918147570961833137822943420725434103155582818669 3
283866836607269163836967793201021420290468133704915343805924654711497083540122724100650394974216 48
669227447368995062528945027771898946913296346758587926423521163354647468642605485613157784036114314 9
026954427505648038478887943295655604844339184060202704514682782423151405070221048519592072312004933
717673835237093085652643448419467734538241329688543063024778255435028195957175433268735831728279337
741010263471725280005510899808792042744778385364274972065430922479605721400339366661597939815697061366
098396405520287669991722547240206396060964299454270591546000735367315498807739083001581335160357301 1
111410928015412280666670587855092703338500983115676285161649242550929283039087709889349460723490286
585602054220670371568046350038260527637108239865979318483093676416536079070660523343411137793121612
020588095146143773947683538839504721294528349865480864837885019467676945623267019987133184554534837 3
608451276718005678754235887195105895652297804537834484650646814695167753813695184510308323903749657 16
214330796386015448161449552393511121218944302382695405786011646737366479565206587250815927530571313 4
383569920048999618043254950205219555020617927799305642458366587216753519281750334499239183325623616 2
650208149035578612440518344040381599135827178433734045297449996405991865666415356124243080016261793
375092142956088283221957057843171697946284551330968382460003698996180592987950660376071243272559753
650882038636095880940038001760475078669744332587722315438325998399864395011449541507700972822653695
839438085091284110416290966370127424988176163441016674234005063816674823271038894223948202530869672
229252434075060265129885763587813750085100568868743282747187323242898477335425815041625895502385448 9
068496767648928297072811584351167607761726048913558510981478950842984983605593659371053202059790443
697353401662876453206371886938218978015732190762998103612568387648387269853601294481607317618658 0068
059683733894119826500873262426696002409088320762261178399915744021058427898450630360149933928362455
402768350998972042185962090201621015651922358421194882020912378392755718560554165620545534719697866 1
235058348962821286082084034973119988107725904545863376610850500958238503075128425964285974947159675 42
592403495586097964340196646672175723723707078518464663837067170299541698329886912472818768027381254 9
629389876072234084657095098943200419639467934671293946851343732630392230933179068730316994180074048 000
687251365978575985994780199496523427286889887178135161711550577839158713864040578956591823213708140 05
871380883652304716712718220060186088112572603398624035420675212769089210815522603293004441018906372 3
659195711953030428284858647825648830052518126081035421351812247158400462751059244487058370954035318
975215236103420408450761376742347300588220343231604746330435062814232108294872409025947644118910322
337404979474085782776220482618219514282179811243726766258468951951069986737402273230026026150597064 2
152746023269994970061582359288282229783863136816002884117306733244962838403243536504 1
397536205509105219749095799860595726941384024267555967486377429308583140664803184453153290815321549 4
345828804429373556800522667016801800094788733588609136494945838526892791365598428817418645559410296 1792
999581260809706454746509024326184034501081240335390006107346941209783867162772161370836145151105007 72
011704214057510295511491370254533502068141165244769178458694354034118791350719472868333896624761011
830170049726189561183989816053909200891172772452827329958680838001073781314001876067250126926454645 09
767337470023676782013523567324262478880482343629009996330109765730571074508621321877968280743439896 4
835524271448757305830321802494521092319912041786298321106456189823450495054397161803095685126538014
922516948784795547241863827862758232782129939783074286755471092498218244686147958081408355004687559
626157906171759021927186972378454724112985575731793747953518295584299136928140588480421571538074685
311302335494627214184400563239744587537727518071466016570650353750000780005476100367863699111323985 8
621322182246246434350103632239859670172899284252341131543432629303907359534291441393387428218721484 1
861312790716268582668472059546640356511332792729283670421533337815648978787243472316577108118905881 1

592205341344776752129774635506551109801811454708921701244106349239492424226738349439407865465836386 8
59700260199154168385586155789670127220023220031686195419702892475742166676680152408824022111156190 98
290952882934227840649039533967200864995696544707521184613434097785777364263165869169876274954188683
133247514531590023354409517149140813592731911461920067579215856331076125470709339611644150880072729
394563684925327185891551688147209601141540566440982102811864854595041190056901297283947161996760030 1
877000729916613487810389918979927793308260333383340579193386012599266354350647100912606346252385743
463526847492979065780017287665968256219468541077987421844550471048251138993654279944593202443898985 1
344256726693278613295048517020426704168104239887877662882835019312545495101087037669638120603127 99
621889318777830520450194812047427052045732125487339039302866808539289855145395183070167372533915679
276903907336248590343351476117870517797664710107502450768161655725395482009480911058631732989175311 8
416036402192503463573219594755860083208292675123788495516672506492270609741203129313574353745218554
549830258041565179862278016468937481723971338112369536375581105739391053691797392934319775188032524 3
525860808275537409997210154008084059927943422345447689707580313149065499764572719969962803326920908
915583817603213989264488023769100827420906680043739925405441223684971940977467046731673788785204941
656447370713254372831395409623181337647384889412182775687605827547211534840641119286609198061422822 9
552490758852587114072134140163523811099124778913757446828093424728231102189843007024439996429064
445084478802766865394636578359786330143574307385522480118057855163003059480351702305291761937668044 89
745519006229814174022546879385980914228583744941429466840567844786299687303736686339751013910079845 5
88319718939840423685517831262556099075164256660914485766068367974480652972403709933396229284343832 6
610413687134472596294417153661683256929874607519349004367548712450125173882289594264322061718377059 5
166566490388962341590342836592467623892154316210947396500986925708950750411415781971894579948516829 2
39976768526050909408476925556032094730179889261822947383468868848787742147478211246290050487616242097
572295178607339598869641860539956912742611053799648648272882147298654473972705114310364153995043024 9
248903898719047380481217370572566371346514715413122205631956995297107448454232578540931960703748062 4
32887305740374143132382158335526671427568755755136182019176330108628379725855115674172305047190608736
162770832629644295804827975636308237643616154555406169800458196446706678102433478598806924847727489
529826204516943700371201912953531129197138017595577974532179706899810786979967116140647258355731385
280378144794618645821634745203985589751231713640779468385145592041450052177212291446692786476520100
363397889970941956779542290004143845487143488528565176308029925167644424768218649062151291172342568
685160060585978089662668582012839650321203074654818211999482253881430040168114450362116720244462048
282967776160165637897576349795548725510809105781339420347277448474876989841921828085630416492602991 7
623036263225044182962965215438562876070374218681400473863094501591091325421030325613511075755828734 7
865626080932546504744633723342240855858163385371536094587826920202523950672274753690019380114964316 59
458297164868632204841795219642449832794880631346462010891393287053134556150378876921145927268505164 7
713559958906322386507647782826901680360130617085698288633653339821664116613355480403703821003445838 0
815055830340179712082249309503856609585571395374637843224042175193426566863925591177433783255482070
386105633012623762876981734728224250946153189070215082050421810397748940765724149908324785285459510 02
467959739308411062725225451569649389236827358143460772759803342643125982788894418184917380268704496 0
388670718647708315647875891178035430820131865658203435437342292834745576965149868391503976141261 3360
789480997559164824906255168553679482474050984649608568188917203699873757964398001165295270277237226 0
193575557202326310147468692847626362851893048492690926409854724936481814128316893832831257956621359 88
355445206674089584092314862575591105196220005030802042573700289966012413635548088280339995694656095 8
85763219926030004685397559802876555831710706399750666047614867776356322611612715224267109673618402 52
910825524461538857766602277960808983082837068778139849238125451717898757790676916513246031087551814 796
00121676201685543613887535111144646445965948986286850084293816775979619127299904591343960428362278 2
145743849108066267372039815968331145831327755737193964762139470369487134483796533672088650760949443 1
06748938628101668600893548372019446262183642683679016223424421625009619198652825018478075009309298 616
878935144048527844852101949729314912293366428383610958359117926697321050328658637196191306498573320 8
66152431989177517561330725336906062894401403624673579168612419076797307215389609926091477800392182 90
966056780515742453948127051582786560867781398492381254317189875779097510910374324314804905099720134 0
093871209967992266732745697219975739749835295566344453243455703262602782931368938896296769149005111 7
916415739641516223459624143879984997239721062591045242665562829601459670012861764153524786433047855 8
149625711139560325150363183745061942587907329747990654033781293234534964770959941597021691810368147 3

3833330641513877132215173398409381746568333237521245212042635149480179537306485748255881296241114146
4692661774781738601561556967768080635428081339262222680573586043957391627387714350848477018662653169
7488864738682430941960189287589120213872770961538488009506565320734420589849785682144810993443271437 9
4129234072975479326476182962040361443641127465240436917542838566140595943326100913231448641642049 76
4947955201717108651706981224160848217072171016494824798077491801666631807604571639525183860958271832
7208657052982558926649231274050673123487720349779982956094106360305165816819038448011147030423901 8204
5758372731652085922539947510938900121122194266654459086779269137115495078966576676546096288277775 19
9570554507297923666208523507816894340032040754374040076217990918813510949936969431342798599215 8062927
0421382675621435340592467202350206425854109685955128295988801679474853488276232260898821426027966 949
4883399735380911531026157275260615166467574723112673113045630201064427562827821914879246698975320 978
3265292168258433047908547833654269758433077955719520001012078724019881349498443843676382704117421003
6951169011180168326999466120100860532094157901928897613978403516511159934642044414827682054550634 184
8306161979946027048964895243897025843417177319031533093214798054202089619512507592936490162781474 077
3224772573220191350456805599978569277543549654578798428594684085867841341145382412407206567559826 48262
5761903033834174251848538540384703710069087650808535086402176210101567282914356736771103511643978 363
4404283023478073545669143817704745089458721178783915416653092472697951952686392823300371685067876 207
8775481783910819732182904787993291396078874176883186531819994065979276782213227134596324714095294 63
0761973967499846349363609758067253661551807859814534953582160148026023317625201506366399391351428 775
1153532124112251505706572311520853765028432210158406189825794391718649072412089171456120249 17300
4379934999420658663798578734606048061922811946433156292568671087969712349623640619373881121802073 791
5981801097590801132722578430025011137880349579204391892288300516242921760033764107933719681331920 675
8299182607848524757117752420168349348194140053916463352182737104891500365804792597615834365135534 94
3843191509214629308199501835916709425302654032980324967615843963471143532471437039221486178438282 611
3866885521598461344505803302636914394174355991753787166688140045296893435198765272300845846550156 565
9852113011048528816939415867063517831922185595530500029864832544477477199550165082658896713964088
9880567958066916065806094048513928010222769761561382608319076033245484652866146492483966773300 8070
7320067510426251414296244714536875097068785066005939402651877610327654702806325729906196897591887 38
6672305110123794932925976495748262551959273944176400925561218577244308889458931304579957527258 670
7145565142360341819890315195457218862114917103453059657845082618680743649773583175770086475879 96432
7445489500780966711961621513676950853089233612386662834811029398046074355342724428104903280756 7670
0337727112094912844844874508135688221560335043880351754108148303753443420841220816836058132623 4576775
4279316198604543050444851055580041167943376712305581470587272088253604731064967931847963735278 844788
5205873182866006563349325602359088989835377250797020050541440510294610720764924401363372278799 46
6397512341178836631250900614162322765702854104805079744981271814676430841410300237525653730495 276727
5484545999787163325331050619024021518146810014651262851039759839412882369862113183152477649679 577744
1913323947985528716530231998698023983984731981788171333103443398908379580000151396534523383390 109097
0444714347942650262857403151815203546507282311838519865802936213522437975431938019834329143125 027577
6675431686988860286567701350037258969644586868341764738783906654442181923585773107870023191744 542871
4160030268283724049464334037876903573326188114310108132188552798589730345053440330372276915140 453182
3618783217199889890550829089662419765855980578341428737306480985290782145941126494992196511361 26577
3076994605802064072391808669002017569564175955272113593375897911604759823155872535564456825714 3746585
6689820373705497045290715846973763557060928012017698053293579678380795022792200105200167689887 325
4108389319250907174288810810708623207551018480041769696826290392399839381162366384787130819320 185559
2678658980709850229537394942175424696253547043954732413924764852103761177731123133850016187130 471064
7783932487585306361991196778775326807139246898446826593605108546523692226192724034991210383162 2629
7241144083556868049980746048371359252075390170144693739164092864863919053757393294556536775435 632948
9530854791973561811689434694346430308714442549106098294828815811595635629933779473922097851104 067
2166448032053106709130375948843457873439847370765374740479308090343894339705830532695856299847 93830
4808179750890193297881964474728134854864856399736790769039302521285919509594533031379751852981 8662
6201176126095321392633918278125632758305911893721069157764388567228422852090091226125140805231 50810
2726247737066716153729796236517171183091817115228052653759337375581282348642969322667847133869 5988769
1580950811504993633735690590084289200705482525461768954164710778011758607143286624044830552364 259377
5798552448696080726730590765024885140814761891799989629290795406069165098627507033091008866119 931836

534781106895005532321232310409943156697571284321105892729075626652983068346126881743502763445734873 1
30812787853966825948045024540899453850626222815657206656259080710600907194741580643428961731315157 06
05581139989607656842772395481206246549279224664410867393017052678406522475041053604323508688152543 82
18840578152295198789560649956069827453289227327038537584520927092429466734689593377789658067695128 59
0449057399130794876253979899894685344867084276328476440980465348855120943606428893738371053515595879
50751036819995860092479045220515488077774998306131379026412827371575710612817362497836474502072277 56
19521267432735816854961196988825831126166950522240218814669306257495384708699586574599887892786847 3
8719864383790480463746222816126871276345113094783166175997075950853325746028493740010436450345565804
49442950345318338129078508883338583786975771084982066510206279570766983344517793452718037691141020757
47743154293290326295321149788262035159874125464228843952777954992895647543471058985851590055084900 56
9690369399463805412744078272079588120610950182666750528291004286440115969091560260245872117456045510
940768469797368274814597904045521904814801154566347833534380881534140372398178819077576306472338364
80766171878875254440731865830501186475632030171398339007898754244110262777492594557872631561608748702
50480620381626062841567542997110084572360794368388317756971160717747601977362998608470922561241903 34
43038680607516077836502789166628360931767596955301493681297935466652393898654922082126132776378982 0
29467995816243987059362391705117507050493924429371228752072100479003695203530541747026881003131427 53
11744562464073545200130335154416115482453063682206203182141204371057333057060956104199977441294378
972333619529368071162946497417467460616194428419577150642124491154067073122138420641412696715456643 8
91594717779694935195834684336783221413037431073429174373437635441590737738077683355454522041607558 32
4500141271289974101494705488647243158991296092298246037064821369314963102100005958713471920979723983
68312801101752643186861118351701735867540649265791513740582162972420188375102977202769228078017323 53
65852486610373552463605197417587182349087381977451996041351604688060827255906104828222575767463581 9
16629034390705475970348000403043333174134614234541274567798725892324090915108730592024279001497 63
7015134772151425714802387818972890993192321188518040030497628938731198868763397705690319074145176 29
750558295070905515128977726034354672225187519472277750347806298883507640883058588732114089935625654
452632562629304285439933282550329502893699077705490294707962200839029322144411265738208956854344785 2
25355843731269337547937659943069910056990821560314508198864943894886795977365202377638052694955871 4
54270658517474445964682325269410568519337370049144862376065979574374294931376284954237476962984236 04
04069903223286254828223354201652282912884434215751270602021538317845218564841150669394364364463390 32
94628692150012003317372231594599370244046654644010709546377862736676904564259977586034142337627592 58
5363126437089730757955269968503132069091830679132654203064003148245598623926575975731775912862530894
654125166228407163370149790267384325301619010137297886469540342569455726305220387629423264806499623 8
163085500312651680544788556819973108967957554426839220485130919026882403377120177863986046398002560 3
7206069295346015363530133009351166490475996904153484428406496443578396273959796970711999596987055007 3
98026714315391239146116135818340680876053466725530504223979280965662210911184778965033519003128193 0
810470647874036715555211403407030398907223233915942351265297171121449159128746969645445570922804347
3384101385887428050725149320183676549863546496871530309795693902421343747525920284449370703291284
0950852874392941278159586475430366953536546450438122955385696018708146303600068102231935356775884221
706627177875228959374973944984606881958992605790426632428181883297682570878308901643540546417536779
75214014916981613499304491104247417299073183796985131245595860636391996596689961080050494007296397 09
89595175746349501131523954054363842477157637057968997809351123101270006068315601347056168842081621 0
59065843854685335226530994080955068645181964310450069852864036972272644964072209107280506565175903
363319425718826190168520911094446230493872762260130096650980181502161161893149917554486648451019396 4
0892424535185862966853588072370252086290396375135442408416767961062540774535439718720220389829258815
04881746263214401932459126384677564387821490032187360528840161581467693409742434249466959679745512 91
5247541300383824175967755422515486890349846758461066315941988121179713345250927530701408561426350301
5271473708797902296635683001787990288084193893922491884489117670080380387588878016977011134533491153
48021065850875700255173632568820097554925748712253571842551697875315095569068985464837948430351870 64
92313573402963136527912761526230610431409239565352297493261018023574144940020107575292488958929324 58
03518893483362326626211070472227951789643113533222151331112813026996570565423666607124273606758337678
35191012510994437030462907634661496449596967303212585228400681288632060138439115353293209115790604734
13936297323492759180894233656526093948313364810290643586311830825965878597847150234490787476879956 6
74248520510401039997571039402206306917347420208969917560052888987367593966293674017209821954183712 8

23393286247731719643862566531455126992223677667770819864349679984152604519464045901639578960492791393
29104349027568381726840470522981408906713149152625044175452710112357868012989368284933913963833660781
14229179455434491680970664913188453781202579621552321288398882948309602541545158301556456231328450311
7418576979799101789556567990682529165537586122338380700692195796394198374261176767691005073575014710
4127391778359634794411592416074096491892386416231443150984337999995741238608456879065017966046590401
11900931064914597645508744169170936159780546717465899930170413753904682544419849306977396303361433004
0322637044138842456485319600091024035914860434195671889198585615655465775089014431777128616456862190
01284594607421607429571045831400462012463901102101932368874623187463906390518460908247466122225868316
7169890636064025404893508750600193538323584773313528261107299033893705322254670042058896404652393614281
30791854041696667832407095595877094232009415610955534334439138543884086108248624289859619741256571711
0424006787671236859353722710915670406062194347260204099413954720131755244915834594427491919291350235
58344041872069743488605383370185896572062254666838991474061713844091114054244489241812528058737844371
43759903370271443232078520464193147559475831429194169719062976979044988213080192587590485857570102804
98900927646674318174193127387987919086670564601741410451836547363921120183127264213529907507531674118
60411390850791740441726580092889664003508561829937247213684341314956295704019813001606087541279574669
64190317973325939240210741676702423535171422118285715161832976814222607309029636948713308807785556663
23972833422527065507307251890309039502297514545692850413644104921750622706436286101757117
11204836658149705824635780075504562645374462805259328415678857985069010580452797562628572208304783544
3668131330317233238135264707525779523330152891663952865431899957317457801678267281460226240318995669
3799484242109824897420088233111400134104400951609308313090546550315955515473977480221462406761105271671
13757998628354396665783552456700793604975187679585004347859444483487473455259996323925882010445288787
95767233391108520813799484234715318952612818751089051213546549692460665576734518571774051139809005071
49322800709405692069256544287992876091385328823923127446269670190717997985249090304609585020328131081
91897987538612770509831127696867211781006094288603341607408020448532441414445794547210546989291664998
194159975081170839975855253125347930707237719482373833676065541850211333735753571160498408863069626431
98901511556298277922304333498449367839151985626850432520084479855462969126299788313029363064633704501
33155263752004047392241570728577799880296353289006984007678189697563540217661942944247572564984572261
55067873599034057238979378191460803198271139248049794110049229814317594991993103280897957472533768141
60617454331326448924803701346264266926317342443574270517747756506755634133360059178313763737592038902
0426517169186542244841360946598636267754333221729097280677212181229450177642327167331091927523377241
24505908085892765564344114584438889532132702856054060064352403401177439426383149269420366767732492134
44604615279222871174737320682073795040775380702487513973697748722080794183627245792671586058756843781
38256966015288450159363605799764879554666897963172764144582671409839302605844437311832195273578242991
23805304609752532175592343974646967178599565547975260104811160900192559877031603692890535464219693617
90098547207855125725975325768780341842823941930116764712780200132449054170687194060874864250992490041
24377799025672362387521344875468680180570026771659045717416750935853746632784661471702229127753816489335814037542041316317966204626016830058358402780842508632146763149384031449261455965653995124411152722520988557631807880862837444395373363866289139439904591869356297993169132074310849123878269670817379838027600732835237130570383538812019217807455705392131250482197669340639428023453335096980545219146149696642051922322293325224980990680943829868423783713594952723688235110647593837053136149452332629060061354420207090080078443591851423810954062463592880074917472326014282915066473564694149075313040411948746127584251725422993376561913228344136159688451230509792290809347780610001202430793367546069567184

86784758902916716281510479109848199687950537574761034383929829818347559135337283428884922853939659
16459422968490216357698460365666770688497906149378958662389785139503019552520711594791624303805713
10441235127977174258949971813208993972409457630504381765420237749372929236666408582635630470188942
71366217962807794758141064720396869005733588378323839385156436769291095321263095302372341887763775
13255857198868415635114346544492134618362582001773911196356597362091748028951071191193121616150493
61400198915406771914740604502008489007852104489840715587249131814241237453147390958592854919261955
75281540455554894860530439518305516386529635114358554267895788433224703032239846296940370036386670
75518962282166849472551167994010237260527619296017504560496637107762638459530053344438789944328543
01464142075150267652998714841483859242046843515052858926483541295999641906383622255006162029851790
79995517161428927433221628069635121629029650503454559800242920380661131224998757487778145433349578
65580083004587905455655237596430899478522941565846898062943112725975546930218879127310353002168642
33661031890511086335963986070974737419552934173502780136533786877679115147383325251300237191023587
67980393855297049983218308998533373533151034580443460273042275868260972398342231501764080332731762
63197576598977297183942297165227761967340857344413747591477931793338924359940581396032281345924785
58756550575159428611613176739552834815080851852071547951439266728751057441386769718902087776119592
93592908638396962005768650996293038181454912807341809720403203633667664994431993626414522714507350
70590760938575244000094748229713623772159263083602213915885590946140740697630129708656969069672642
61836355204727903533175309360779778798470802390218595874487896074523749052837474189310268062804817
33813001982212770609218470078152523271459867234378543104978390436493058607645357560259838281625409
96399831811297188143863954265440828008619305729179956888798818257244092308607770862635131360946309
77409970284726682966685429084540520229190030343224718982049938248060751638727945668940849736665162
81336948828583395313505021705361118175021010169609373728821746838926180687188721171220320673364983
81254572692763105648258790106857520808063392875136481758375810969596993384841991062692241136561171
29547967778123862393493414008547053784583782851512327987308840937357211004586036545449453018681073
37611086797828280343661333397805386148635943716350087711931195598480218289266777976940913930849426
39716108751625981364642795191163033917961291900176095322165573491103747120945790041860289922155117
15683036249471608650518632279742956136519830351971442760223750716059990046852270284824202570099299
27914781801540664169179259696502306167496772478419474142009242977297138883251166552227303389644473
27049024147727564715409237680664165282261122166055190351049532169538299957017411465102998489825164
05699093945986820498552583128764456838984342109365890949529671420560890861769828831638828382
52297076581896118954572982581073394540172774978345406877641107780044074292966398079580266893129468
89150061861841892185301461433221694927821339211827719021675202080596726274946300280538877794596218
30743842989256463024089063366760647389704968736267734149346437826952783769070841846568389278933623
16036869658885994407170901438576500856237035707472881230042776474740377946320005543727473658472402
18390250818502039941310950397081248122107761282492399569407303784228294941790417991353653370609295
35804128449019567717436326558733430284401481499075465103281813878210907214338837454150957218087233
63931411848854048824761331564993540303113131971438856668338021766683608295032360405951367759271551
67968029585973380361344069307813757301161300265797863431943620466230186872587963057556366
78289969536349814522388660073014771879198641606614390805777255192448707082910976735549911200612317
61847813176543955729503854529236536694133485621787892615470456150452308853118438978335074806994480
08158304780229139502742186786919801765554681814454430741910229272193184644079031725993977136
80997117617146890001377240700923484996330832030429732196758278940008466523507115511183481032014483
47448851886324133039606763958576623927274335864765533252926113916010589720694912160419438436276922
04083601834811827158553439255537462362552372625331435969964366678255933821009175987444402718
12906425157453594796130527189599488847824353172256275413109599850442774753526388871189264977071705
22010568232530711543895647583031225511628763968835143262728629815240755887995962098043946596889329
90449010574191419981498587900005334896120691631117546825348582900768395376626414520527393607865138

7224150689718558277653895215135856589764073014881111337386924588890982276022973395124413507510038725
8948206647854453929055116049254655683017922363526377554626840904947800372684716102649508826206935748
4631643968978962700833763037430175619457389078848104243092855236310298355174517454466065297670819947
4320599516529159600871565211546129673541395732775167844348486459833913758485625055460175079220988358
7719338601394896751483376380213903415290428362645376374580878281537904708544575967614210377236123296
1964022920228951469688114470582893098035700150410369484170547286922751970446946972920349519697662176
6414362032109874971729908144000371573928189152840831974229437336774778258709218717225298406930555693
5225836786253877699037108329877979505169169629957607026624877256111532102452287179703093738005633540
5929310890187400495751330564575544684588192533443722717048411076045834505257242917684423561001394566 8
2456604288004740729261916575440953365000544583331333486658192749276001242323822517184689306940857
2324615542392813887422027661693797935663441045037115598236757714312491279012811450164528880440942390
0517346456378036293953277878016360441042759186420251677118235910812494784489548807258551257454960799
5691801297131720540384249096087363621351310975286209862242624187424357837677291264179188013767520032
0563326189241018616513034810246079248508060110481129427409009375274857858586840922973157793668860861
9964429073615749053303451047462192139213935734146937826784053313199235443303460719515520570060110 11693
0106114645564891680500914500556137849404543693134859106184193301895454863852198600808200402863322268
5779286594749900693607510578023474201505331773730843649498916728935090935421209752074727250436661880
0613349590930611407110464599244759717542401996540730654084169735239250415635500390264400292275155191
0862941367236000269412071278113087636531615224433516226691901231017136295013316980770097670312259233
4099542352764898442089109243902756481617004072173927256602429583715571067168541418713003571310230544
7259377064542064373028381478419102186882184665671383261282637806975309886082906633039669846839624674
7688511450649131518615524629479248111598731109079771152980588092092858162277066325765777195393 11203573
1943345934347337295189941572147227619036780043087959047999266424281962016300988837148845043980122446
2455602660408616133199723284978761593171268140040450561896686909847082394137085518132619963768897021
2415231378187331300156011219956570354141065353563845243965564267272174345053170897086203476547586741
2814640719792280574469540684927959945904582090187317658661625178733129693572624876301827480530656002
9462418101514313186902408749384112421582150837307412833731003222668395769073506988176827548481773049
9539131031846532783838656626717476001806278850492540088784039279856464964515527892734502001 54100313
1905096271080962889523768982872651391408194511671959166631818853373127982279478096384186330812324934
3682732708847168484082306510680498401989966141848682192923712432629328442483023610179839100426 99040
7744799190837610211112396067250719299793136517706731645047986232255197970299256653151015966045966901
5088706888298252728640598951425047655646438613971390930202571945855852527139271981132775888695544 6292
0605202687675221367966827468775875428760778634491382699563480082544144131825372049480142126543 29829
6784668605437790613389102060765389746783799090419822642829171356980043472466696993015751149537152043
7403191810795486894324062290945862326245220966757449528566016465787368842465402656045769732900129583
7874201711510057426597492533286825866257024583781152182201277576078722787536254417678516818491979949
4976491000369349095508194502055381122496479676649650785802327123689446228669796319715390249901009 91
1767205329659201024327415836646285103519400541457190714863882469446903822456883885007824410392501635
9715374849056094524560531254037917603310165347750019986049580085536614207169974117331055254042055 2110
7731858104589461046273558070947146683528783522244239519510959648019339972822544123729119753352333 97
8820050032094830778066283364606324667100180087066628897715761318039445308517785997967916175623642457
9913187479952951873675602067243360786278316446550471333425577456220522970587706520846148146180327 9556
5723112891379150610787823672417063157427908602758268048328204825305959448653553053335736089436683787
7887790883577331658156656404633363117896557755386745135965474379288244327761776652997753788443212262
6758789612663833068438490058005776137309040428453733141597876165553722630161642353345100237463536 8298
9424782425580648076643361805237741563140378933712699900811546084081424058692844640874238912457751936
6466994637359158441193177950085840652805204513861789723299109646117709762971698805474148640403588 83
9279500405680966882682526783325875358351600505794585314848377702967618326360649136605647118508049163
5911181680573568625676757483627962595423144408426869444178084654590010983008324701273276732518629652
8101198756674251237185471917419644610996381436922538882116040799710488004488801559100647 76
9829183364432563837983478922499242473474429558557293151861103453413733567227462578276718755 1285229
6157193501863251721759999422779441251276949166596411764533113076783943587557015112683397880778230893
2767292196739065650167909884959899971836201837724669791646815888400401508326413390170244028639070088

3106649068349767628800880971315772643341647052515364717730661392722405632571001397299899095593747730
5596363485600615984961253518310745042828059910113561527646137187323074054864438709510376239129317441
3926799644747323618213633118585804069936583777606558414953328326602877854696894300229268531019343019
8737058717358218098006693891250766257084746595062899184683469499119620505628810062352434005024075121
2565976218356834552257668404916525157075841461441328952097009306872902271637056385906105921696945735
1312296992925835675318834452109537570173563261816644245918307191732592805373518481830987229456262172
5404441898640397503845136061110062107180886892905388556538032123197766450079788089229139071971832155
3376607146881588861466593708021811848640949124415780158696473723909595858031173549396393423239812188
3858322269062273043691547964773290362031023158462282118660828589608169094090006189644213461734468252
1433863060864107649130309630386061561226947756727056616419832266128295594054185267009938944181452669
9815121965396719051384313536550321313806824245173488947592503124192484175255740381823511390261635537
0936864688471015259866820062966604332671588470284672528273675136369158934985721514957696957393793129
3333878685870155864384721219088131194713370873382327500056239923744771721034792168999587001504698058
9562365188542682939856667127230583317473946798938791798447572639669972565150933504944962393298941183
8095115220273859361991620893155937352131938012702984818829682456924664015891024522408334073529472376
7660187190835662573439468354704836224454619937129219945521607705226537983475106667694632555115664949
1168070523052817308690882682380129412541814673058459343581273433407463471098101697337845113700036146
6614777973756676762211878253942360370654923722566475192700250824888860406223409811545113334223901773
6841135991533723731846766340507156896681938103585480799073996134538888268575672435045918997400691044
7041116287865267920106161324719998448237152334997836375230143135132826955395290108649420581860439615
9053058259754001573475299749827230953870577210005396418869704874528973591568792707994416581049442687
9390222782002617388424638959221139263874954114195943300270846714237068128137782298487438922580196067
3229557664622256072650083204373463689206974257310148877783281459700550621125297094395513482069706780
9204578900900556359931930300746710425701791847467995201645098538151015396961735455277804306267757948
7710979913625936622349370648370598168419144090086928384175813696077026265263739842179275186558553400
1802494738842479507635936962516581598005490110797072695324488613743499388440836614685920902013874629
2972909384539568930915524703254564948483425584353927500268398089195124385714727889228818004727979105
9416493716641756765094433746540972890144063281301891438633928063344349424002602288104716999725533939
5707641070678950590524163290221291761570720281337963064985988232242671029826426454822793271548045876
4992124132126818276723090347559579303115848248948301417317193431046635619982934226608452419772743000
8947575190644364250704011573813177948095995339269155798840054782895365239563674165729648804634630573
6737117215809990209894453733255406649244555657047977207914581230450618880669347731611549213528598081
1109640356420103206503138783298144430856387206579389407056232795868744608528406980628390128319940403
1753698172910119302742164874460186196321594468453807557098722129647584261058043710144144891074881337
6672138354541424787136666538718207128484761707002802307713986200152328528467498051716009417700848306
0781630740674129158570458579809143541609290613494597096889257105679100556752900747504379946338211192
1199900912215396556317263329873593583866650018970210376810565539125811274256503658921429101919356774
0079666127138230714081882841864932545670050478902357998346296652053903452672297367971122296475763842
7953370703079415632893117466348996286910518604722726888778758797953654811330971852577488362549950780
8962383116823946505116854708626136402178204452762262185094687714584666765889994793710284570278582886
4945578192102470884098054884049428920275863251351203276836916550933375756877423110361610668383215808
0256433346454271722024956218060593586057783683982546182236449833541991908181754923962168710528049514
2246375891201136159799843803455389874368637941643003051303788958312792484549868399065860064078993352
8127851940984016719729727069932213390718420955178247520680268463616539771651234574340304432466147817
7119961085537282430917112635195011915381033226170096078197922946035526018787669236212486362488512903
5442839737923251389555064014239130766546753811452440247068376528064142487208913451379638599944935160
8677107460143274772385102847494666363346194172301607736297628897725128300258084687726530151682029250
8730013462199231565387199041060550741930363390184442397874423384998260696760570205353684564642727270
3489439236648459002459794947394860416671133571702812092268052781568833531326433175902994653857485218
4710972047718248056721561923131996627637828206706279778643822558087274035538875576372258299905067359
1541471494743726498397870576633115053342116121745340896541521554977462478886291183035260403687328220
2507089353084352345808150719569588924126052875718396496305507662860091116726175300728173888458812373
5985372692992626426660021729769040932291664578008028615731050138340599605215180202337467493294109576

913999676638521753746488507214642276836486091831973323639215924903900069678881210112974635837340525
868785445702221462087368587279664145301762633541558879405907321222539467073782654675608107464960418
433957953872113306464679928612294857139338563297616178508911558276611979023379998663577047496379682
399350957954508205505111189303477935702443035283044307024105904612246808113753997074287434351207241
798271000829331913714192887714098986370546271136142170603160388771587341075626603462603469320575746
632653061205961474109678643663281284892462172769906044003546813727017184326110762869070629628767824
833725218167849520870187388883526681180068856155382102917938468812597592238717575687377663652172791
935988891248129048499965476445965554595153192301986734214396269052453374634498603717927205427968168
929555879457553413146588128331024557479280502086866557780153414390677074688444371997229473142
096243098464505318539652190602671100605662171450565239616762915821410039307333892918625670333714472
170924079448220819579349698115524925573254080088316481951994849251885979718179165071886497535369431
576366026242617229242548005605957217481535934092538283243334477794234508946594682954801561640088402
355037323496549878662171076680106251027447234054777387228237063244223465713099833553636179045129664
535920772793879392700954660146105091802732697555135713654909405170986914333834373538622395662531670
081322123467368781442761854788305850058107815556788076939732421220873066182620090830504150607987867
207800863874831471046796221804394757556309908624424438280907075163603921360973967193494081982005189
084634184185137758694213859700251923572103523597814756562837006498935806194277478737367365686044012
425353945460837473646222496082982724724024018753734151044388427140893032903966317059852727435757224919
804396406893690833070460690336340376113566927200801720601652587016620924656531832178359034831846684
633623177354463039337934892379583823380148352466207076888417756468257271713619148355289440361157962
682534709995778541481648466735735611338031920658221354967829629458379489925909065715085858992403687
724709560225206030410594547223573430761992020038703424402223409496718095119479811812317662161328126
574188879267178040238578005598529232561568824676516359048340588004483845823024199841762420397502821
420332378136469561291816090880705226927447850235794371561428549610330999701394772146061745007882475
170067917818813373073553878679601012421924341739873289763228098676229374534372899811725930082223246
437598540018372660873836247120725544913064336449951001947825445255425611985444689633861923341088611
023663625200616717734072684448767087078633992885187857488689069559520575608065535597236255486657680
997300269614499791386394913764334395117818656169724575011955527139876663310241993649615936734233367
593518995108210508545586590240454243950149586570975168801729800819922259725289161528326432871330191
072026225055993021055200593642720672067436580819591983894686241507538027516566228260425584487236963
231527370496473601247293747058235189463777287608586271395235999069223258703599191070927536077178731275
481509403512701381707948704002794636433688427716924012826404447538300216806055597399111532756743042
091689664936534610664903033926454798262450752752970355117029389549392605026116732805080636191135041
503872255354805249503072592208321299167699393875896052191904022653296893201528053855848832676736575
685837994286855431488484598784043199371078484089337419779080033863696659632700004800753410733130286958
286013592876613508856941307268952706221194446570901350002850781700817329693606994478080116508997746
838327533544622311789004142445613562629619067137782184982961262050387138637461107547138243334326066
111911249960431197487300355784675385580931940536564143872408715930702800223362024034209266924841036
413924603252815139106025806900867924694784641513774253049081133374859256590325210843787058369018030
933853297001096960008724054481148589259856963459890827237127625540048792707640107020870006758524
443257752645790361826803605262387899668756268684575871134948261702787264207740532779178396605932046
053761252878362242163181476420476433345658692429156194617414729303267274533198796275905105825564390
641279605996051629410558357700353656324285671397243309359986178485440971818172554477791409320959184
501649984386128073788718816754788756505663196319767304705864894624045942697664532851091987443373512
156644888145132509782997998568283018302927181265875997491525942142606638449347618236694361301000788
347450454438394059463825316417469621679637540394915211600835534045873007034167447688538635372417591
919125730296287576998669830602845055125425577813049196573537081097538898051449828195851720963288792
975966878585576268728363857714282335234668548589195448742419522854402758101327257258848404
654951851622252721485896972726328951526610074191959717832883636543759765705772628478559544883724075
162883631484906477914553487265755850111942264869962439109009594842195050118285457021898741034571838
817904863646490829677731508367769973355150741700812202580538852451953639834531876177812318292338454
149201896787202174802452815919012422558651698747259021550762497491226737659456330761660211943484032
397991440702488173434330292725710928657389894240649561810909797654985118424771133900728809290163018

6941142126117034372249667476682598881833777489153001358002314602426047205527579931989940964316114429
8528316114869897317486423082626493416316845278016222869452176524878906995560991510960158794169103884
5956359668529361259912457292837693574949960000637454051029331252353719421156501331547537362809071241
5773185118592763310014484781157730515727741163632176299555971064374278716404082983071046305419155993
7148153916125547811264374389039745212073357576777750742115050829810085737523835183835753993320975989
1578248805041207059046440732276884830874353451226454706269409754445136975725708915057302324257356727
2108516847067390111372101828058046032247916600738326914541593198292432543374648604963396534224817293
8325475111403759384378055810026790235933847989586548607874194141688407304341734969042406917428699629
1133881573943227710156196247763551902402171274686278247219799676266290091917695564381053856926401859
1476166954319407769348965559060313034157909445197560296624875454879091117532699370937126380672225675
1463060740233445983148205708770855253816934364805356807945204538688872214580520228137165269820116250 6
1657297379748075002972339219097501220494749417006593929672960287387671952225506308643850360228641843
7662400917147190328339083995367474686131010932754508537010324881645635754895586036799189361129787619 0
8356737312274823781827018531064263096134724871436054931890377026133291202074118570203965492336859086
3272900347872265989418631895233975752644826232284678147416783941758787792841341122301988055483719
1966208599312263979830338865167585446227913405384476083942355534490331633000124757996527616852631945 3
2930959627313146726192661819832194564480004891272404216573041363833042226648978515562646553219711447
2973260582132148615280100972768015104029896520206386068600459816142852374991208520934729230797773503
0153390597764783428907474877815173481573626888588142554838517329528360113779153290113357598117475908 08
2955904750657658456860498951986993050662506017097078329876063046816009521097297787513601863205458955
7818005958757171911723235091578713751539125515052606959476059331576350909179773320833613680719455 15
6407495330357188422563693171183439732516057365037747321453500563595638262474936382247058368475214260
7279199552510745513042443383364054939700333713488002997859465754494276518303412119702102998733619 9
1177647930047649326491521069549849847488748616219526654749176518304212202594729578364241569518389042618 62
9319498502398088279018227708670870719353901835763828047217627026149024918463402523612956451260017 97
5449613122087284837388331938574020189082817678502985052621563352750833939410145554721256376368546407
90946476538165508050017967373743140990894748416914306508118210399009419171429055442874348691778082 8
4127163283349333738760189805192382363730219765007029919984095395364081929393544844337867252570777295
9596138710071579214721837075804150054413498600492907499698937903488102082792506930057423601746771263
9825044479879477551383688753888207757212063531958650030089106544714907549279711152718460901553945 73
3251863898128582468776959800827417655549925565263787184749887063291494390803074172603675896802548718
373999996196829326612241217067133971378809252017026230147178208006391621359820529738550555820959403 3
3264708915619555223566380626412574791346382537492599128801431261443620117181006104722585841850028634
1562115688441856620282727660065536243416531861727054704601829523329536489605733307453064730077394581
7405562218010296586954554296232136268085193584500258735730586665195617446371811134477563610293164229
2999771284847399247914997524697574616765240133398871189935029199507259403541776788784275863173386202
1231433122324549995210226464191705902063721563648704410426983363333138216958848319812093693683591904
1149316234787276366275921545684107024174052979256942981914981740639527144786905118423433719559261231
9193757906211785809093205884794836305795612156010565182075216489529364750499783642596878804760925999
7018653611113136048448104343137267321149325176406779591282704180928409930214809574578663493779227121
3755371254949836461321910547901950805481637782317553180540483447456823482925821306383035954647975
5313386037131657640783340885935945737671967408625251809781817880369860116613883471299915377112383415
4528740489956469028260300688542763451896235735546181582222140719678668431026826557381151949371316182
3492530436547791887277303957729166760359890682984979275326445793026500419821581783367975832458201
6703344016375194361307936066877060596155074581873007405885541857077712937635395461112355201737452675
3650277123601026263711140850249375451997238811849720048529540760535775504863349851760339343036589529 6
0860734480553122955335688214567118057604758841942058349633845421653770202262887320328142627192419113
4698071535062364060488010610176139643065506664667145977479275127501313346596764639606994405703118605
6087812280632896781657653727506296728357626397482838464729501891798560482419850079916092323996671 9
6476783301263846508088042883111098502554661296861855650350012361068529743566446561984920921101266375 8
3119546240112661948930083828438659999992833337948765982135588393330975965394351687477025420380520337
3382317893878282543047736859273772357478886566858736092568691056377446851131559478673651648492178603

892046920573921373965976293426617993875988610557138473901586953800144003377394259635248692636896 0908
70539526251096127290887376798222410747678488299026259214172065135443271991645998333033382050970236 70
37918912977711390002179646455681701380894182584629598768936359243937980370126043700019545375320947 58
56566862618691377693236555385533736640181142604011712634532053725124468808392504553860642547662809 345
06039108615119488314246477393594536113462632539790305310615515404757043183580698889116850882783575 40
62647008133489427756419881104615033910819669743603856073267087150877665885891060896072087471582697
016905626687199268158483351741024109506049761332210304683200931629481966641786641089259633540386292
41301524764041519532761824707235278977276957745431491457204054179953185788374981085050571576711058 1
58521670552011002403121471715798464585433248907341098761099295649676156544718806244284933719422274 7
40449837985964755841334894107426083336152120775019298011529465672084211550763881645988966176464369 76
03289243245105302982505122426807037312128083935122022554084151322047999805751376864933655676167849 9
47133492695263773907848371824828315567929819728778686290361305600858009907222402153876614713644801 96
56148112453886271654233428875619797920485573019299750041758818622035508435269374224183477542357560 5
34672554956154189883817856029267690850613136595288039173555602456879717723106032174576049759503229 56
29419637906309379558104490958762135916778658295393030657653092307043986757062576067142706385260554 75
95952532130478006326107107608083216210014579464097769268006913907193727253192285262742895738950413 768
54774592960335922726252666668531270703189496282724565281854246063037280407744779985412946353979 9
92464746913355243372318304533548980808093152518135768485272858917328591746450365612006882947050320 4
71690415376867800192930506366957785508855054236989012229808779126706105235629735806022201829431580 73
55219093758657747362647369928881279782933393499869773523241375993155463631192982070653727478607258
99731206930627210401572394384260875603932638706392902219030858098772201985593853726881479322882922
36982590464309339788165229985971114388791916811255637498313161109319061156325528926120586515985149 39
76127055624087676714060590627593678972863204653894075319271591295117018443755758535236978206034603 08
11140856162220429042890528709348719387536681994211967871603447511656321704404160535134139017313668 94
63887385553138636824336997598597061645706267041304591264372849891483568904556009934810115809231807 30
18459984087990904615749310986143133159197840606356831884195057075962103268508407539511046071367743 15
06318655681175045684291098593609486346869593672275807730607288379881424681003426858744195332034222 5
92225911315687185512988438399771818481775752765826872247867997560955981443326979802324269251748008 48
04373540267368844464825094568371986966198330889858783525793232810047849800001659240729031460028150 56
47241103452031576527657717145051080460305129759639033690487822708390133104005385149373537497295161 34
89722639790211988963444866201881902957692950434647230578452652005806799064539004955427487393033311 15
13343423239392815392857552418925427533689936707673603270769534071539778317693299858002902473809122 2
70247003014973214830993493324188082111825695862329465185756368975416357468959866026651728710637317 82
11544073283080495822937176862803685645159152570329027569036857129883127811874743596074173100978847 31
56283864948619310435016618122663037695937267645885383809430494530230302680142109755025038907214842 46
00933987543991538384213775459724640986873792660279416620470866328438766273660878227150035989827765 170
74454770653839619602884310285238409133872378563679536825788370583048947266348134821317190888339633 67
24123153639729520379956140542026523557331822605360301516107672701613667753472021089952406019019073 10
71671157213153131399108734604994855887930557732907486675692499177914777762752572153315305919154375 76
40208556243114944537254595680970256475764244230904740701449387200931485566126738641899425494931363 1
04759618933034909499307284324090098660429647764160636212894769517265674169221041267919762026291755 85
30596160588359815094381398881554647395390022108597871859240596478027678893293422804077323241680115 0880
99429075130067286149727378504160015538097278691011653816376029956001998756771052874341796486349487 59
02284345081024845232428506194564649288833802467453143600766539393253169069347153411102590915595098 0
99960777108192404340081740090499522416993657870841551263350446837423540829126465380354941695384687 1
91594786448216907197188279045374175897865653963536341749642113833239127266085382956774626422043748 61
37508696560381441154467817463182415780125489762580240567221816519025525646655104178403139931552734 97
01282746047879672743103957500116764350123292187217369395615725612096294658612581792259971229360156
04832529324660590007467538289113588769660502304327546441577272041355354310692302099040958828028424 9
25456609225504736663353597670114754779389512216395039174883700606929082143131056511403216591419 71
60545033152608756244430397512016270447566549744508291084491427532863512578843201433719161950742434 585
42671276811026007996977327310910874040713888398593020568547705681283700324106099488089120372337515 69
16771294476770105736285175269226738673324904110576188363433437399317405736193536907770580699187001 10

387550682586512339634192984733096678757320329048370056903353621683728691586822484931645864130995561 2
807613543158394797965036457984422529399803252134609728622695362672407628997177963276334614112070415
414830530440196754581623598606346657273057403342467568253998855700384203956509771995410026837628297
511970715692877805882319026171014758008973737834649921004305707615859532250733610872957027150743 1229
79203137211031512057869458182420174183205651517533812845799817329613000972285911308282090905331 4760
11967818503836753030474000578748360997590909129630344182765505119842942611742125017453108837615 2772
103209162330883357102085772162595099252986436418206894396569085647751243182903018353395109114675 1371
853424630585177074434321613691304544562072955779149889048547850294942518692301564204823672996782 08
543277081713993729713647285516236910280943949049809571114798735326336108615449092136210719578627 1826
589846454595870090692492488205234351128687386269125293356095565624401533344756716240947811826571 15595
59791791785828197578481237247802296273427132738070171213159345402254416861464162061854955622017 5802
7171741932960403072428557591403748752441255836486847826530579021129301504600930097911328939110209 2842
22126288743972398792999872217126802442695704364082691751239472885809766317352190347740207831082 5008
23068674816599291621420437855969070083963431749157040070491113309702304687661585748313508014447 59928
520207278604062469098624581837105663182549206666339286894164223168139785374174558983550239814134 7627
568661622118636756113454018536236120145050641466476620025479372737016911509105700588058385528775 1553568
346135550888143137449856363777369433473077922369202328195126019883348531930841391296921034511566 4615
581718451609186530489711953801102485257498931586472339992674537252191487877997888075626737506387 2378
054697643526861306774761161564008898107229900613620291385538640836842458354434201742490652694313 1926
363064557919103281746224652305086811453922379034699935761819228384117831112734266093171716054723 0274
85870001047866059835368762042349093563146793544370070867604441608093430388964169122938462935021 66110
0210761640546614532826133025098992955319192756290946278978983603594306705082646985065096507142428 7984
810793149777110353425543816635050218200475598457194707642967827158772684836236111806592445159528 2915
230181808971672271763496522837506807313174144453350933010558621571973367591051672048856745415728 16321
725939792701826776592787907269759865244447986276693949146101776057760367110750866034557555712
962345406637758448773140658050218144414570121613889442942543012726143996039751548809684175388778 7099
771053156896057795536335967007806998565011955361699581910918533374036619990661867745865365937828 95158
619216835838537205517181966990029062252442971964776076579212083499798148310842533800664605654626 2844
1059597587010538378376695134144117115765801529197239328318237419072418270556211429248125950086219 348
2545185655397012584064774590941610778984486679878798360359430670508264698506509650714242879848727 227
810793149777110353425543816635050218200475598457194707642967827158772684836236111806592445159528 2915
230181808971672271763496522837506807313174144453350933010558621571973367591051672048856745415728 16321
725939792701826776592787907269759865244447986276693949146101776057760367110750866034557555712
962345406637758448773140658050218144414570121613889442942543012726143996039751548809684175388778 7099
771053156896057795536335967007806998565011955361699581910918533374036619990661867745865365937828 95158
619216835838537205517181966990029062252442971964776076579212083499798148310842533800664605654626 2844
1059597587010538378376695134144117115765801529197239328318237419072418270556211429248125950086219 348
2545185655397012584064774590941610778984486679878798360359430670508264698506509650714242879848727 227
330336475959713294583569058759697058365984023752645595142841527430934760028480597374451154823040 0857
745381944142354918783809229229783184414022384436112322168850562433541858843251154472064328496208 45632
811941082705883189354288454365054845356330448266856935636428960227669230848663361182991429872638 798
806829980861239497632951046359133826912525187946694508941539649332734549729944898362994739917547 44416
471971731779872683943602401052166101498152655416254038545177952158400249587987974104952480047535 5816
454411607796967437476734422118358157376370489681657618646648473995745736638952849565618957447866 597
778195075225888298702478900964065318520474237695238933550012184785996600740896503838595147018042 345
776878385607580956164533921688489754259830599175376101323206354325344240488600003090822619003730 6341
848686143876373649417887401204826095051275986339050977024247252980175882639229387079367325221116 705
792644140908543740148530459025037169637477458607191405425694381561170144378884418883091592292719 2035
841298716228668505324603894356500203073416708375186459566326697226922304866336118299142987263879 8
080682223272341214084695966733251615657580665902431013032064115375116874077567874060359258788617 1973
634936771114265430484708113330323186633985509494314397480484078764776783277053488015967141016984 4356
69780845487805182319957564073978831770271153463924204452033300760976436796999004095585495562013 135848
058753749472569340330909172832394183692139249151868723547739392127561179466401851180013807501027 7721
713064204253265553611432390788203509453770750843488923010206936485172849761293833257931632804024 0236
622477073584885055861960214818950756889614649864710858464453732949655233372641883832621271178272 4069
322657157078641755728961453382916448918652049552729526330028104982310985733943081602256698171115 0564
218030749436110781361389682204877365185667020919787109427227650347063385085500842117094040508256 9924
575628282627813751332708052945523221608454057654378549071799081270684360655039537497522864067146 1534564901
126938742671140362151382047758749428565722785336586487290869174951010235874976607230169518573650 90
579491818691542049514818950633136723233600179192443795940164167719835945106934272172934837133152 7082

5228587814764495406616826606632817385906468170848098019563095401910023033837721074832278213901168208
2582389277936139561206216213391578640790409627777430623945887116813593241244337109448308742299489657
2704969668919097678729567856837491826622807594707308763909429179184646728989350381665716032383413004
8221490735573101147560439107642307049971417179272249889362511853771844565361124353668033415834710999
9781275045931072949201640040438736891084890000220658968949509883554543303440863469068362642692622526
0480503822296566585644546381725787202422393060316745016053977551655424603074325691453841406677000933
4817262533785783695496880181971420758304790250454493294344080654706966709208196687180957451822379033
3116866601065885464616222513680755807281783990499382032540352222147912787357337924050581704793436111
6046575203509649920300943063385151557010396543615066450209175408368025107569627240540070613073911483
9978215497526962006777174612537517747408077042146949807246566921031380365590139144631933785249560765
1289588470395683600524056037732266484889767598647222236870457260025131465330278949073668317542852793
0436416844913090148229779444145397767000504764545539441997442534009022064970795065778667625625790467
8795171932282160484279042228145745555525850110505111853205128248170449340850065111058596796611348054
3157990100271163704146255884514695315016137653098634679351398306442172125391421048488401806995555893
3864698447097220729204410017446457448578985219133254971330254820980219920946867055130885041123159
8940306060776407088621530225283963061061498449297470451281206439250952683933163016535406892928056518
7157265787411940217478091727995418741181137373534823204924028544437285424144786673531720397284099921
0753385213734849527547637515508038238203451410449033687861055113974555644534413352805893131495072
4154536504253686358765114645577638528618422250037354433860841945720257808362467051613544121936052124
9265478557979011265815919933225542147336102522035640035827903578835431594674179374264974
0947948944779573166096230217323972884026061215508990745102462967183685916037890598163574392667278295
0299181795702806863651012454451544131814296541845245197887305202002880204338955209521262425068207362
5164648296888315050959701000226437213538785826025335789428499024259849382698655591574552277230447
8367004512926203259072844700707182646394299397105796504924027215130090020163225789293646620690791141
8909170955485858170999693984582418886230434638646853709469201908664425001423704907060547944016363622
4484204946141454073340772061365373799471743464186961441635564294715919709591245729889392338150010041
2294395852881242903163818939118293640475674801320054837776422413083227337901680551345611878652637873
9084602983248449677676526714460909842724092219442087290507772474227128491998627528849054536124426 0
8122367302636241666463676956582340509347865011435452230172110431829674611812712477267475584183473918
2964689242439083589830410778612221646674139274580844109344670914076889081154804269904644766179037069
1318643164487293481162475314270947951218371189584301606136867423308652026806838261961480478445664749
4832329837112783484945756818482357381296729860250944563100213870768049043011088410435606595632913551
3636595379057745086346584183793785502138550730606062032361892026534379655424091388667805176466023556
8680102444381998217408186830806326579344501366069588311632765901963710912216830217994317817811 59756
2569334811817590163704539548800254386919502939484969333878802324540268683115920771476269046081472974
2564135237707132655865672926093521313563269738633451392323794912727416044071653328372766636069920782
8988515818900740681788356003383955024919154384029528975768041647987388754419071010073882
2600250529371571205988217997519052515481351289265070350312953887973951968071463129797393988552240677
1074781329661125142444094254620586560563864841176973765093222300581373898885989302233630809521934 26
5228150675306773116834992003074978449533173923562877249889011049829135380994323467387064792939183 82
9847365091741599344224180136090702185376839482371972551488138816352825082378087561773037185933102376
9015518148956680264510669556763562703316375504282184693552607931286771716308172910912216830217994317817811
0995237587821689870722832415540437859493648816597106019417011177530819779600610206107580954184382263
7717441589308934442548077635898598386460044819130632918212125220072806340890562731361562825142597 29
1169096962116740824716314518917473600695966991423080878338768659015986702232142869157014142480704 5
8972191054200479042072618389456591675766243374816523343101319777787506264814478962379685449183339325
4452263282389839955214350864723998824618234678333412034969693465231029709800703127298113002987487 58
8451556284431013156099089461587840584003836145836793994311551940672336880332618381
3019065159316862019183963648218697041164945876942211365769814951731860439447681922394006701455 12
79282540565303246423524190837891152091652075345011477513376176131603034635001583043241198303450 45973
1115480235291472675565285396154982517322187028118914755821925109751881447499627018320123866466654709
6270322119673520668256883487375964507251207969145168739639987295089292861505745093918352489864171151
5633710772070437194298978525854106512202087219851152011968200668515495090775699216193168057612255084

107995644735723621151384426059118785236111157667462461676058949088473218825118818916537294130184 7563
650836229040968772707590630759517373446538123581672056998615449337441355115808285999779725070005 42569
584482904215703296329695418372061125327781850782435323918726737975390106042189821333568001491762 2276
358973974915103361029448548755412659458830826273087297415813599878050897081564293241595652057224 3886
015842078104750426281129044255263505482966134319834755788519322226718693036456672710264959940051 1663
086637317274044545694973748748521103317754936462538061133447431080683263084662203937077310524427 9995
1374501935266142352255141868055104005021438767785929901108592518674991313145000872583711669369824976
994084161606242840630833289799716187050576519624049243165999551589664975475039001147398903189687 8326
455784745372518045223597268776687624285075381661679248800082340903203480714652289022230806149657 4270
447722125022661692371423562609291226018250583731811971039075175338577137807762131772452879479158 31148
432273147350683717788157985202303528005999986776669370082267088042043304271761036044360211957405 318
323977508253762435335992587448066952313140950826729742008271959187161696013540654578147571012432 9470
340498901172403145627070070858913555130659474830510926753310504767668510068727953244323689649387 243
491401886858021766970655158850256174152070315092726514587358530217166907411189566762941681340578 42067
338866529843358282099209279600025605373161195748651729717114043583683023331026924475563496301826 7857
351110563974947335708175806329870766803421309664361526144710364355406583290195474 1126
3216179414368623878244681088510060879820657196947315316827265582925484100600262887084707264146369 81
4546760230690648480001950891529208834752002948330118357071474860460032318036646630113783461481020 8801
040824162464398628580272535254001411787225784949244012153580882363111576793883443994167425526718127
068704857905001700188276611540259896645362826952840612570000031201513414621462743588188113752159623
550909618693482530381968085084967571308026522100175445215043882446963539135452229483822752193978 1610
063081571394734757164331002885723517461749219228601261284795060436152564421703635540658329019547 41126
321617941436862378224468108851006087982065719694731531688277265583929548410600262887084707264146 36981
019310463020408836581232828339328287752748597870536473295156141142985324610343025553130194964301 1670
379286563766956985479637443740469514404752486274673802558967408496302725388581738320957777270442 2659
676450234624195887257359338615526808120477513640278605971489936812371201186212349054817129245481 543
023804103650148753567454311180060450042613078726821587544267302962084048226136949742620817609999 350
033446197688418790304159595153926411196546477482084960353618894576122048571862646143232749719180 0858
417216502492556122848670444079452809182539144469876181366331943960646378224508161381778729282783 9764
859110463455622717222178176922974115386786214605724201588982175494554749486363176722743647089021 546
200732501302570572121626662522005303961315167883101300858016798771386008087441449608596103041041 1974
853698311136710708247974741971708082430169166617707713127633313638154531589137525416839840847864 317
750667503948846636772146792112185361223631672188803806160698593702379096318692240259119146345846149
741712192550199254747960048460063345981846406801159374470373166319535189087920564810728118772402 0397
440246021297391101349926966489897822336465536512949732934154340689469433738182663778605034749343 3270
290837561801105493469017933942873990566379697634781069552896198764618985072208634587457757335586 4468
723357249179047654807751039237363961854667533349597089174705010313969438090236340457990307072485 2963
285143088878668807424981635856363393141947625230661525205658963070371420915744678667376833515582 2444
226371755529054939532882366689651533263314935839281282245849325405555941071950713799703563742340 0973
130986462139379530870947165361256508033157850445730000941413946001474525441403816920993604115965 838
005063036825456630806282500948802003418002145584175546348018765356776441151647710438436690085370 6116
905032530314683543713358180929240076805095818888803319229966040696511923553334427159951307690820 852
662967740310259473022591776820132591077731585784477312075886450933987756187266253938362357576251 5880
562030923121386657807216261161812700376053446226349498386252566652249232344365139697208237825995 7626
108099849375422735675122410923244793072428280291762353753383670876387351815527482112244800245912 4640
511151114996644626198433900579254635394962288892436232521864025248104905595540836502868935748905 4320
009125338674343134073422651959981448876264483185527327749412287856130622582187812001162857352133 8086
043652520123507908301505963245468281892247598913287169435985142267573258150924982124899051846590 7278
237639649232119042056438491725564318734416229620060447190161161278608069159705072338317990240010 6211
647477584390237574678931316957011822646217903641268587186863582493271746562706728075136743 1
597507565774758376406338044944820668352178332133327896776383657446762017288395723672110981540162132
700681687402313661948332501044648564640364125317413333237960756729373305212297457933552566168558920
043759625134203063843294306097158474095380197411549350010282165055959259459194853348227327155448 8735

2136534472942394955964530478805317945586293418901077793490276022180849918514125716531651374508750314
0146677425197647620461669311332604538789645165729084386151944311401615142307022471639399010043790686
4103416236790741850646376825660389550334773489673113343136294285431488760312473133541967098000845264
2740142097631369587622585910093111299737936001355335292074829853672042761269847640066766986610534552
0728721873818067910581629074870107673696521668734487874382771997327186492554248066842383302741069609
1855007115354892417444079433704231825456068386702420523393305803173064778859332292996554662168705712
8180663158107596988037954190286710515896821839986172264565237212159212726998561668843085968396028717
1538526694147931732893545844953150218593008668911797136649492410539530174013607858891547134085003970
8036453811115720861295639470964557427082387312687498873097059005337318346168969341709300000861680278
0058956741522844366300229652650701385626568435888629758589271228973122504501939753988015999295859466
7444885279234641037247334135338390259480773955176406741476465081414955375125878391526600273054598058
2800834158675087820182980291241797731523538577064067711668452133686650109064439918466472914384415228
4355957780524178692213439026209703590303502527032839798676548711129716415065768915393509094042163002
9212623423471285210839542166491175188768489016016350794990872514594248409076951969961803771282792923
3063139463215096579366488528671853658985428232404638733828178481530209203088315697267343925583364321
6320660898884580711362776399966495706481333243008044307069228179629683286131639498341581788714262196
6549905140499994905132275832902039739248205425751366407428377198389513758463568593319676365422978 79
5979675682839983101815254236665985727858888680648518945970716203467370351680456789741083210206877691
5310505668766877329334920023893505744369544516023429794578060306718931567679510089580811282704686785
6517949494253179098985455846351101662924150670161176221975729255773222995795702695142731341258703 60
2132593747642947677233855393949608034943296308145907993381594311461023743648260905274892609114997817
5992425233969728695252416687315009238204121285426136165332491366251378662874417287369277732668533899
9050914428805931696176825772855927778548891224880886696290222200907105319867273320350125608327 61865
4686069004612176551141034532831271204435229510016794790313350534235556783869192234312490521332794361
2569046803304540642593143348598935298788225491857424881037641375411484499829522748902796950898 14986
4690716144389575234356650649798259415250324263255294411659694055989586650761215339929748641052808309
8879197123728761697290730295301586338095431940182026691046931393035266362835832196293419502205582156
2811510082783702191422318615775289443074012512069822362570413511621279344747993737507085853449040 2518
9467769147420649139024731524047392237570356833125539744473636977591310167248556425227049855871329918
4758438211851524915321086608709389477465558909768150090915524531843711016797043942272006065934727864
9237655946958471716429025786327183436043870606152679931992517807196060181997889618914413296815327355
3656553178278789877045484925656831540484336866358934827911537849960146294330178535918922268713560211
5638066888736024524286151770771110671285143971739462566840777072585891951865720028302687827488064624
8625804514333344541330861637868233257296257953800673509106053396523255759682415048279519619744959051
0082179623656701477056459027478980181006309518889621379037693653372987268128208847887010630825541585
0421334101495828542771806949463381388168245190344482050492243551000331414292089422576831348019510 4195
3956483428383168994699706893612395299336477360596737956301617803184226182619920816348676196602758664
4711808760325300708745350853575490894833166708013253482497118067652281580236070823339041428117022941
3525360033063026112451561086493227533897653332750883730873546591411189798341977081211090804713744235 63
2419974361958142327674056004446749156949455787149355479222541764298223075736651596039395678729520830
7621299572905646333279790560873601966838068415216005340982287176820543030494829640714377958967789178
5265134420901479656996958603321761028398322325242090918749795628502362444942356873501034701847 41990
5300293809698609087614945672871126806871959924240064653277115700461234695506725963015667229090544556
8896694903638197937468465866534067955971944629775631645824343862403793489804730057570983951582161392
1444041889422681665358548954143282061553926819933381323414313987908720655644117610051979103079211 59446
4124822986954039586697896296360224807663263111856093817090755322596581714925458095004864281930723758
6533109347410268460883510176552329792792588642969057722571390829119090719614708538459454433599189629
6182581379576619523375377709395930937558695979150585469590600816003435570792205728418485855996164771 56
1906337685045329365545474742979308224034010421477940049481806545729224483426104801520489332597893682
3575947758489390796539861320097738878389002306649650738152650568283985821962580338070209708988714 41
4621585654462637525431393842532127573407453319116295517118791369927035391723508149986623779442841884
3345714929271033322663099327159181177798427378975014789433268497205154307237560639987729616687253234
7099071746405402407398765307649992827255557333971022446852281974406356741544233989522404042548339769

55371473159903911519958106094959851210374536599442439645586621895120731402017735567818531957450015913
8619106408997869328313648390096137571062723478005228242118426427552831612858697601566046431833533610
397233746019991538893157302858269160920494884541300922625883777140487965516015543593745110789847180
8847009606077890762206936840737849633609634250958470825725633681267006429102982227999157619394123050
1066561932438529131227088307156747196820218627201948474469147750995873774866029631262112393626268432
31533917193569137898919660667127709734322808251984750619540602034493330703784267983799417718823847785
7304923986255856611633528615279571343531452481039163835170550778772229762397920840708871158662391912
3319336495574109949375410066796880142650207310666332190372968824698040807054186317885193804782714122
565417999942520847288328203476858489725525747181941141110041741566799996419753284032409331219639121
04713467023378515181682298661343846179559222892272724792295126971190232496391380440439957405009271208
1861325429437494680803495274028786638624393417108857657456509859476694892184500640546563007857601863
3790396114271309657046386091763460387568116961674247700175012096224159952976060385534885700148140313
7001128029694543163723511250880211913858542622105689948995183018091417190615926369347364953071541759
0667880722820148829198820515570776358329567219112203577042495168506188295308898893133774280092605574
823119088319103131939299334559231342822908244952580052392312035468409591811803767004110412429526004
167497605558227538402785572289944290970792203734798808735001702235402887074872415687791506214652489
1733255247701844863336042379174274985534336281951376593862764032817426362481472009657057617273393219
7137016249943760722325613278742493777785892693303359640162133441364984027114932843427470577695437786
0117564691086194270718291744124265444598136378594344020432286589754638643482729148367579090612462084
3234390391923443343496772773556111421320014394443227823081369085729795736326744778943865774890385918
099259886296977925891374705285779546130320543303677522033508550526418524683519492934683524328602941
6899457532838210307005971426445390140901802991823336647440778847072021623062385605597582213448377296
29959883211943413369458344614783596937028326827114104848145288290526166403281494081840243768297808314
9452046334013147931875223737780641449565756210605303373736314667499714281990742397055859815350366620
9046505844835829037062788217951701095497639603291046554060692645863021268740270333376287090086360775
717231275916195076539133776329195822156023557434299446871290804612180268971042434170908330991098588
8888352540859422769177682881207561794396901190756634524170061632008101418475332908113003093109758677
07303631842545293345309766152917523663236564742169042280616975160533305992507917682502236464599957
033774761084147501885998326520406832253239105872448491232044201507636197289000040065927204424962
7607199929997651568985047882085190980357331157415446555005241314901243989950767377911479714212766615
553657000299806435223585594633402915196557447372577452551736846772411482287637268001963584486242603
7986498657582130805125486775367180449618700159104473879342430418785461787045785436649442843850304116
4819266671849752526707365839930254006188659463004425934986421887367466779140102899219351903419847325
760225853194848393382061146480703648997867086531405317348151432465185340564085301928990763601609914
076707687486498786614472416438462254922859816791081292052188229151944743470410361926198224968650183
287881228655261494487243355986406705334886676421607699601535508232824118270715618196314343109228040
525693801721006438745609358563665337540961520993684410900644239986258652717374982980371763864
41554083399332473281309549009091169442676470996060513667034017441183036622504899102028224010449800530
6393922465176432819632004447864310710645181829249015547074663013665850277505079676669447092311169507
428457926919864654796897674967442471720250261993627690191868938537969774824130205607630433892247367574
7538314713417546782974962447706654093819818294053395278667728938884828299114239277363245716014373375
263048026324942165455657671976751934720546499442516009891508526537508002510756054326553727234223071
9696794527224661597386602174168909312272254713382591553324526960697281730317552536710851911358
8765425443579041298241035431744232764343271370654209963215706364060968713845246245663352691301220789
2078038541203760206341195539453469466949309162079581991165930757419826929877866650365908258531021071
7015018441367529138484739081922356470866562195031986519855569037476710947140876135315487181593027818
838207813940008699967045174005890292947204951246680739005172243055169301040838278147544641937702694
24932724336812520246015715348610440607590563320374174853214395722778823496441861406872134325498
2369004837382182638400925102155997628249241483239110024692789253625384807769987524168277515798144534
5592180912352016230923561872620356180637137437050124625681248863511622694756689681361908739138611682
78104224664184881377491637758235371575109336365135920783202850748781773294567957228802702925330393035
6296096550908111234599045006409831462600113397660037298133881316144986246073840041038738952334677015
6047656476774375309135303602773064948548181815798555845871362783153768046482215248418050024360485920

424819532883678403638789956319332163183177839752991937552421965960896506553739404460898228960830886
050908902496512647211910969229403866059091378266635979448407832676362544382973826316123851271358818
951107258099419857239426389659059498272417809235047599580728228798338367066100204159537645690875936
082090530464645605498351090377847790876519746000574937826856962268923656473689664002376142139140408
530228530142292402293423918474607289182440158403161965703700511650374832836121305179276792029495849
610757873121931236279249070877470494027720766863951289959581037918255527573701990356398551282402947
351343047014985331634148827241470870511372210732163267818677079570442493525442402658784910323099442185047
571047629262215263799117735029455404147197973618939164136467958250810525362210095673087070599535110
328225540688081842424613290550341124636820625956442929175740192009705746751783787094973834620003515
023509658220513234951881288097417013862800727749270607384295478676545651223280696018735927938422502982
394526545661537689095003761204162565183010735370039070291502047537142778943688059173203027118577899
766663425716426953716695933183141768646992039329287314780654549961055635585878580355988938253252762784
779752774869590582901784353170386419677914076504812980943838768811633599534749783496325840042566556487
835230230971526389610852632841399355173700550157924331457133926350649126910328745743368016847088328
101983180572589963564174994799914117646408783098587388760126224392915251351274316311424091659579854
231194074263914199573700819368632439542888918921590733557117772516588695449446490515695732422360494
910611398818787978145692300825681635089337688536087849150976140767272201765263270063040429882985323
041000240401829907150580295348667865854914752310391765304392444451966513425148588659357253061878933
732908416340355221647410415435261375821818187912790652821080566445817008188462120953276418236193739
371584541654500461376347557269167252476102557802111982212191616769247994681485102211085346869776047
650797023269791794466405825458784123513783915987686576580174717357584005545002169915662489343277570
316234398574651211255669761595794165500430927839580643678562017610936953432744203237277829212641072
797273315388054265718719523146147608512412071162145370705234609873528525352626939855173759886215228325
706233171771057646834424071121848969716263292124969611887604178683965335208313403993199974584851
636867649088685910452680787306216050214958591937822714926533322860968536503039861403799578393235926
091077890548555861088592428224225927744773651178182700198138853160730568033657967627178457774291699
791936962962907299726810304970697061753617848728049157145532340248970086518250571841390970899811443
210863274307629534644830106029176031739831629885580769714433956772901529479249489257305310362880929888
571097742043390389424177496084967853115875752446072106263522179995794483282496498179686087770356649
069740609755815112095162501327709107803913461147510049698677195780467282368221758850855512187378823
843550239713535647653128488751114558439441307561669080219404705402509256163887305799593571007095421
524240238973866144984302696436519755358000865252066344823250934289128159468246881311076706480727
271539213380854908893217446305978858112744253448813196217550745390469229226077868286365875156680947
504786267227357076953714897264860136280801508442263265972211471187217154458187742615869707938869559
310355347744844271027727918126541939125547604846310893436796646334042828332733741850620986549994600120
905668609109495035208441838991634030696334351997137223404510183936562839490571574119917388142068649
188564896816333551950660009288433325248067355841713374961715055093426371894023250354259938439418771
874208814554354356164303489103148152057658869444728704649109953352128432519104912469054321738051067949
418598805440128942512325899009623123240538773982101446405849655974158659523205814498852510376930649
748931350603293607448181499898201118274927781520111827492851001080547259597551216746706569
229444385710758293568651598011790199480453582471723450301763989149022144948902160198684151758737919
682661098385738453765280414989033775503234876758875765835081680848980488994613463846758358275894500
648026022470795960731123470870190122939638421992508786853711199854331293724294847758736115174083584
375331090665942701325803295343488975837756220067414677801697322800545673449942537241384582367759963
065641023993968520341513173309251386082983961034483756434854709456374110604561666832802636976059410
786005301485403212528232237251732324935578824926593959508373340095059845300844861549376083772932369
780539020694894365228679285807815810808580649532633173056468160917851471254000880722573937135989196
020321176985166181382057266644897145605056476417427368418914506734245675641604829030981897917595674
479970441848154395604702337843568126761771579873748731652445882100164106192876715295197730961257950
013279951251230446071737653304434889758377502200674146778016973228005465734499425372413845823677596395
995722855459307838519140395047441361758910074146226819297696949886128652985517880249933196635638248393
829419247431923558426763507319858030301534307486182437832522793357999383568537811327556538647300247670
430672375844555706643322396705837897501940110984584530312039741608149528633651224839511514265139521313

6199492804776145672284844312856596154497313827859533076736960149415863707036217565867010430358696114
5791714834458205482295971166547021136277282493540794629070601403720169035778923993263032726072545060
0403646050283809296107600067621093582161548809682798180459087699075582797111496748587103659798l7790
0559920461992108621883339386436767545357823633698908816193564218210955951100939837537477554658607786
5594330622484912789787545081355800095536186322477894557821672858215655834857416920557822343615032535
5191306945196005289498694046865586452883932391961240439959477905755435190582258127024682573216026993
5312376217316256389732475711628596069997082939495981464546812429119289449321657589363587752365870831
2626129768952214037121333371373636570097549574389402162548668414635249814865349834637129932566103
6903209843245244363745789283274532540101378735460870857849153391330184879650215888109299037143501149
6211919724372703633189011799293100919897206605891949918385269867800580939230917378195429850851684668
1299233425946670761777675588620801261412614640886156063703867564612888143788618168840692105737310071
4712755602825523846104942873199498380141927494375100694790609597627570407425605279204037351322564372
0532009690271261787884195824392343316525246682094254662729793482420950273277702953598156498247338180
6163938715477491975350493217917432066843409206201758084778305188754961244239520118964907047660186063
5733321398793734673914908088123513455137740715586822235458845754468634333775403138713026260714622240
1171706024010652254911986468430964157219449244602828173252536670353723004242498660648053112750195435
6523225687382635156061797817749036314750495732032582720879015800739472207847114408535302162674040
5066505125616695005739908327325068995212697526160602725247283665244676995469356947594725756685581189
4258537772576809893976858806496441857538717290871623665842954600064572836053138758138636294410413462
9953741827762718530615919342612177132010310021145256576690287097455533109073858111289551403927166875
2247998498749917850258926890214824595782590518644548253086709605415246391648674819969569196375971963
0398096105803354613278943598184350867945292000548590030372961683071957622685435364173118174509579g3
3164776907746402740259052553880918629368604586739521133104685547484403817107250663691014557314738282
5053570575661393917552606951868964925034446864749146526156085505037913942029819942221994735438231783
2230368471373302474855942982638040651298489197127316269793994244638681397971930894515313010228207176
0241132296391228061815709537618452022878633617842610353107341759203978291654370239534309225910850108
0565591877252777554800270019294614441537672275682553321417372014074713478844343631675916914387225594
3304949771961233842226616048964396242053267797041430802411640119680891010920634290380279255156967953
2441619283486641083828605441536996436593196913777787005936030482022913026514592296134558029718272438
4196876724370844926367550705633405028398194493547326536320662743080836995826335167607082354952g856
1583433119524392297003987910675268314944224875870597119750135716880807708801638578432778181513027786
8311685891946114301089589218392897133594139288856485450916372593983597420767157074601749752460598g3
9896695746231577768604879720148035767956484589369637907288761612309559224148835550576272323443
4842642601175332306182230356850804701101189919325257172054996292664129773504285043702622897235g281
6267563789630203898474355948061217387390668535438453039293129881938833041183423778361478059757505840
6622541336293357809319478196429742350390848059320069789917678833968691319748258864747086279971332
5613717273081653340613946256855050907275458645068646565277682555342972140883383727882010289029324013l3
2421020026106356642443696612083041768693220104899345155973211746630090867120083557242052922510628503
0294066927058050440068181922735142563465843548110959320734012749694900025447207973603791646697031950
3383284835516767605831036545270857655498002823947822313718870396521642078414038632005016875592892442
4891643210796200313711074626069359189558182399883659153109700423581742946007359612474329057210929097
6292410410656620923503792443139268903030620340787058475213684434981400664399682817772883283068089
7474851072684228563950311923967939970227828083290403918794270125640317319867054809038172901093826770
3276181873338233299287354251791214674169684443841609957921734925475411516955036329294606721879838177
9848868362782909979843021720417532522299672743257163080332626794270088346799312372277892804g072690
6343593863344827373494687180880694508882406899726165871343751874071244353589993574950576391055026023
4884831930109776287514383618192273514256346583483110369338721304909025418488426977514011626937613955045g590
4730039876222569569528522702700707002236312782756472091890723661453383150645086601571667250304425313
4573076142482529934735508200948111074026427032879613545589972387692438810975970444457279722559558214
8318579221168381920223766601470535503329905663899611395020035953143148531999733956110064596g955
5821496162158045516324961524984625493133866615566130574710730660649476125925134739867240429470527g4
5870057114461774359248919999779853985891554580117570754584198570746444171573528708831815566490671161
3720524842124067568833334632630939467440591539281243468652741507636710833294679930796012132262362971

922889061129439568658906746885822588883989165018835533075233198157903553586855155782065468218332159074291034746956756633924854152236453715003886217890263431378530266227448817999987385332341525005050759944529160103849242964737923144851996764003120426193110183900107455976932457439965196822111570172250000780185200769092799527481957223522490092455102100832943506047090382176234012352784838737727314319812353312167350741624784195463253446152082891223780469229085093862807526773733648916752751088671869074857315117987191127589737172122200697902686270153977033376235391168573023532778080515008525981753295550807877886672815650966691615839112721698699388759112688648485453452898384500172007531788096127347744030045241675032393038367061707101305504380587173067566833533745378303685599937759086951306218465528579235933917419171205417969987256132453266577397569709321705621938004614828574999893752316435134770736588209810605778865416514732489817870094630138507925592226072971522620388194374843914310594099258
43344656576817396893261104598701003727543525116377441612272999941018619566051421596941206355131448597195452860809748682548745244590363260473138064839379734468186624970072115471060193500238648389343756227635012792584941732643662372023278533594194930450011152493701147634634157542640955747394306944563542362081212241176373576970867776359301935638364440288936305078333228036674743942486570798950852508724183268352719951577926527198763749979076208438946347212620360783081738142804787855497828978622747244177003016325501397053724176828153235161769069219970255699962054642437265337754725102403129943553864594831470194940156026684943031837836936554661866566254708748948972825155891601538553495064513844742211882756298620633135692134350535417532546229427385701851422160479791812391358185702336381354453571127711719432160046614310154741982155492904752610901895720806062349088029040678545667463724177248681190074206557848221929501065966833553206790875855344900927132510175578131112328600410529188355048282568212439318097857966641441641974384650359754316704183852145907794335773149648457421486085488674529131457458931518483420505854272116027520701053028812182044257185040797177351938264441514303400003896503835476069521126143515144940969915151783328517247989474052420610045984073638435113829829335370285516415328184689878043592175819760111037188260115715212198992803575460838874094737522040639123362898280661873195323552920401422000951548088070610074538656397258970803032798551240570967529948775250348381191484476396069023990808588751011612900600807691194381030260949480659847619690480593217852139982865901636139729473334245297578429975902328892128876174536434315837531437849574608874737342587958758219901935389814229423917941515613153979302514641379860895988767541369432304048702855541978092295804469890019290455890684659783833799449251271604941337790764865785894967575994050617557632934756808289222091115491864882015921461776544992118272549886765689622517063614832195140603084486884294749081714122766989529766528467187010729193377929282443532413828520635706158076925928260322211942076877904292408365323232151023540753423210947605321017167804788960416851071973969399186187946346189679713546786722440290516444396478293266946358491866150401165503213795823884640354533706750014682450893963508407963388339316440021557629487655451496229849453730746556398464035396350205035984435100593365820526129449163112526931176182193404052804977001395818569950450626908321844220985806560360396650520405092652944916311224741224398545523345939736021584889595645760035601123947226002900911023235818327077603181928957893191200422829722719276801057856446673340203186065785199897676300451553441212227464921178419210429930233047545934081486957338855853118789257243599624701958104940834271300659716364375165746371024987053629169290060997979798208147147131128995085184920039864046146530250994914143035836955688421615182000667253985853203567078753447410182134499703959178739753496211472367771510750644341932607097848101390611946814299656594694980301501505043949916581936434061754712006023253305100568566199539885210969917968103065156627611400123939441274050406560022170985477796442464587486319694618955103513339164119590397189387610544264232046441278596632014847955273227434092863462009840598245338576386151933644320983391957496295052527170415941103294160552007079700445274265503291068016829182150548865729790830657332005667110403931664289460739742613573588877664840674065037969090673762491231815694613258421465112430180886283850187273208293049320534883490802257999620893153820434587462062523629681224077557116676147083325307435180280156466152412335772666545968950153512096407409879933525112423683932555080407795389063135984869314857268748994901440853001062540369844042433985785216277629454919277281927072710075950541945450379091805163615808362888700153867243945007027498984321855676474032471439236648434116062093101960182503178068353985725839135718344930361449170866597972333881453509217403181174775203258167433894582649675275252036116132601226736721097645431340238065872012513451461172380163835947726875228176383566558961886132167299893940149412510356465833636887760906588376967418291919203119456469780942498386090416033096376452927942341930023017400543432258546250947434551708835436975603565501992385147371849267059704

233277579791173815247435316334241172984589412910750455504288587776273734066330416039180826874172606596159893360778633070199222318466648889304527152405517461202230160366192193936615793786736961581625973005821281128255764679594940281466745760455774706739022001976983182597002938195414927590811337332360588587778716100672583596232604960016058991488934220473616132710075452300484394310989991637221886232625722472307119791823049443514033574476639708361069860714457006927639663973492029218346297644381860189376688053451277770384815690854061403280361502803860949033534893230357925117393530415841133265471290567398884435930822283033032221165929854191965597971848854238871580891403693016171725700155706148369068127422955027934635226450068693430774820746663687476147620022750181551796978266737414595043870588723873389632912139573039946630543402891327746816875466950216412465503700912629179830290388734841761397234339455693566008380160943556137853374689207145442337764671963138464652631570101713235839748746654423630277928541904501566645788185997947897125148114050237769026172897930130806565716312121207914290705421508889837954536591643551233417459879480927694175114903117460552245578545813558670215309007703195565589959974680574161333836164169114009923341556438683622586644280794033626701052266693619246747237136409054289852051883510036926818799746564705254506826839362640699442231179129973336410667817359159716289832741772887230205260980844287577710069881961926403729127162845583584784034049042436487818337243230371618788149318366321324242242201471879866012908295449020987399595428721390667769827563089167942174016882358765397504203024489864188963690963162701205576819699291549927751425437881294676650832503512671684664484445472404101245280642178327322277604369141661028807835887184741005180840179580141083528163516360380534630763891947615018698673670605014755654519125563485474406162027393835035627856152958894681701699940143323110952872124482704720605460258500667040757911141368279069786865871177920435611489299687198880325903495462586850786451560737217153995339107054574208447004989811282899421602122209926244947274540561035820926425126780398190545265944373751942813217137033612591057551698992847294695342429807232562902588836268426784470298363413294966054416253863479816732689346993559501317206641132108607604861957063566245230487204929701679457814582009036192806782139458937433777693126987686811712481640849105253884239333690894635409235802310817255763499799694364597544489485664773280449886762357891730215026987796454984277123360252396013678890263912763167334869009946588810286310223749553599501716187794059542720325680750991723040604059244759347558781923115070860386436400166976935884441077368702845770337940928349414028221295864075270639353993044472385084396885275777983552083175810709482686545514923467711645118856722380760062998781844878270052720312938847998209719432022757136353829360035987007560979354968507222738190964275756846644078438497623595416437898607166734860499536429215769092696151709528254210860266888128762132282888701236914111235086499884803033860226167435034880512115199522213094722311882454739260808544121534421043454311042835336107232244610950475490308238497623378772397985764670714850727011553035079176889428553256555785689141110539376812300764172733235555559698197951617876561127158802381737058125843376445496393209033630884238313467413254157835281076214784672273660353329814219889990010399651663985781627083589624813758112852050274683143462186542100287357984530641972173311190325200734619298128722951789824511770328323475986403956270619085455073580791658971007776402290351977055165146315695428841424375579757709689442236988750346546591323568232485632308818621486917525444204425031155171125209326672093524453853285759307857205196311267715965633535956406638121569917613427105035796893469256097759229113557550044954680935994198807691937528886502489711246859161951191180573662336507492183673283957490669396689948638812546588558838303386429722354597161408639132801696867060967477934970251369670949211851826806837103932976812879349090488099268529794978557333715456812291190828899996496173672758296772254271826422328664001327243273092429509230566221346977560274971311377496402160451869335895994334517013147431671669925355352625191822960685511025521061763913058993047044013055539478586361684376918286472343532485938877973370002373444054223057833850423369748670050160028663716354807214257274236347165982592000599527350286342941390667926697237987304373539379577587467043870950735671245544966030978961181945541702455921930094059380552294276921735098819503385424390196223556566505059811895084955834758326794413719433477706441743068760728732386031909376467452918921839273405652449125058695651761115620698125003931533884581844064908193055138220268081023933630856535953832850811518526024907380888193971974192664563416144842651154131695283525195911244283826288101030847554548973690254035882364831424405950604333637217231136979737662537789832914814676854754118971023646587793294524553660846298717097667145211

```
153935936295650841686793887474517768464701970560120629111965939271694287820010473842269120842037473
63388386274792663438170746008618165177012473800268910283248614546728946437703394342046484241967025618
79164897251838674622230416516000185643129965411759820850056332395241632046675535001335372968491746676
46319413499223744247224633022021859547406463788211882345939408996899586667766370114295285312707935566
323783256196678213665709220602831025653913540119066214292393816411208069617216043809938799300327913591
193961605459067259657242443886730988394948040500199586995408776106691389068427993564695024599087865
61048152626194880291622037728544043617619152330967611523384395176667069509566127981248500848626425884
23111598815152501667563739574019057703412613420416359044808390765374858277752596662854316298833142077
4778261209504077604333883663580243089244840348835410287061473396328346465783579669774592587401134634
763321608104239762225168959734768174286155125377213488843126429868916767069631623873420014694898521423020
83551071071050255718846277856440376053415487371340570353047160477106775293200799083008579633505891364
486977471509376122562844835142793387526367457072220025676912748342237943660613198626760944062105152
7198485974737929740617724333077753380254302218943943957667695095661279812485008486264258848457967719356
14664646260149649514634714900618867260130216748107466054111926689184063788353110563044170808357801992
92232956434743032959791358894379380097270442582150692797998883144672532976896720924331096788735373
547040496937826853533277996781762915186712775413657266836934910872925660656228159152633504466949756792
229497645839604031247826096808076324572917963135706380553018179506155893461920055250204212768920472
6523519590844163705976227580527335399057237372924589843113466208946935684628070879593423614342618356
739728412165260195438481774502442968737870447818458084566985918167574593630071250992994559021579797
1267979286814183617945293811474383459113049494906254577757396574482504189366105015672239114063379044269327671783572823478402429229040376747003937134843855464064270710261175336091308476127357563889304495780143671978013898826534243776067204873056592069332816973770772050673214000573675354498089554053868
88786715911240760240287649361091464856324351392282896169205384742208960466080590382309965918933425890709006223704080066997920297981944092771735207012733640648208673831027079479355302208227752154460927356207151719553874896819084680286066268052662617307395592893243276656082055892649228114572078932587782368082793050500307417743535142587643209181854326940690676007919082134203963689530945256334022131073020986458629768965547248652624284610473665750904177173205323741407565848993239270868216794264326875694735191217476911115775407999719992668288850793903934061031042132964825040770647705217690955724326859664717698638291441153779769670000258192723944692010498480258054700148091808108172665045679671866880646205847880930071167141907849713393914993952552454520949465078434971981036142877818403322057069463915147694697276774770642486460793923519565436503083070252079824653742742569969456544174202935844000025812830466592005258
9434532805415082762099060624110177435844486270376709248843139673126563568059785342858198446090082502281501506367269141887603978831977318626572931421807329055093358562444488808705158512055619441373703285405475722146340713736932655210869322270942390754299408994425444590686675741143225242616723435219127828545438845595161679782993283236427374574525445460529398968062635137335872148508088202055186599580340814883
297012537812305067930508188185685057312332575555424196054273583194479764324992288226604355585233496066809055029052163377847463519347497130223294939655104159878397401675166185936051793389503924662052455112688373111207852572442457996232944501683417135951402520951792646811568298203136188273964266233216764415246954875581640843582125850442476706996938037585730039057901105154147795571791693127290959982213641159815952014586136789206666035321839445791129493727464246482392154779756157336708957618405753220985047485835708917663527278094935542744825251137393891237833518441471827848818093776259467255433420690238375597656846744498857297100153365702593838609837888370559661656612261881245463878074036437755829259340134511785844624554074690296162202292449791738901424327249724654957283299442767967831448193467055175708350294265732633526490651414121023786109329671886310371717046176289311616725902906771223985883659641492455308120728570841006606716854351666353034138280113381967791228997412665524495134833893463618128225649905341150317914141076919332677687742325698034291407998029191076113965307761804076221944515194026040634703567993538832743785881520110804064908851752700820562380205128642184248230026324320559979983469262326656447019563570067953905724415039816423908213623513271771458619121032811235726
99330876625534408941512051799027314738681826266444752846408572464085723801535139318504918912595893739926501687752743769741533748172422037707128664490771162603154417119414108348606899525507444772203362742266818471195636157137724246154560704796508783129001334349111362929755836090601759494537968610568179080507607566212738100117918293076118629911635574502602021275654360951138560948154244767220734006103733426127360804485531214575788902375590577113174550094118597486529670588563917389715951598898701417586960
```

48654185324863779433780506989345553880505233124949841887573046444733144459850552473986539970734623381
19398008577304356954761698282658938100306024112186656859802072533716561353350992188595601078815219551
992984830737114161764839950330037898824790345410532502054955693588016154598918936886572124748963613617
1862818854644786179243581710112551851317871774504307353645029761507292301108330802551534951869294941
9716900991730394769733378956502295614877878048366582834827540230192303690385819788534303828558273006
721561304247679650997367389863963084595330994467366005278953510077510623540518095062072959121477879266
2633854287925897759586305806465044845262391335383426270504308670094670362204063397672529913651878422
23065839667022625805621222107335411618502936356416166557792377663958604946932445508050590361798644275574
1294983021046969816164493137010370277508486015396166586451285354530481559829638298598154556259248656
9186328817630110149973720692013586987741862165578208788502897085678297019269582769523940825795893466666888391835881554906943683070353276320793494510936539945097204283673067035144196315528875321482218917
3259671737078127140513347473860809636945635120190184391605573384080516638291488624793513794037131979
66875856259482942074632416148196268288849800968875641317790265769105550802543228031258589984582872083257358894763134926062496271832200731813542439536437705648192953995700144554383910878449144193680471
0651634740317103744824580518578818068662884417079356604269800316323634912030291975370099601066619389621731876226707182631485228441727943344684818103101838417534997349697901352604608389864938417085293462
9279158345594247787414758186206722466248117722498568622987440438439218402456060939019123698959782488
0644631955555593083281673460231204066700724877475998063268452732025570156216876628405832688949305051
93900504959049510740035408345427760426244588466362385974454576164114198430383570036397426666703855509645239035702010736523283527692067722136663585746080761599482575890261556442864496737256920804585117466
2670246787668603228796511978576164442650025536622079972039999865614691551199659189260998756919572198275509506475978615626474235578645011389704199350996670065571208502958424115591449729075235534992741005
1294919385596259403263820252498822492144475588270029003679518705235762764423558418333071204601246
2993991548419581355125514677093447144330924763732150118612798381856025576131417442644210392318412486
15613047098148024733881256960519677269438321490104652409981501183394145060083461319779933817165917066489759621961933076617168027996614596490848571740823780571312943966103687727269790434903189674932321665723319003721541461036471884246356801971257097712420455997271894016308075557915318038868385226329349122868944587124407187398513109807299600005402969139086326671417923649776297562971925021288399097084846804390717619829838625897603127381810275493426101282445835103972461726002712472644102839306036777543984038462374655711776604274790447110253227526070881915658696907606589908547798087756486559413089027045689772974195596550109221959323849781622587517646552420925574092571769546886051901000316080128972898705286108
54229730993968150775009659717137146008611522092622608527082988364373624387798127745117082236808061077
077413663347955743533574250663440979289899184082181502006262900581367815452848577595273359535974840872450053882741039998701952126233169862828034388497269141695862925036202722974886898490003974147161674575114133460273449742435505878072186655258735064125308324573880356085157662659100847907204770453688975
0719974356650630663167587611347516441890509949504410711998514991673976622942694451662140808774913553
6734530651829997758201465753081579408167503572563130826897527686949131751660314196274122716209578299
74512595073689499764765315903304455391676187931636640409697787317115800412265552886370914062581788469239036439876794389944195963322773315106241711111175895820421382268247158558623159366153128943219165
4892821195976227665814359674319046931897070954625498480235495501869231129366402929099667008638784004289044208624836617790064302063395013920324363516079432570246586846668977153432807721709879801181485175792816444921354300152529961377236010772921085951314599524616594227164157476323657025718806117063487
6292627323600831225256996543432189374507796744529154278947127228947044648131474412422116659008100572172330443870087373605331646830292870055572001906994319987064544655062428212711171245920681242948105504004705924105288357400656484547245607487562647634725962019554163080869913086567869678755397008127911
7686691949683135150988085209582767929487854818158433903895764802898509257246086253006148886286503065719865793656157955982572991894328947716189620569354672805441856350184626344267485715560888443376776

77518111958796316841853639123374976612377125870557536771425535452801023619128824660846856736084934133
33119579933540423335773588963780531839093444280492270352162230871494436067300423117979682863905171195
15750520976559027309967099890200513002263326473818452023997691129524606155729336699654182678756146444
74369388729088789425992271475632620666673290809469862929534311107624328164327360863086413386486466683
68334034117417243361379086047880568004597543289332721406080344475032843441146117190967017625398428226
66864683881706100253649900743173847000861481761634921460919937381887765482706997939841539389749096
461030806089521056237233733955299064854565477711132351150583518723974869707635229334354972561003011258
91267832849264645292657116115146530034496144130407078693714179233116662476964087635487399017477530
71020182114281421448264213204890136655231442441340428775298118356673485565936917962558531536751079801
6714527966374589942103118815474548075246518531702182499670582009281703471433056490611030296600778626
186439586203091262195374593191550111691331559547339411720861353588405201458592736046321982720471542
06140331096148629907590810331330659145749539436538701843065303834279040143059829881096866287936906842
13401058668136877008625504106696553622430760974869206667440684275559470825940375954543932812646151978
6010904922100304662393810002488578082153053964123626303586044502330247943343213441880843146928161839
2368201869189398393330782579391518765988615852658830313054820647419923861166216919045975656253336318
4467689507529925867773089781132205545268932341196377415807042979172961849337651669375621514648813841
72662171532362271080278418137745969078655725653492678708800091880707514451795591893207464840761990
51735585848891330328063507879705231316767693157737318795949072123726379925971534942241650491860959163
92980515375415305601083541412442663508841170889542644097702742282321287373858481377393550974233555
1140624466436894204535523793022953609025688892472476482856987927777177043958436924724006222094132555
4943292326806265100656067112487799788039988221458634529591716662481653228741155327526411248966562365
3627179051708201531002673539588247022352816399724015346412203202579777808273135512050193684281552081
85549975149151101699141271160844540090762083005141646188255296346203608737167901905825189468389540468
2662971748668008393029512607769299240694352257774386318127596795069437001560625056278559143415124
13394030327129532531071186174802577223494892925219801297430895212231461957666420719348714533896799766
61233291059527961018343106907702032287162520836119564948117529971327497305978355285212857854784428618
16852571073997916448507379463019479486010937938364054003035089249948913801089313227030643660409213615
22517503364759125529933623450874620625221611521345334465090731527324079559395600327487909738694260063
6143145093479579364252820760576736682245561277978857985090507465575995232325768019785164732223573446
61249477999064293351032029241706181476957105077280191727166542272028024548065568292656244517078444348
09247355832405957279281370093179495842802006781667030234830107405474219268605401978802767061773311698
549010053252658070039193322183255176221950049560232954318807248709892249330737590455348878518957734282
51250967651971856799652910171995101746478143027813353716956422319340757137678346086967122438121730799690
3831217042049112445158622120573818982602813253361650633270961268112735445764503438627183739199389437969
58561167126683833937598558264615427978133179120578291237899892276277256159512584275400014463204457910654686673241405345801
61442304222582168827371032153821229010450845809849749707951428828742756657075814204548242022106120376883410734
87044616953135658473158464995233288974086138926037436545571031357309787030515797674186488333083334683061776199649533
23342340591686783886452470412755314395794027884216137596849182832860006928911605070761181598098057229676116423560009543
9990228360118255687572387881258582934211212064353513623423335454800037637353922844133746644754648997271532487062343247393
9494074363784904172725426574267589518279602033436226061843406548292910969473277581063150058025056849492131837057106195538
10369925100616045006231958956852776338414533870921568787580321274603112849248871469759238966121654100784516651875992660729902045
8345627963403979156524456500393332697584161527418689052810339622772868014570700318962787707753172895137492853891601345118147909122455428351145074640526615208744053271983106491319508431939379405139356086244871206328233097256310656806715935871203992140966633225119104450832165354362199377758584328122723097176497270028523353202360334694516082287284727522818468777375072298833913187683690268249450862677793289829297493102574748519154758212367871722469773414700303377094862942467836229172179125739788895723049380035859123639968963121613858310464837079637662679929761566821198469341159333916744468862003355689651840618965020995787949475034213451064682941891357624099495577188337647484493614890337338736408448766512857990600569180358580217574328223720982964054139849176861425411357801923284322356622012533756997103821037145053611352158087544325

887517731498123415979007748415485246874718698284371642775679661218822589836358646123372087316163958
782993815527341580288062228960322744791973151341958948838419529290567529135828470288997290468242178 1
158812544500275773489756610693699383060028442488304085568975649116156938287828620459017159206618355 5
970557350218309269119605068711363792198916388264700303239855998258529737206759685021223259479609213 7
013315436900473475800526697316366280876754686843154412005445181096396331779963270733270078424261594 3
287198367100185305221100049935858980934727282613245222554744663365234690260799520188298486579293356 4
334105861920635765802134949712381542333263308182496330203863618060743007893628480494572474655596897 6
904796307725843589609723556268852771769509575485467415631893654443526825222687331658583367174104535 5
186016897390037051140387216074925657286694144634234819342201078799444793152890807044167837208591403 8
078719202046871489540429657782742327726300675482683925720474289569191676005205232153821140887324067 9
725588369972297703978174778655445133936952804673097919875464405405015355984214901764939708993362368 8
179786182637137747761899242139647546815218023565700846506246212580093382393758939853525532473703072 6
876131869326125773333729027491969501580404817998772837366525506402719361477732598808090814946394227 30
754621137974252544785730656206162327584566336871604105456555821963228444258001613092292561169521705 8
561742929711699372987985526865736798162258594917332186376150773517115337805336399472531737904670 38
575527223738278135885645323766083898120229497517958499014168963452187860835838411893184728325276864
873474621953538997800875424150586749780156015931136540552070950803525500481212312377181521072900332 3
101759183786240565962533994854471076202385234083415014218901838963027669846062889973158305000605 41
661052112618332456308874942376132111738323599102671544333398090301076751921560686091509929757948984 7
091340484776037253316486633273997745741707870588584989036478250500607565276677666730181429783462997 8
631154724719046381308270269502715524345837771328884011332285612327642475805491414533400430735136820
016710304896740791322041732936558863808199024025042475897990619973949424061393859002043745081712616
036278391241147268209085690526837442250689109919376772277687371267701529071296822615843757149665346
296154035289806984981990238158813249007282842031664545864516877871817177277928321252269683297664124
549673971527879680434765895761265338524573915134383137845005187385915329634140536894844397225508011 96
079269028116229367043471158371953865778600341946713096534254355235613503926374333559024877800931 67
585566502026142451755202310518037979241601868165327213490744741879263046379357019572546568707696490 2
562831139490830659813925877165753432905185288830744220931539456267189136509323778527585614188691505843
112818062116453383416546998610279900887831599923320834970349909096448289736199726084130305016138433 75
735033502696791991003945764850313988998040534762079799510355628009427119807714138625374689420067112 9
290379402110509931288176786355712128822058452590232988278444899728557676433765513209837208453651972 73
566294540752078683774825993769508547485377854401518668703212703251083788575535253237422467456165530 1
752946970492860349352376619377585131269112157125054564936628404613157549323436116143868941551917955
211604032794138704059736596828772355549369536724926033527449892888202044886844365275158954685598 5890
718317292891292345774428441927250527684755038702706328297997585388259938790678996367663472636797709 1
137000500451915150705020857447053203113428375039645068373494746515254316164069588396569604776248100
769812576232402765632471455867811665356335738413320375632857711457947736117758910978449597487134549
945405008974943123702669160022779621516016443144632155674657986969134302091737537932953736310293482 5
941848515313457700649437390976420858957317742314576728829679067502992231525732869833026341233526316 3
490206490429708210063264881567676324254446703921333767894896001251362652354725651702225595569986284
251088668968471078726001673324221562512429272130805593262213072140936864354996898778430352676884922 1
231834492419076374715747446252159745764663572427527952228915040642776786566115191933191817830567164 6
530481381010666734245915686417445768839062419201865410225266970615389099907254998428548419566819245 4
519747093061422753151298445309182757715136118167303580932146032258472352811825504706062154262243245 5
144689645726938231665552509598950410934253743085999797137004258858340304497267109629969763233607776 7
437347987883567301028647138454592879163749014546046751939489935221236247436613174783040886884631516 36
592243576762762344666535958979646879055292390270201075721891913821483162685249049584867543293124182 64
134662722820936532772839766755726728973193812934193405723962072329200718638674667030636460133111116 4
254680251228943053351250985386012012360704496997852530935302771622679823055107676926800022207
418849030165005038534475971018301637826819436124165696392522947141035743185176583656034123276433900 9
565118632607917338991262720713516175222255241829612433962825182328696862544411862381233060340345331 5
560164069574723203836514566355749873441116859941616551824960425979839267816131483180902534507164666 44
267026276118597649132476829527278057032238343515063672177066376374024903046590962859602719797255378 0

01418201998101813981259504234866248344043921136487236662920206393962884531448374890102608403614840730
12006741562291596696366940836032643341496371209854547525017736696019714617864645155559941672637397085
864959877953242158328218410939164052835679070686421078034660757197989148155400542005107300962796234727
24997011221778165679844919433222633415033856753082446773410455032742856115745538742140007192843017
744731423009836576075151277796281014722053066817420350596794105098046656313637782517247091409925552
471036812670513824675211720052849429521974886284898527778356210600487812711440634990881645924451898
010442935708329047220160726966046198426077224783107171439093449289737950750506471053802916187491886994
63530135729350187320664687311501731531091296794862549795815121682207571231891909138338353447102365979
480804712338827444035053467999529133546094139272844651391190836076226579839815642463829159990441628
52768189353279135674740322735150689098775472181557499848834669469822711943435139572756093318776721574
28578303301830422257104963297161229683675274898329472315069749788741400219267006720656772921310849
729407635928956182812900110978472432830384651974375857368591230981783603031405282300813026631330413
925339921794157647985347081786361137720014085708386394377035291834974037418351162237004017318826093
263087505487566452909326566503024394453622729170800381578513252902365105680962589179424001638714961
21216946992544239867472620600571311538783883388307801653788387752115931194493595694917939405788486
62395944418497289287230849557926072113297712137238869698636036829162225446710880621094479132399015
0667816028934694282155062721260541798291781738248919973382953016826679061778013533650471886339784
5335856232791853579779226627038024456968296862549118748685305498757965989184862186237485563935321563
0489928348655615415406495122104661037654818060250676549134033273862941691177626381383478511836410569
9960094920204489502629446168466855106066242088314537401268779478139859577769903987707399417032565319
3556130005281435994560612616083223589903248956595675275769853245744035607612886079681857619771788765
56198523752744722359927202060238916718790814708867068302793989768378023337576683748471679204205611
461443508423836973785948258849782592514143167689849592131938941287506959614919327114703587453366081
9437546971429195290310193894363718359374914830742304491340295962811166523899581110600093862221626552
5297661065074527135894973344740728167392634862262971347155532936244457994650823197908769074458525
034570900774408153416786387014809967412400383780852342739778817469108053533050911944331387304083804
0675060306826953212452290166750385631858585931743776049941574010572749444840013991522952924016806
67458424696605569107697596987310507845182518576898000939428637121910166980788517105711446695070312
370696200473003567536823520581524918682390839740880926485270458168006391534013493735254715092352704
161926572110042345848085322398093081970121586417329130532589871715885516842060650340556996859371591
6219395459555855700934771168117983599584279819556435636530938905094196464188924341766121771175457371
44294027293771776591831074430581515319609482635063365572386141392081307544146107405127413481388906
5208965174728644434890201501872018366138417280798827295820189774861263383603711094140868044146381899
7551441905115201402418762897868823386652887495647401107245990553799217515564781980918495587675277828
0803822629818043941563979561725694090929525187744788365159447872068267859636945476370623820696902396
6206659210812781832191274680814503031421779867353368489380826681896912998351994223212726387715975764
28521321515883717146485428124231224684028390561577968199897855625102710706283793994319073579797536
28737199478452103831686685214208220192667231558101173724423573548599051489386343666322615794260371682
580693159057152379480256619126870887647506950850111370257880233381801903021002975975592681821635953
07064188571900594744467974174202521309472461919502772132372470257029616316814647621846436446517993
58777590480917246955673967945537349710322193694559817913983406972537884155020629583874830961954204
041546990222684347461767711319737486600879354436073024336632808653684733506687074089001847030676
98214753133731542822151551318140954149797246706763436976645830928679521201994140665404326668344081
9686918622917654410364920807857292423388755061809865912226537972884111201306910185760304983295326
2141884259428662146952768806320825719648671342246985264194190222362411863391302841718447248227557233
79969707048200243758037179218073420208053693574061876566416960773912090981349470212072519721369694234
4209305478465069237446940208887326302261563579196063092369916027823649300034497471237794559512408588
23970994657027536675981330477750505053663457471551655837277310078578178715303161327684892535760784621
1147886035180402976569605848671756763665930874810160999279507871789131042038449478943286084797051504281
33265245718864231983999328563422686078834437453092728931460925442990607871117367669598496330621775148
48489933778676785978526528057054866121737921355212470239532560819067885280383242296807554417143774891
5014302315014696122549489533836275694486930467419802292225550650874297727580760951068798271091938371422
229096826872856321942836727242477443909060036804852784543854819955828743344189095523099265929588482

```
89777196750543920577166893855239773609258209069343057898674235729531205148509038465249314006899617371
31735816222294455416149357871477506270376192498036384400160913611713729557661808926384647940279365670
0385305779912988573944783757639092679443336505496770742285963808721870399582714758000440222404214003
30359036096054800471884730467828680774098983222526245316803203408443510937431949938029908124179211089
95423927096542582195848586679924115788447815219557498322583335674226798960098032009354865108549467671
671340531034349986434975800021528683583572136597820843573260466412605704644092005206434748808684041999
585408697477316017505390253064903620449458476440882040053860571525182217793518019414711660086532294821
8106002191594469278346037982926881867778488378271314816066812848087479043023420037713089646478178559
946183751062068844135862845063034644191394289376235474277758676901467822890700606926832522503246939533
3375667289976602542465979519632609027426151574818652781929779836811013313396516257933184194070269649
889513869239612612753695924066022969008742083472084433521252427433494379023824949444578516493612703269
94047668241678096352035664154332541509645019779105414609437498159904457928380288801335624818140726111
42327597289482414188702595745493425472274698997687716231609932288504202807083810081409188735263331838
584207407484465733978384290853471060023742199872117626833349092907386533795907492808928303010720755
047245085118333467630475982066178999800446274480337019655502132044139642367450695370878169737996937900
061637848201169796270721270358480479488580658308963231288673402963848241128765952185362411256969974910
990574782803262998612317247930503236377058456987857745316103866706755584406824089105118184292580329850
85140961573315387563114385477921529836638382158713588240820127738409736232647584435263028166475607993221483927156321249990837098930946329558955923728443352125242743349437902382494944578516493612703260
4233909454480862000283535262617529818355252978804650281353991128471612813541446103897031654677395258761
65383844457461103515616418092733462541422179033101714720310599294953895958436885773489495225982103831
5964206232730714837166917989674445418414890372511272853300592982739374737571099277652356370360647348702478483968420374230978899887438787654284159356597358834506093612992449258746769154280459813281582587
29991103007806315924817220522132060107714923366010031827100667272664889495509423368979354810557964233
77154495413717740799511775046695465741008015579341318715461713838220333287263136997808093756281698575353292539023656811435865539828482832417000516419900517643835120057569334304218029315236854114240598680573897377172090328164862383954980500843602353585825546188559424426129289214341478982670941767604522513492987299743533382062762240633100488377452738118872258181998221942829367666600040379914870018568674554412319573518712379305995148215954867701054047820258539083335640618262225208028648666809763156107137641890902360603954124553570380667535676524726680367517673845564696693596022634258001555720896238403647713214296692193472420798729629861675776740082959712485706790346703915786585811127382574325190397829954457674305752992830286341862425450224962419791639827349043594119589840434895757833246482216524972533181193055585614050310507654858991552542638525889258725889554577367874202977037846847563664249476708485430735413284036163491347107446832985890951110201242544884643701227174886586964367223751257407663807577968593851823215803790138851324567042252785387661013519568286523394604020035673386025205513475307900744869642641436163812466020943398489828998572534654355285404636104131214993769216302614831518234694209629715494417194660720665528440044356575326641438934277220905575184236912080347379886707962283986937508881614607383824642000815393674001886257307369534997308367252810149430436456349752135453195190035076482370361845384975636163397442943098863871989880818086747479583176022298426725019591837178700154647194377440245879644193433052737786174502452497071499070005187269292834587178630
9174843878506397547781397976147102974805258939069221666222523537344901135398626602814192647629370976
80131872041406668762554595422292493849462711775501758620213788767600298051574112378095519278181590820663636540356868332445566200951604633752256882558545829201930638153387365651794574370258875626473210773227646623152269973958253816250741193599257543470320751896392792921623091299025904455121720931896
6179934694954150218683377015220759113008886890238579915282639867824654608874627852622681424733188588
5724166512619590003292244047284089619602649237730727930328698385071995091733622206904262113793573787
89633982192711117924375186838175762134729227304841109052893123739756654644015991083095635953955062268
4903481778340163888074775859106047358664560729940094200063120435623081164981455149655513005854611735
52405216715566604133347587598792044592175677563283767227690115416492112246422360395403368455011342652247448989545996720364424296622482473608799649501574046214825111167401638128823705492676679728007574632906194561799730944487237467063062834619376926372843710250629430239838747180411277944515182108640
0015584757912846401287399509776297708262263458825052078183457605053081571276816461601267456151310391
07176973845578732241330030005534719511669012581135208015630370469080930979273536585649135747113509
```

441275907649029919388200826217393959286123336572970664641027058783855131893465796268593304795602601
545035967710140057993336889004022075384825139930863716343366007923712406457617650036410612205435688 6
881774062530570060230189829110915340711775171244237036436371589022011623171026356501302439912154042 7
012730391660434852892171767800544353796026814476987479405571599377835639966210066927419271468108962 0
407361116347202589862464744081961204033687520897010880633542844369252180174251211967856991105833494 4
999168309449469840707806367546667767825383723040528489291173054802989310613282285240139744212784010 8
229799225637499186161909539509229235240387265633496244744690348057513565946504625030962501118599636 3
024036541878244570740245894880605074168390715058032424183755862679604489403118420715618426638993005 9
683519608809915500540801911609426156177994903557363529335005602169384530294074153542201700885059341 0
802153774416896976552390007001131094629800034443560636076613103027287389274226652498909098159012376 51
570432773192185028448811193320110357105719444387121835232255486772644086734045441353674039901046417
928811413277329570523323399878009160267002892904670034550632113551822596454563655802704621531470603 2
147678038734544203988775731536419729437456867827633623111986467460831716249593805163179101602174316 0
036372135135506555681162767164832287962390037143316348095868924384711690483078965100591104965015992 8
314383120189325251667689558973105180207091561282127947857682315030996548701378014203423508621889445 1
130917415520121250377976572630511758844557918166124319147934998793718897466767778272433292270248246 5
480284999856755494526946703275037839400366514426856820813090209490578996221008140773669655662797895
875993816037392940818983260231197906051459780384494121855073472344404641363331714829781976698696655 1
400518184541976331055635044884971342236033913005897971734678237347232923051738850500463025681998062
728258112455591586015018439090409864180917766345078434739349112735711271075330950790361979461708733
446648052417888060677311064558841428743120553686450754131237892050164182455985291702855298234917568 1
519817495336504045373588004097369310021016197409940885723368139890685230580215225783079858444498849 0
026722154928886129250288528135271737803182076280866589198702133918612113360246187362649128598385704 2
460547885994420824018091973627117515404746563411804682864398751105260186007632086640320800588098124
668287276915828885145355599297214513431881771664556450266633627551714226121270282902358703146786242 7
302335998951338331069080367912289759223209005353983610528084879743470505105124297994696958773290081
207079728796535839232426576733921443804703617065295956729932344168693092018662571582035045922274601 1
334917847686783106363023672435537093256269498230726186313109105016432300468067916703779309404069
607135447772041240171387152541478713374566022914253682810092920558890079508483723267871865956212 8
376549304312274644597738111563966740927499199030967831570443792739641666751097892640931174682418788 4
653928794391428071913722819456062011996049420141676575141552265930829596399900541011164776752925649440
428795835710036845090703458019087499993092734233237906647410746289811710104027788338214509831606137 1
850584279038953949613459869455343321733883804422922186848247101171485158347106099757869761968160124 3
733023068446927105578932616600129599349859749171845033446105624084001095249031129151310207353660669 9
142509744167108918044279263850255766220625664347056888812091343129654781619845396751548210810244160 6
244493185873512142860108581558715194193976552610624780925408142475964662701919437855071869834968769 2
657517135017640200359938353017830278176710220244928865565462010559567415771159047285830165422561420 0
548268513719162768982527266000770336835926768927117466145886443256295441705121686083735716597610278 2
388486067014463296368213637303317464871763201427880067424934856844572688678255255509250061546975828 8
549210812222476682290277511682236950254398732456344861209996738050145752145346770108025915298160421223
116328760264578489200881444254178235178772946368491686378710335598802935287975131660096503450213500 87
861481652756934254915758254475878797790042101592801135480971581549323586490211513898577566392705820 0
478330810319358617209592850309837197795638466498733455490133565606295899331266703542551795858953425 5
685222167057206373166820932241554656528706208202685332600866580058396609069504970302254534936941843 4
799181485403175216153188936016989829712382727329618815135404187049273485262656664081364863788716802 9
974341992184045267003615580203875004096372188655376610564625258596762311209145558061492374462248655 9
052594146783412301336488120864513178140504641794164567238577509045217705499758332360916182468663731 1
995974256373924319368360663346878883664893997708709923975176942932704315716340505835198994772125986 1
246595675803136402007793328797865113011947679012284933455937274546777306994245626020238675493090223
357398303966428565992346239434307543557661485851861284466173143979975977684470929792778276470935627
949450937574975809402297195543701438592212160580810042397438533045364671191438712266270914012615384 4
627736610886518271556640204899738718538427974087178039858785748721689263629340793705516018371405087 7
149628160787383362335597883713608096663152189322875105227403710184125482971285689541641949279438506

3945483861715452863298700743447464614650341446025619364938925571934232096238572840936220720551764698
2530400643228756038069773146999660101861018409083474528089280983391290914925830365117302996765473925
1518450277244844953768047638864010906348729677479902124856127316639984427361862308855173182399678171
5818320630969964851472957372369464794425482501448372786430354266996443153981527716867984468577777317
6724214993063597651813595392768068710323045802519156036464184552722886148251459740929971994529105998
3347241041854202720851360543073574876227384079200167634661510906147191081330087692439890505428382858
7174596002008845764482519031375548086017940341094418988372652319407183137053799835234437595489813215
3424084287482442809898880471971054529233998476551717751441096350331443841574283608079013413016396157
9445590873662789091442759842592763054393408666782643140163757170561881345065363728887368457730018975
4353864153639381737629018229633304944189194065973057538512133986275646249847032791841511149121135250
1046851190089611707902188891880624882538422836411906558740888381207312323141344233353144433609656271
9210824764039272060888862628525885199283013330589057652728295714261949791649958943631773247495809598
4149163996087240559405897409518518453701084239110782354479538977220797522617599737993180176602584167
8345852154531357858420969913069952099187860988612444010607411986374471530993510334286163756809485035
9275704744265896795661933828768847466730735277987555965494014662899892099869716485407230339888393
6761101330378404511307837997043311605332621995442577030710396843975279691973081280251126223600777540
0051308597498304645404951309764388260164480790484267719159856451260861870040257247742745143079701
7361515617731515750059883996401418804973069975506691012924075303749581557846276831148373516100826421
0568687863568408589201192682524370390352517666900923840826467526170926026971040704714815310205739799
7681579182981289235304146491987593615632212451682746172277968157330253257535223022968339827799416034
8264985693826397360590562321392948550742764853294267105896994589264214441196000845353331145040686573
1319571484348541504151723470687159665889346879477616050652520532551887794276200067792917428629514803
6393715562492149219289945067840972054346001956298474409674862465363711302087381417544833816616561518
5111911346847323655382485319878581818145010538694131580428941053108502625828157123111145551238854904
4534798670077621741380291892762345238939142805293096864556020870074750296305685666872397749985
13562083485942647022385403313966555122940526772298210771698748906831236566788925348437392893984
303927063104601678559287530601787022213306811299142564872664971680328549439089540115982149379017032
676209298763152947610229638640039099466517428602571606511721401325095927055929483973612998179810256
7138533177106675033131782787325121501327837748650207033135506227558130481080029460517179886418646938
3014272229157943503979917780164905282771302956245709102784944459005025001264756232514016120398203255
0275269695196707423516842119109820914734553452473851605349134565784736191391309740060117546069520710546
5949133473564878481268188018707352591838083903679072721987136512691128737953771559527413426742052986
0582672777608419974697064196590299595629560556000221763788281809646582429441160434101540324618371
1241183341336908604047381821867859296600601526097920430269051432225681436574695542007161049260710551
6213629879300555912132665254334472375154831796125640786774243070070876622029181406550213601916368
858999886123275152903529870349532107528969061140401659802188280376805348714902083087191780477
84141065967519604324017985891535324434233629910033903617526189914046681027614845787215430327572435932
0503117165170343024273760823271009896214950638496910029025771617136585044898207551353446944194185199
1214566815068435730756541271316654235668122273722233387587767393622874603410636815651866493228134
3042420400173053913958045034059680448257513405554904641633578438160588686022799179151567769184838857
4581979819813620129705387867266068895188507220074426102970737128356942669779338208780912705267402281
3448878106828049559510794308881004695618358720934323233044096976123771968962928211991687870983396113
5262017695940034586738339782341473219138249974991406589677344743402835803153747984899676192496998
2240176819305210022455108578603846905687663641289089751554366506561650619221818558606395256352043478
9459161679812323605749683748234389055964335094229547726071932198308253807155388521773092429941314
9026355802110535798865362641510146460711985598298259490755148136944093607176494262910343038171821614
3960417698528173282088210369061259770468314059876589814268531700310662742574082809102331168159759586

548561781614606434476519302871734300921800100587395483454403762764601382427634053294655808884773746683625616908343097270069974817824245741942626077709798914229003500843321192397735909559457468156646947811010103696356946867890309337571102762086607087821065555377926088164133752939159615391062381538131481317662760323198880497921677961104910238832275063964710076965247441460982262594749729758448101388084101452159329887353851807306981009460126167868930860243784986072082802669244510981539169597360182132879440792230675841298484903656303436908701425831419899105413985226930754669501999027720119934380998096571948285919876724145591715959557500060243914736499990949628072089018531774166215707333892838863916441375939741779799619064527740965797692823653487828864697225365575452522803168847103926942994401774415056541083924818538097224626551468903000207821275494505279154369817549666187513478319186574125583553397407737334160156114452815017116175117996399406119110086304704795713409531582791149697550614252596618790194047452875733438892990800297698778930869873399027324722936187651932972809463921588010581209173303560696088255232179700576000415904488179392298445358037974607129470760820065151683365645121428112940020791146082742431095203360028505785103178129690111967320860994900367426065788332367599890317467841882737621128226150437357826128239235832602350621502580812050380876101057792341102006638835964616931815750428606621212402530812757970025787253844905787402407677517611182828020570070680331731437332224972089532223176991483092291852542674905187716587192833914512722243910768803814676477682560519916242899466641586857003358971005817274611700131317272015264539575067017238873314438527194969975372458504518511212453237600924304723995439893327632584741996602126122291357665482604500412057268302788950129847903701012363477732672039081071163930328926896985898027604285309812579195732408053145359995068028164763767861620499087220505717926326447013802103274475785095981537692794373539955990692011086845727615873747414978132199221009794636168368837698006801932672463563313936198022844660290825747970887611612619391798898914147203605599369088389305805359369339114503166658376790682533810159946336850527021605286589896942257096353454924087953244983450152302310368334930834082351682915189641667157504762901953467655050454331891572657051498776384149079126728380317905379403906551343242579313304132494807608810469731249545345457856264329245753975443631106604365289403443842934131029921856386196903953629361901016399352853501057299327718394468786490277192411969477667967432169166174018371906560463900076521196114835072075559291017853787705695420746007253475463298759180083020271502977497789152839894533255407195166657532309264951394211425540451153778645696623468005000105766568622256559753200069485364386223037984856936822387490319549004916657833974366986091833919987237194725884528872540128464505663054723627109926427857024582922373042200103989251437607418119767998004961158488903136574404814727769349793351969079124128680495050177445358305674042673285789757264025168112914401728938893096007862203398066619650785808534824907943715105931869232064049673865635312813040791072222375665482187805198585300198832071946026351214279937006940708565595872468136554341671216007026774829236204014529805602122441854833782595541641910010698441606111936134157284385573768224370273680210549049859651658297294455519182415160406551183970720272020846402043930729863001390554348608057272087118125877938449849043705292103740970100166399815194947629499986428493736752536317521883313087108880797883924177046278893607737691470138020578895049478115887563990450268575505617416055899462503460092102109352130947675934350822422873652738883742321134710601092049395617317488537802273146628841603886788153423753915600377407866832869398483480806707192360015857192029231113417351022174559411995983544561375619179631107040180474480380943839754826744551977505936659329500786951398347929873388781017794560817544738135591808299812492315003735066265433776452183166172959235665505036298871193560120410793837252007715931419053192724580169449393893968862900011911705588515157980783219758636439622341155912478451870829040220207055268885676776757208433019621579008529471279823397076704667834310190431373909056741849317948755991990551409619689392255731938718224016540438942429761659128259606455657678962695006754576610574970349472098549641722192264151810279891105903306539154665967022021495454292522568011997322331862993012889772600548883051801907365617848724489615573216482574547538161434670840182571134649999094811442960030308232427627244024934395610559393073782790939544010805851085381144126655161542809528681117050960782891078997195299893421677946200201699898496514405533694909314415663747898278927807417117097798317152252276910176290637536782986869272805898815004823069790734921618995536707907034793754336049453072079648770163350380462127167988940461720890124337095817160070941224936215496495754923913389052137928481325600654070872192952041174511465783621081106242854180781661949418458010948482260635786400953180380553175259088246744194408371697512638377222189990351660818503784010369641914891107162792024978840785035770140561634787422640000635581746899577459816717176542473582133634390557463367004182631314361141181603329667629676001679942

05533403640351866054990897216378910193314935297881620939541968206581906428364266241323705903925680
464546366588202705763264918291587136360468736384505449748883936325563446902584799733972437826866792200
489425440222386948917200466558257328802332499453398108946466293862106782615852789557371113648254919494
666999004651483304786736213896107397991573499337265679173820506794890335785170348015684871190573311500
307323648157543719277078267888498830184864653982118322884774599068233974462516125853826623722831969860
025204337862773557515550720101605978712174650697369997748850582368618239740596282389990617184581682390
669939404234038027152149341645806650609425516633060493107167197330836033118091222672826165364917781500
454713336139513693578970790381291008172086749705669639627505283727439791628764864455768107697904387770
885368984373003601431294866492623107727490370595621425875142932668782847880787628478186245956781686600
258302682036445978850982950892525944172113535585941142399515351241486610058114813371447620574893722400
169221906391313782811620178786086438886050658708324086963946519451168837952774635359778005996442250
118812756016017915902247521397691067986320928338406008610218467139811866120510377179678586471588911900
197808206511049872092732936744466455522783331556127984651988348361976080158313179067455124100208677500
238220653464455939748978692163232155531192906025885276633670028108500043677602024625577852652669770300
400695921577615647642596347433564716021855127675264892216757598015472591183531017790110214847022673250
795852527548423162615892967491280149800575414928473240746443811161096891279255334866504394777647016000
689704662497021733473368079477321086193443442264216085801303502463711108941668104226838203658743823440
293369193737263789168983387513755579102654529523137287857195672725035232727250114908852401112212315223
953566814175636082796889472019932324527491150005680661906710073056211312564018295049933428178361120040
413870830454117779908282985239103165217553221739083863377240702608516018650975472284511975391203976200
759601990538382949498226841606942370746852851185966687687972298646027571845029912469206489946948977400
830505013519745922277894280494588893626617557558501668491138034540655338404509254779564836853141322000
672805295322177576799732975000077209030258510443246399340674510943312435873235986285879361322862400800
278150765608155394815807462635927191062286452921295489912738899347689840630693501530580839565139440200
432325066622429876592396826941031130834151196755536137839854221179219411449561916538849185979570764300
267369745935938106687751104593905615875963496617956775816312907314395200211716241602363878096329901300
324703785938655291405189485402021747743494118074693780365326131399462089455574906039769709495500802600
496879083392220773063033152719944794489978198182895663978726235995058458408402261607128870691179553400
834127291556345078392909554962176378094144760392676202991209797852610126161587127075358678660701332000
293741480098053490197876511723501392603202828319381375304617438618548707315228789604380162602151932100
727812501015366095535959394826629785009647214769630414209285608367553697145767484658251377926893000340
187673095366561540940181812138145157189348521141376323974759124935970455118111351262638521822684142200
290576167233622174163859695942855584152660325817356968734787152825385486617083171156789798692079668500
79385203867327936313110970175090915363971577478546692389841880008598498493111591372836081341437393300
408772681388157631645299040033170061435515871321048490860408630995545829324985126941508965889662019600
441030335698575787979191371874747941989023985576445427531759127821820679194009597944889802165519494700
107670219496596100713430683605884398062502933339291805043787495845783955975219184993991556568420764400
401577026373497360798043894373233710357794629495956975286424169076671625974311702801304335272139209800
959101629315031440616760330448713108889303292520953246042438715807137341133338496752189467842043520120
051017155888810140727903370901390608296232573654349022515857159734383964325496828242539277712422353747
636514746268630421600367373905728097415180263325364778806297783650316962478876630494658904139814536800
349648376763797240301315465398808417396936142586717938191404814312622118490104771070020651363401986500
582459549151893608860207754337442587932350378338502501167720527206099193802611795015938187100434945000
478642611784251461725602723804763268699796611311614714759878855420378960304851193694719210877495085530
862899585037777990447987976819174494142900098035975024431445721638973259334748433045739408125512400
793772083829886046552487144520681203144328720218249171133324123894130322035955727805144049560581365140
098616007905100746445574326533939453916473024455409044935918950093007018061723337167982022317326195400
865526063765962406939391279640892608416114880330646499405407464597314824039656863432158312994393707000
782160027095587125926393806265309063722715903021445408278903536701670981523898327042384001634810127400
986996593833187719507936970830836943672885650611025094535240744190595424682093089791961069973226213660
816736312298905624737776292375576101165400655938523727391506690645767946392058971652286497743450228400
140350888080830934461816836667371572514163826056268400920108841378388920760123000086402723310587403779200
368077723633760999633454914229514026340740695000357192016355410390524035290727326982838914645854980600

712197168339517746845727327057830016504343852267328906604188841138250834136137996198101697052736772 4
7197959326117762234453981953204724407339455212937548499646212828779894759263556471124809844788635485
768440400796454086534311784469866699631550713553346753318278942174905295408470023651559371913007076 5
320610164624284605754459136274080249442643024742038723113681403501164916738803397962812882370862631 4
77737109509252116696677284000566965235533123572734471225058493342100065438046715335152491823178461 65
102580881801646604999405884890852354061518389404360215804758131640671292367260694480645599978077249220902903
862440744586970142059022198880684609651517094823060000964670143644076680663727968756940713515678279 9
0649079063277209044524503248575792807082252032296833548515860692931459783852836169456336186314956 85
751252054275834084337654387735735811332332244587419130442333288530403416818532765079629533256719 68
03046626222693929242338176147406217694381301816446836250602887868826388486228440768649870505438229 25
8065848704040355625732567495201114319375477540770919620795371814882506166306925483188887162368525155
484810807574535347566506910899890257900674711653610446354848625845329601116605524812366581334844340 2
303383876694730853114682295290092852033970729724784595032639730470969287727556674107879212706769691 4
7191629646778484264375541857569851081658718271514749550361531732073805452127386073713493283299 50
6794838104667216961316674556498386410665538519571147797989780405313153130469535595601250122307301351
2282359030897413153187246065767650692731207540535622805396956670904045543310016992009340160301510967 0
070870330896286586945481117116447222429564592624702284383736316827618263814545273819011571508952201 66
9555895255462206792074276777692308115226475110682443391341650025240406933025859894563392703641944075
39781200821821358504554735215748043870509653774461478113468716565558883519727921318504550683553943 07
60503690093629623873364086670129443989385444187850881588375087629000114446901288812955855677523752 1
65986849343242206433312659157488729539953386225175380682153450511244744094691104511632399585150575 51
231258294012367477151579026785466343783299768437518167132466395464302078876838451413313112004383627 2
094728966779743948890340393339347714916556098784406261235812235129549256259585831916362448449112395
36576587105078807396708810976691210513367202108849092498348141080851471197931198325601491340711816926
5377176694886325677385081130992939242797311269677458349060895901467971121975453286879491003849694060
70056456723771053491890644506528029557447570185785935333025343731414420530358189472502565974199223 90
0855033384792966237076723346362903759950087543075025796397415020398849590375857682910017344100301639
017484856330642201175201778857984274575912502607278147043373118803108254072233705618403981464308607 0
132641399107350024418877255635131802012518580814897540973697308148730540887173734744284091929164342 4
402102449642639287357398240381058420037345869531070329799502293094662680690179272417870980625238297
549267427174010931339084400603177150398839717565171566425066140863519754573459747328546035257708163 7
9080538731580692280533069866107176170472318941722385413267567686410850693619772850880899912059322947
8701723659957911254740490230421735611643954893538440386619667832273836309911100528583707824962506145
5188256938516357639930307559070740917791768960099420162268639945930987166518752762806121677655991791
02906097798876029113129385895535018018228228425127176741423279437724908344684215709467901049134293 9
7381565935133360706191251839663489878904947087286401024139652538971443908917775225241179826520
6887498243016237329016542389588802987575006283104455394872701249308315249497974712676481104179236 90
3256497908621475914072338569985648993207228126831503709899193130760222768091775960194196631360653 37
5426514541707897265642164094991277672019356518712974238974200774227060081833146868926602940980858395 3
4529813264337429483971315782663489388817585285996432152468492022170504236438629715317037861202578 28
5472396855010947264868652733936132705317091849608428679730630043616542134626766101017003598757979069
9862232054880264185248692510961687963809766535597653653614545744455400165223914248148929729381427 9062
5588597012238728348902405738552464234439119934502720657717152104991279089921169924264097040941620723
1803949694168898542656153032807224682554245811114270095732327190155988537895755711161924596312339001 3
89238727215278612420384168148964678214166675876698418258452443941373067714640373433094041364476929357
83257567547224604923772545306631226140550175638115999431970278836561469974535618662519921774758789 66
80220466677625977438338995660390403628298614827021386190536066366845791514514912966241491896900801 5
3987865583853781157034266034432048225501319786040476271115191413296063961296795675148556053596642717
64873387754842168073267934468273745356610801508605743399198621529578787611185592447235271316900900 7
2760229277857204073949284081028003889566540215556333756222914589826405817184880903521959232298455 91
9169463929579675309195498710901410398738347924936289310579711504620617690105468930136691256496076 45
519105336273179156006459648274765480572318894713984109860130284866561626662957625009817839445743520
39379491631861623245084104361645553981702333968280754081606767892350510276470520409569714819307832 1

599322556225791336901779370937542504178257657070596222970542412067164187424641557566178175183211009184626487177650911903345723077873178804849371765442539452471494240914793370735134878763145769851002496749829672571838957837846494786398544023121454640702316093210360555946195476083184107815497585524494732214389320523373497582947729369785424473319216585338317552249465848759397461203136817692491287900551784037075161061508632843344567384956658915034942405200789741383221214924671798085284634286820447027578369882865730447370179875498833918216443632049838526113090044610037477990279490761245993811240516191996059651390077909634293583119034305624356715734095056163628748278205876154899881362284006351119520178080967490670497658942820319324591324255961711416431669441618015524066188633117399506879643778155380972229297459868674364343772446422500821701314936991814024542091576768939550131681810740342830411268652549868032645793231845029509774520138980553581941410191319848338679855548199401716615848836118148504918646756391772863305837466512519099517627862182217783739216042843812365855035987768316798876917667864037696003396728064045734525975201932854039038432784751855614753102263637335938463979994511977345685665446742838958191103508750244754207501175472655793430406416448164000185489617235736987650020914631244006805066561821132029336859567547226664660246868542008039260740562982996228278866733064506550328841629362588755409646807070658121980508578924056678203730191320067069036421683649165256313537762548308959484205360987229555827524593150094334190090781546711225727107985212227375561583261030239205316792788614118329822645975576534045423461049029572239587533119579619502289460503083536974031984824793750389739827953899206897189129627097026816017999488236851544102490634503793304638505980506893688509215960924934546170186966470225326193881930168212264843687779523961813687773501783987601428797208483664107732876428869253587146978397261888103384503337118151711484715753572829111377136313197782120245684609697774921963796847374969109456442146235274675272612853018007897537643032753550589002760724171082422779772273273590062666386669601352230256085097299631537578243379250716217207440633313796385717144392662381145539390048386781842471758753790306636660939326888319312103232072235140907033605416575082209060337201668113885031468464451916950436558866152125950693828434458152870871228293140727555933699681210990351591016421256071107576563443063511705673167435289581947549521611425193110025890413452890075077583181812267074861693705113747405144794546197953017476006106793453483773940552135929988183465458679825875860431734040554601182336493553098322439166680621202928592930996275725403127489439096429633208730774671500777330083393431588596370124433769577694548260771609767086154816667942389103506090461460441303968713689486799835087804068064381621772406347917811916200629577770139937093439443217249722182319521253791413262753367456855860884410590851123027060655379689486119033343118293391087619618565414570968938743693871061234280197733557389624076816315844335848770733607207064012637246841255098300951381957686615124648660191044190004053873335671201538782626114531444010194905011564171801462355300334608021767589156147995037146733274581502728112721182646692255544413188398595093341962398594556118494767478653220714920141404358734838912071052581636449069203988138792728992853988460676469997338628784322251003743280666592641993060846936167567748471799555382249785746556725055674894930960088811165025990005993740173866604706262123885284817010947101387683522002537004909446671047557900086274869986075801005598973977524853207418346619393789997610753993231144261568925197230784079438383123586478168286347239705070337701551080372168639415071758912025235200309364453816100008908813050203916934115910823275492996997841354483322967187542418224655200379622790431069770241676548293949761640495002830983893994260224304616904843558047477224038717866934915392885786302298924314368417304703015701090230606750370244720033264134872856010032197236565201590949293414826212299982317332073064879601203797276473155636303700929383734234682091833203420880375831996892409927493635290735649847239275179648353646038113180744527184622345859799497228434180527406250005783467406841300581018485542664866184058788046412068103591989839609872713115064108184549045557992760943542184006717645354861510528247568296268598180602937728298792442529438708541207310252940498327891791277490031521755214882526034714160181953845417671180625218368758194154070367681561576618172047799869233146403361380334652040184261580390264182536185722468448606128873686999272027416268063766621120692903461969545811364344741598714018921166046622658266159054206976394359312366204552876034216500347360114934225614914032015794171185171542756396517256864538467095452483713059489858255974527756437837209390373760644875780538089666661399183963055434635153154818588677926291272536342628898525685446469814497461892414958636636719814006506850858886080602242673379881276879694064970299154524527213257642819532491731150662085866520774909529651007534040492273565482829570256906293588816090463738695903673809054306950261654828295702569062935888169091777724209554461302585438178630485080605899063738090543069502613842422270537535459859099326696733216515194812534532747333602744725852724539478578704905484758633311

5718366332359132347588259340641521039072871936326796375928473313161233978154985650774595742630192501
3613442181778657326845949803925741969699987645982495670409559549064514319299753296990292901811 3346
8491893973167337404737610215349790280131722337912799863914710105736458088249640377936691442602252243
2918220359694796522963241504625930376366432984086561602312610990271779794048244237437724217545327436
9030749262617258880652233261410603381653209323202669910870847586819856399049857501176199636905969925
4104368753291819072004157598262345466727015736971133357041403209379345126606070799065586879616157998
4934109540903212436544310873161586375727374817450178665573793984866922911759920434324760064859760054
9782806294187391474966456601976892636165982896565574458040991426890947249706735220470116191536009452
7363253066644020100232018732278197614868663489891327347014448203242931178410091528333303769111970512
5251189029708294297488398137149977805278164934337056043260053532698106918985886898161088992699204345 4
7815535740465293822554757923965125781698498673421821853124073431152960821411209199994061601015821912
7373016501769581186190366897792756904678571018105939373143881192914794435652218962602863258663651 90
1745365921218638768107774209158364690909165182739807531030664998062448492774756188452973294727139148
9972684077858977868656048723305752422857172473443666741818123270841591791478616819780032875242194648
0119315937935152394170490419098412578099090916518277942046727049153450002486042745530730683647222 0758
9308219944523472145218428256209291437681438780213969363997860221263221098220773571441294126406537652
6428543482970693655116806728306148155350067799273428746717408366660205372922048484087025230125857179
1456966575923563596862706439037207268058794398134006676701411765812522334810483868677174058797368961
5599621784654973073934095466043145860154049561577645616734472126427468746140830206387935980428496622
4722325504660952317681724586362611284833874076507582889567745895887361095197420723212623335561519338
2471760218186838953006797502120769043883812835625814501500711902715651443467706481495059019699 61390
4056079067372607241129244719947702838483531863329817619999473128331449159687775042903297703476193806
2955123840138422910035767692969859590544398289266660878010344059670559052076683702101595219513945447
7311874107235279456084435494667607928826819356764666589161362140424785642792681562544850631668526832
7524656002747765241279427053419348280546267028159924539124873937558094012591977834653365573562593766
8087699752574602702166964922598377538196938462547085188615104752264751364891883357381891697 12193832
6810041957653770211921182725665088966824675048666996598850420411915332089858567238092601814 28117623
4479558294288058513698860737275497491213068743742194301486676625969169519536885691174938231969755955
7940217938873722515159971405447289708395510365486662819650364868466103927455684143635901 777440 38736
5120807714563648777284438331563572496836756314810394389680921192823374145076186305657777377 941453330
2578207887933713552328193066778103474487881453302276711598246301467739131806469901714532318195729640
1713829934458665281042390292860440402050301085037228897072266732041640758383521162554729354209770671865
2620762946720436336432735056632041125212872258941499762804689148555075973614650511764125793028369745
3203604650615692381197213251056180613452134381768173276284141810216441341702491755843656811454797775
1958280662844249750379174723620402442450573630609111548402426997564336003530143666597511828032803953 4
2105404919103056731421075413023579348772342390739593893004368913281485468821970534826806406133447603
5746590906564609800971717699575204375563545216850442243908278083480721568003121380513744755738663358
0661289825695253557232266307395195406369794691392739195992907099291348801486760727149785168270269050 71
0678967657653044046336262688872262742931202875173849755421431504998907456461215695542535701520147 4626
5873588291544869367168210972638770366929876270333269267640511235659178773632477046116118752837864088
0382801363490414508513182955873063526115923755404374036600943126624937447350183694088350123180570255
1089418469366688303806230043616899868181504628145439937084394672379528195303288606099631547805411053
9798317391048191798793376309918240071636935359256783586699982565682846179990244849215828682554266066
6948065905588375476477890063030763977320119162644193123137338228364181193438305058658554982 9986899
1146681311711942189160179373602575963218532104868749902047347029942678712761333423766834282 22565756501
5748972028034318032062448495723097390057150931453889184344943828397373451579980515625916432262701 4186
2062369447593021410508508212036049109939053684780560216627046368552572741329060422899563510252284751
9070308253273849955553304495780330902592753163522180988988262911598033711125701721767 6690454568064922
3051574746809155717101756724035418935061128887302404314431986958652186676073303854903602774 6096354501
9525296534070301597032409851150252930588656719011250832847149680648943813007837186239244681790216217
3569122887239480216484647377517681894212220710356555965079794849900827193552814791434040371 38717208
1760903198856458746108109910593691774379471287936895032477418648580648196799561464367082 487089936839
5131507203056530300788685982036072066991673761641475665428771935335910445211691676928163963 64562977834

073706478827406183405435707741216138320287378790378585893274553614956446125505055478266874545509088
688946927185988949423449507483821850184341303236200466807001917504592628408385053643126768698034026 8
115807098103458986041308417355099569441517991754323348063307326133913997978800038210913276601454961
157042858028160675162338135386305242956383309503201928781641324922300476117953582495605914530004246 4
478806213024689786559692629257635740287994013569211675539004002699664556025683927236950495909960271 709
738166616867800483272929905942810296816157469630606061010622671702125396674795138938137485389531757
832945136426006936137777440005669930174020113667731787704469450601292260749691106157620763789334504 1
373365119560032907835235964632657474974335286721116298075675851083850160699896935867152259645307009
139087628741649254170816909684730954886023329828880010165296280897698772319690742102700934880934455
512188188336519784318535596067476350722161178728735639465655434276140685594124259121417078116303080 1
019258762199380989589430509396825182771303303492664885329563187816765900929250897532471669126815818
237760658035402279669108180932267251426142876850855023403639597056214125151376204672244428979978554 5
413190137205600294567063403824299178075125724484463577152589347223368452680005049057940765926118340 4
627646998353219893312711732710511293877462720081317263208671247247831036953250571116684669339199938 38
341653013309247729429358470743882634240070132171320972827494796116356678261507156628250212520627638 9
751566585134060445529010926112638166224236689772760490418862700144211287359239399825005658006452506 07
509458935850072180729322082461082581858746713340673344326805154756527611229009421546556183103412687
017228654162269075744666374025785058198390337266853912834251138897761037015547056978428438121104166
749642361936984063561387622795954521514689288132823943963661294501387816765900929250897532471669123 6
834827515460782728044518243453759650492568064853230999281451475345509275552779627941424557440525521
882532395562508572117996353019567581177443676255097319751155651484513728542407490483081995588091516
117939219104261909284857816139051482314744553613615917841459994712239051659694411039727295741158339
745939062050077580709596765898929461437343487815637744208374999146183451341178572135657651002882165 3
437883890697229986296422686851956288323205705378942178959367844899972838082251295857374295653148704
304032195433016472204581397905745288020747285881031140858798747221186328648984473405311460440048000 111
896474216571999311588903720590031466418126179209119408878500667780109991854619093489269185091189985
328175801561496344252727420213230832626253782975374780849648620547437729805950490214555010896934806 4
510974333005997985553203313182691751915919500655884843655165633523507949744148995954593468266676174 9
841931923239198584931429070339724907410433315507161857712844527045561514872082178790107580994599078
190340768484559080348525612112446388323887760077256405950585761450617292910176342106201632276313357 0
869841611333843351915955219934239104565565973100823169947997845631959834114305637058884135857232343 94
599937160845154432247333257474326902932111340553605260907241688375335436141287023423782527089538235 7
658516596417328110042367022479426589495420024627979730989345076021364247137031065132759759613489 92
427004704717253332642277426234756632022519842495249821272655951139456814127007662315485158257359633 5
382671792266999633391806685509480289663970163358030635777920615129299586818160233278268287561386953 4
898338580784048677565029162328102232126286237036471167259009111622074499444703371546717193557960340 64
957218399132431571758689056357820119356227777634216236724756676876172206101521338942884427554638039 1
429829340970637684665116996845497749248421623454987901900590122307110248805880499208202673415213345 2
619482271804045426592288442955419081544921458089045704939272983216883413429953682586764410836728468 3
313287432198123007451303144400768357402446278097701300214218712351188443907814286457148671303162743 8
140533885760388529469050770691124396402844273531601124684855885908711120433136983355129831770447543
952735118094562627365072134238675487816235096639875898907810543337220397544204986189921453439457001
076699302333404507062355828321979391210716339044784574064959683768359996102705598109343327545718226
581916273764083264919446339254141525704727893001218140995028817766321849854358566679007037620570583
827317495612091752324760127284885442298986799882662839508678943577143881580967550580114816329510134 1
784549495324847435024616682604908605434986260707695220684576930888101993638860205690810166450518721 8
150057434727469240045662801352442429043237968476832649995978595418767960988824064974266522958440697 2
977619146264897831510287400669793262033206040445354794419819989475253195717052085531617791641077276
461392157117740299135550151670976619652406222291416099722702986540871469079193291110174604012062 20
811903347933507409339673354381066776442516210910640978277403934724092269713998641932401833676257875
166166922114857084403473110985771860414380657030581140896522799033707067927777173240196063669059902 39
660061747596602743647723341713212564069573043007387189697608262537784076168904565036964121017260551
105063180587830717710457167891720931555100513126248850749712570882776081846297351566606413813185775 6

92321942216169819818338615910911121296406834764541498748925967693915520889997348348251097717173347 7
4849042415704476573657457752887031370873919072182505772993177204212596617789094627810737489396727533
36694977597757614013390964005994959912477424058382026276743479140436597750071174906872762693456375287
627915386103348098692957428984900870413755216037594678764398436269591372899723772007314533667335 2699
6836629260856570583903882359383118961476361331704366529764436079401649697173956600232484753 12782613
5116274516934985914972842578011566354011257488383449578205475651348675253533930494298777856682473832
247136341278156638245905230733983253608300387302483963995418402866298087668996005436067463747817597
385932001097038400943290848252148607855800720302839262481484210735676894365084782917239431 3537783
382862045256079371443479890247754480001573891165778949114366007935436393206776263021105215 210135921
546945449970588876283657933406061323911023381234789116513396064823273430276115853432570782 2596745669
2639944506549281999054027881507062990362135050452317325013670624394106180766761378661414563785 76604 4
2074924229269770349845012651492167176963310516767267848837295490056786572397844276311347324977198906
0076187591408960730665821513844913456155557115108451321597382911001114287389599461623938505836032 323
442082881450433915073780818631937207838031113641758373487519765079335205354261064839646800228318032 34
766267824389703438282856767409938801224558418282506165887184191773971348424955753553615428351 0624483
2821141075660967969951048252279421587069315186868228065990544542897676834078428468663516950735 92900
5944652517186518412045443271163745245912494051031774974326143304786160442520357940327121038408 8446436
33301371529014988427523779498345747998752815119651594344029887214917065555918493963736 202353346368882
18131437208134936589804188526181461668564638974283915059558370394712383217345273482136 15696128326308
5165038215295650842082470183485563684152892779777776536659828621565022125074611675903 2116542909654147
01229428385457567214173899297998005246418168473348180217325219882119503518461705828084 2927115925 99970
15375097479007950930331880565580101398081229845654681618715385799281286103766006844083 08498397279763
70041660306524614826604283117793289355871156163438086838550980361673411774679552 5054156963255175659 30 01109 87
103438333592007520317718713211694297373666504816162281397436919436021 9793318911371477579284604851057
4542832874817932872278710961589682604151910321603588677947479259298850999490492164849 71385441255 339
1673798326983742448741274709620633713904740483607941575293633639044817 95741117349362 02604851 20123
053605543688879386291448078791005255989328252773354359247067 3999726922739927756533972566 96715240967
730735176129922142025550653142954908742617053358503331777460 2589152893788654307666139718694743502046
969668750567663782500133665443412014978039533409484305880 408087138695315898917544323055761484459 9297
128725620764700238088330825578637544806735453859876380 858476365069526736280986119900402679 8113020461
010194459803592743530574492962422773793955201691580 0533206697551301973113370391256119133054329794169
919294829390557267975928177726968793291142577732 05002014179680915220206152451794238984581852682978
97974405416564805621008330286054730103160503201 45740518887619845350296999924657956500190448 27824173
86095401791037025463814362445250887029226639 23366404351967800356565544364250362507952964892139065648
414018273697145021452643164662100373809950 4185588748963082465740733426300253099497815591421287519705
271001157101716377498833787956568653934 4266119544077414439924874232060422661597714780065489643349623
5839622113531257289820135598481565702 04591737378662802539307044655436889747949065845 7794940959 81
2244787432218660153009734548024696172 67129477879519441513440173011883236744872044780623663700352 4258
6176568380848573688569023709229088212 22720834170089791292565435419407826899305573151916900118070501 8
9588596474048139046260970734195449422 7067754033525710269483606203275561731021591434684415539095659 097
465499279153763329473599602008980264 079468292536127789832058530508561861189451406113222427128611166 8
67250257033634886315857173971160328 8212386637491874566260429531898520879176927985320596870827320607 7
2049087562497623985063918737880636 107372115907573362989705665746883773515985230620783485667267399450
1972457061604170286561431266850797 5571689508067386196133576130779338566529426117899557612822039627 86
16303669765729274631760052262488 13002620159344695993323013044439085141440696129490951435113240437281
803780305270116147143696626097073 4195449422706775403353017029483606203275561731021591434684415539095659097
876043952080736225958640787345 8191531059455852524031075377215743214571344778184309844749991891988572 3
48815392190452015661719255542 99875334540023811091746286205720053711471238979747010158547538626802813 2317
0286201290386585144236448649 28652535817137202938801752941010225202856911871860796450108765298 42535297
16311744186507520187692927 4642106131317620308506790665882560616162451232983338560212251996132 0728581
6407020906231271834484822 3007894040924391632745240874566781367355700397551427807392003435331 32128228

732868954206188188329141890423929582934929305948139194192101543032435674476992430684895495202395456455030650470971258953086410011569687327280575346288820928949980376942767152372469074452768749411737611248907019198792964232474948371863913292204033404762728529048057020058877226359767881747157982142176417664349005485153222335130502004742660842602758111430811587735785368141365683667133917403160705603790852843226782164643593015192070991234338444761974897013637518951684986306334712439357199345332511924340248672268509671224442280954862096706420714603892359340106840047536963864975135973583310937832600190925157443192037632129377089055745436284773845335137564981164133768920952528886685199915916299617867208191837173473070068228127018600502753029833901466999574794461070267161116027871706500234454852655318046152798030135889431096643822247543977162304676463531802819964493756623711511605197875870834292146749800015297141091672570579693615487561571782358311177101359012535395568712745799720175926065461005937960897846190272218145072358795484271499133150396203051241109165044196697558251321818617835694275540614559727035523826735711023180819720853948601822054826898663602869566818348668524544614408406952182633280487604446919008266967629645490754572236923327441664913195564864399458899338775870094850433186399855404932306776159182859287438960578046409842089740550683296113972392222697903646977687551730396664474157472658465478065969395648953581955700357971668912269469927151284486477227398117418148866331928994659476000892113189894296771965704861852768613423688150000417238002829796700527792276554084855433348616889849738718678861898732323800424009638640679843517162511269725924658678721210705380153194957716494850629815798946941714428204216416558660579410549436649614964229477931498122383641538557038089789007613901032349717969632547196564912274558263541323414236435745947492979278569607763591484728012121820571237229125443324556605340748495181446769058959806952003492300124986619376210850051236442547826435733821329660966973165335354626473073809288176113373972062983640305054086194062218385024498547566872126006763397437315325783835487482440978097393361487310202390453380947415977664560313768110629892140901661232700390050502294761351885912410647065603129801460888994927862354781233707563735243212718006105308551717013450633073763018711136693532176984282601761121860063584896534143606709144199779246409721142744954698914635554828643401104014722300847400589719394255567755784403993657012630077092330401772015709714022618972549024996396256689084858977504157130429271592893380146276281780424512433461156729170872181169869587131261066558109715515556963348198442249377278998499001691334065013922558374525364458711315377496428451540053640421859769093297900720083629620246732239658837413175290586626269446735210442603791931521103560615613271779583324238941011378308624546295140958171894165381882609858136255070281472074410103228385697701212667932146472801115968243771101640588292038122298228255054488502000903159908409669801425015995974256102202631836172557111491392144361101853384556828607031576644860576825091944215850695082849408530180664499120714286759898668841270046539483571531979162968219164287626912567216947228774367226907463108534195544061133608487163008323207815371548154354645837025948503616747070730275849967231328096016293612335740850516758670470322859872473980864732679403949139379130849736324141390294392845766383519842186763466430180868960692143383196042210094720984397666522822544368342220547014325651194268703971992934248063911831147159629827659016065848116137229441994332637690168222434355926079908820340023435099058591392977157604947270770280953095758427070913697704371357590202672271213553449733043050674103705440773459584309274840395638204885926470438636321199354222550000256114497085052705009916289296049847398305977089314120419837701870006385807844261773612787580995515950384666207484815725018212354082543379820275256807457417795530258946819457764606537499326993288820539515184740851474436356821092425017105258634953457915902871522129553659591580769737340069643981356148346862593974252934937239229911260953527890147415209461916937523396079180058881837850668857882217408930223276707892649055901783573302904098710470476104756624087404030114420916460792765924033500521369750536665803340501312807124279897000627372915259070166674643724566787432560019555342094453633439391956768045908713309834479621296632581163765466480890071124740281515643434631081445327339715432334549268619307448373532903424290227063234450908155379512377571610492712179818533585099415538718219449427086114969262082831105450526017646944214984796916249383350864389979794372572585034947332112364201564544595832321025867338212082720110236171813291628134769361336231630005551854169451237030874717693475306380949148282318114601484735917650196830571470471397160854127760854152779694148408154332773758085413065135195193525855496737687537565590189158920767435261488529376321078731238757206484082373753032884640234882928758674575174119642595547325471299887844133769717447828951480606297501598104109651792487372542415860607092634345112218051515780503952320798390703844559080248976751242811186831122353594495362323280515642845091163825328468276903779894056097304965982167749296052290642715409075963049075004578669480644240789068

```
9141604249897363962238355342656882489249030940135322759829226761802139944680189960320675803429397947
9500581123563398697279023670197762898407331986142970899455340903726057682234474886920102976488719461
8758629342131709327277691651871002498996665855083811335678961748113924008069204414625665382945223339
1551604594824870277524026560308024160840633583104991593130853947426907323472088411918202484707573291
1507212454468985268553115998511179393810802097734738174933999888978565369940387595253362174823947157
4783480058946039366591889289962175210471630466038444412373450910382932483894560012983649341732042243
2165642758286269646298730746054943708647463427711852938248043935821601986070062171196591832591807574365
5660319057421033069753707322301404429391360729974231482186205712797240843229742225306478470222877289
8891720445413641683018620592669478106500157727303468399829551105996427519834420909942982394136155832
6538840685298375019610026743279608322676608982437399230458381935994455203647109590825189916629159319
6398638609984425245591944423740980765055611750578961464359199255642675963119627783751287465379652386
6525681695700326964287850720184716605737922725209523210990761127024194913962807481696514943204319306
4899696752352377801336017153557941527226744043835477478346154171361080719710473088602374503121389008
1325624317204976886802282925502433493959917827467695941155443098503616431316394425732596746141758066
3424492506040232202802946987375182931270655413778298861958481093502663645207509319615250820150239512
8106961883020837877917531424737800661366107125561538095687549097630224818254733695777442501363334650
5272962419515030700162982343399091000615550249467371288432197235339945293806723419621701616343292470
9375744591466577156266828511079209801382360277908524908614061323820470260336142123967891694992342320
1715288832979939112946652538472686405318731979322677702760178715157112713190221683641694532904519520
4615033553473446978714152244708770708456141208314980110666716520746555367187872873746904987162762468
0530857583522804191995607327665057156957730028792070628730635622829071093172254124410289965621943930330
3935979312729824901885059982075302805818268734536262076988428838928963551772699775527089280711983832
7126416498135850566092974668194143322037636016031702864540103804193375688745593512383982760565229937
9694631115736236765803515756436802079709806973589289503336310275094784781357000647031676531179843748
9385931992844679705050145842227782696067591977431422848983462638095752243529234359223513044120349
5909610127448589140505827476775911036197187481126259602724586676557147275493941205551336228916904263
8208357399520615723946454714449899129702110657095944985564756710539101364046559061659117790624364595 7
3934607185777117061118451700154545080998850040593155875095120616554563146200738739443327439156540822
2265516671149813613507373995489339174808637419664809327817100263955025910407751934720526936117215523
9192894891128512312563105277097234938670089309885624797358932759808463423218516249853050362731645550
8600244801128794870890218752873453929413161466208814826808614162015154912204198602598488609910010 0
8935521986004274340573101214273409247594356726977628542772767597954067832154999870802605831328690 2
8183862100061003376423791980019442370642063331998932146516974334576991282231826114090708986144154081
9914737474336816449824325266081666966973361969533212776912977263579843015097108158562779524101403 12
1972539950098548370069915726381749334231984170867848596330912936969977383562887078408002392235782310 36
1229231334313870871375607264795530687856767876140867978538841839750850468350928414471968343566924539
1453969726703865703172962418389792354536870706291053584095862528817292816924710463959137654339775103
3038618690541627854067967188564523146253468346430020306636243007728041839150504839977463906523527004
7668220337015691585237701390054126383476484663841179107634163945096265761345248340913898753793488871
0844082251502479447198768883992003573792607365768549301553430268438488893140272196682038727684904 06
5078641498395483862399144327100354841428571446636758414086857564974964249205885696843745946865744 8018
4934228027958235637756438826823262418776221622607090451984593267734773501828543606939352416589601174
5073761140640689598829299446453818686066474728989919098229601789297804735791247963218691536870365595
2944433499542516058090830492735940881012512804591065050476694626758224413393315802709420923435482413
7545730590871565675676701092095054671117837732104759766979436357024999172477640990099618423422593896
6846991544188772096469510043371831157287057320673987595782140794073633423608496383233359179022771 26
8263032769487753200468018475770539434300795119666775243961591663082780839590538322951127233820755074
4153078897776286771662518810911875351973300986377174786188154764116022239030319559678981537333332582
9836004518897341313857960092977458801612401045915014782067973943636291993558227602367510347827556 48
6627121882555285317825358601035108225814502612047470924017186025646900068461731767990573490110072728
7689261945275883358758221947384320834578633052807557454938289523900598456827291413464323488171485846
0678305948260145359687621596708124955323057637815649344565257825522938249626575117074949886687654410
3827053374089892104036867736560454285845169503140223236337570264179657785208175448924655709924081323
```

665578698526453533888011188919328625192592221355667681557609276306155759306626473926089832783476802 14
60555713159391575134197362381437944978888581196334372829232196632033578012611307701085721 59898202801
24527014194405510821109126282616707080627427106208073756237473911790188063039151918982517 18666137575
727597103898114628132094118243212011578288178755591908312841658981012959939724582828850347 09053022829
222517897243132948789737432150734138953199236450940330089444177994978550611469561529485655 31212229462
35263060515601499246416469416535817179065975346477447518944903388637694549101338475379570124 3435328
3214492982757320649460057035879741372847308556850024140578069949444209656454542067040671269 177034205
58935459214751399465865637953244989394807296345359598977325503522425366975520262596619022391 137446470
1515637749468173440379453288595394926777638755908647972478088680068172355585313143563297543176 6043925
383854057568997429850951712772848854066092032655982812701957135642813925406613098387251913882 8743230
385070066019921570688871565312564865871792836573552565862718814401443171436793748105969137093 21216
80624247162372966171539343995336805414569867938971784889101598603095412935298746169109192523 80974294
45514885583184994985947959024263664255486555633149346893561503414883748846312946530056590746 7697736
01945598152863062761262676635577359767581138421612373314597087298604713841740124888891879713 32736362
62511311334676537629299840890377895200008539399763584744281979857678072104896345990778015426 69717456
7542617238402203277336476535715491665338444272733685335249669156297692483436265047461988233594 5593854
238873901364177046945395798118752712159776844251179958071694546686174982003891413675742529519 9235363
028640995847738080667594169715580833542623996679137191811974565095014215741445024655942830866 854834
5477550417532809049243969757733834069290051034226532105842116831153608063960003029834869862 5014 18
951961597709853098934159069182647462124298360754364446243464183758912699382353801438495736 433235890
8803400365035629450989571731821193875360604033750257331182525704669344702757566572460202434 358051761
79804305001576612206859107119164478367877552875576514953864629003147784335242018223228186 289983604
995567100947130736251175046568288749334511080043705700857207322750831967368410921087264603 08626987 01
21760440904028060979491118396436052184314923073516315889044454805679783225559628577325030639 07386237
0333133937948649073559546517965042966618255310526459824517353756325028373943553101805927205 30901 0514
29092433527312786900151513055874053955339526490302813127589266878502638802775398087841969542 44470759
82652771476368013445122704646965861601405000598135542600445530412564538564899316031280559646 9709727
717647751362379318695356428514304456301271610428264940367974802933019013443998609284058688 100268883 28
778606907005752229163788459542951127123641629041724925928703351812177245720634744414071045 798701168
3486538940860879346428858140532372205220333882548249160655075142849889502673047338689414665 7715191706
70071546284933413054595326178257624745577860528907055681209392136360207948400190453482271750 19919985
03517218198291555373940524474881608401142868298542198154173994615194467566539911086257026617 28915721
1616708612807864422596878065405524080476980926229798908974388648718812412152858621071441713146827394
1512454023152718464160210458750969483759431807362021929374001190274750271765673435924273670661803986
7767305606401858990535751230023066083708711686914757913363840862325384349267186066139046684337879651
679007095096077004454535563136240513581616173843997008720780057279737476697330001951815716651 44821 56
3406218186569555695230599732096135279064360849164645440098534029608616745334380968998239436938294495
50947169541676412839424517141260191624738270866976931180696097101557258147853194257461453503976260
5546512543502145241434329824924138121771320340323718671520627359281633744103154710463963111770834535
01544825945760866597757173914576609587753667249840517245700825958212454852196164378931310988248 17003
84223320632885004325420849215944237347069301246933391887639562414254068324429613965686240316412840
305878256465234281833656651895855036909943259530942506080824684142553143703739066510989676539 087358
74993770334589530194951951702175226452073203602754639413137116116222899004570802847540361406381 4770
890413618963981467936090582948205418506765374568452152011414513584740042491714183497910852675 57869 2
36460840510296405685471629114796051660512986011642189235847067744783697916343692203021988896384 90152
41726483510873673134583953442150224626155336633614834786432335613805300749667620842875753074 487476 61
22129520464026842561574021989494369036109218476043128296926638081824774162423214918051745760 51652
767148595672360585448145440213290753231193378970542137669259894146244375806251615321799734696 371796
455473353549249001405607436631064741766799857256986302528399444346379930518242598412579835864 9590627
8906953651854959181602825231129648862456642741449878023489619904934728190652647147577253201 58683554
7077836728257562399243554364937745447797338296029223953910136392484245799089520465639804512 1789111868
36846111736749565952373215883192224769664295949694007368505028403776890626580850024671 2958976399152
88552871279469207921220672013205280939645226322708682212314758006603178586184106934504 55257909777395

661801233418741794136221578605007833903459125252454048575436327708936373481376250658388020484570153 2
917287753658510284304303738094648279446317034769134486828805493874747420783607209181966956077978078 6
595574070906443029786093090797207307496070108155868501594809753435305272934116171483174733966100517 5
217002301479901086903452297704877598405728629894181021529690074555659724334836185037771505508603301 1
150531761641696187723661381272278710986517345203857378730216612722258062394563423895118271063809938 2
819394680890891714268678748870323697423698254355888324608458202840234836262358836434933266480175590 4
603428215172195639634953043008481662178271514494595409011794488525950494726556911945792036793654037
611386749376191352883897654886121318115303047160619686704364428843498822552331296875029163545865426 8
660889460469293705964951284489740485678035852704039935240846809882215255571996993525794545261740743 2
708598911300142906714002259432746902189829517955344274871811164211742934346618581259577501754135341
180155914001252399483939017661752161192306632039293503074408005564853217364811681583302031705897646
782329202381824767644909239971574846690066268487079269797454361050266597917187256545818722599567184 7
189539689635639827391169454008286772208397356485201960596067264555193429252306818637594677165747163 9
851037580102664513149058946532011702593902980921762556318811216870595841647293227196915373233155149
878813034739889489625427170463108520020308934976074748096952314381448595233628759639315703476429000 5
251909087525316574079244929463176141128160500604332367814921702447181040900025235729074509400875419 0
944850132342773779028230376877759883889102242894658630718857836325840114004029582015157277750552204 9
671741806522968128144535963077468399197436150775608490148304881526622616887549680346283304006849672 4
884583144896108984111641853472467904954298423368792950285053562273808693036290643496106389027342398 1
644437129896752462213499807179098325385375182451431822981701498047147440520820677173482930757424460 7
177847252459461857489049305096509795390544225306921236070170383791085714698302257724852517384591456 07
910558853704706629305268611514962570167775660509871522189311993190860804095893272714652003159911804 3
637406495544985222116311079240253084120858743275083025736046735590502428762009605821782540707253588 1
942427482290612611565067099069000096462286669193502613356980448499030606977708791796420344947066473 43
583130498593239705958907652120593897697617999546099025575012925295051756463328193778481798272892162 6
883979150390284154892848405014183293430169393085976918820760983272889212135516982344564447332530729
623985792356457676844465574078818475328003206204091248503790790336969679985756985481175481183866884 9
282624893373134636562096236436076406756288482557468798532166892032758120831192672738707762830879194
416406020746280318221576402945658339747608798691752555031704962919619171215072124527733136375472863 0
499003750245934859600321151449928406621582574367422744755010639122242188903912068857149902812503322 2
930101962598793831274820795145746636908690111021310530573870501028762582480472978297597037886652702 1
744112460837370072764091503713333614917140090516021354287018659906055371259092989698875726878000679 1
586909108457407802739901018725834025207067523492790845564584723383879369483932121937056631027358110 9
630944234629357335874393546101715097484176032594835621751671249004828578694431786367077895613431474
653044727101587308359186544227533506094500454270943829595234500617950815499122606770536954034708723
167037735800385888201853606077407859203091001307366861533205143094832957128610836602524555929733726 00
103297614411917437427678278974751030295465030810406048421266292749225871319580437832558381429792820 6
104671644545396678275066337611956154718081141066372890445086071121650660339928385555376753205387993 4
685050349218586536215611625607737850783368139484500925050947034451168901434565620772431780451284417
499021187993090482889896661964477429526148689757457320468170131393090511705612968133624656576703275 2
997884256367261046845139557876174554261407949927885159419323455730645855363767666277904565194687523 5
905070702926423593768921741125833571439855472717969334716699045245738576573463632340209580112235447
644417233019968875948411158859193880265208241262541577592395355713900994061925788576243834396708253 5
985086771745203064771259716871629279811087226407167316203119950749535335078557905805522805677944
003588621450841939451102129664180301025071904143518026258391841696334287108392447011217284273032774
701343798411733012446913775974881728083780863283584806041092422086576772875220996324008042994492930 4
986884989845824998371385891669131411594805379770420015970689347111831573389010474647987808156521926 4
411241753662668216817707694323814663364194867908638258471341439078678526625420255079875005983442086 4
335320340338540716970048589542381941646320236449921869693519762514875893564475163444940641619894167 1
134104435014824843798743619160009785800714886541351357234604662347929727283142415592080025103467895 4
427521942413257040263069769465401613548546879857144294868030301018441408638904144811237544371285330 82
399373283668196231302956918569585662741137703388585366462274719316725033611040733156570076520712424 0
795699950151716821900645117887028746352292980881877100729033972992256642113056013757759771901399412

36326728084538940031909615421499319261336411225553601183662732783852674019875478187635393573394928 47
10295825287103809975654397325671294875582247836268074527390349037453906581151941957264558587882696 18
85994749183952654963544757136504122860593117832774704317170217555427338113164461220577791460736577 91
46307623015698777942799470800066693339080866312852037258042871394552756934418643828321630754249357 67
43406689842924817540762456348435859978947995073584089721127260109801859131872698582043602154493537 34
22820998321512799675477151086725568882198976906794323199304536465975468942090858651868854156500 53
01770434777447943867270393095250807174811438806676944040880337002762289229403949546456869467365627 65
12157444272755618547272970923160771008330327612046440019501088255436661183840175064330797879601849 57
25640922702645363838487828264437784676467889452186137353554365637760647667817089845054355114691231 42
74141648367976459749610075175159580739164799319511126936601648584829390733187973916908788195618674 35
88313735155390613140986275515212954448771045808977219105827633989474849102783391699525437768143940 7
41866823756712442332351482346586764996451945247633087035644068714145683260663976945480193094371008 67
95751239811906085980795618977028617046714402026390040755211696790396797200717133559771467791845713 61
114940796671246292299933147763421654122778357482586275349900679211197806030788574954697328419646448 7
24815492388450416874884440326562774709529606774811952785925148170372859245401709150186522857534294 1836314
12370858732629402433909819838680897918601274620818939502098874883195920209220241991431102432886184 04
386721469844721804586274751188553143354754759491570811215472430488110982675308501879271223606726544 2
72255495116778349951376070493031367598922164576741761560889488299509311427656818795044573907260386 60
15581213816955577135842043549347895364202338974649549276685353620131752865705505880944097716682618 77
85035656932488370061916868813257698980891525770164294154770352800561194842247298518788774935545695 986
47312837668102316847838642080357150661043760802759009924176268410207319104116864752349250603645636 67
77296180495366105618145387474536473562955760076828385002653903382204235925539293284193944942050008 9
98872489918062112870490391300294856556143749543445754248228715834065456324076678494274930963470015143 263
1241612982110976577978086463643472088059155364232264831264515021605196502658472066706130120493 33
86969922060721205507484698313250445670936197937511451056109940597227061024553164341842315931353905 30
727736431563672645801322676377266861863347959939843430432040025449854446412582181492
05601492188788848500428184182953837747170367919128937630870104272072102579340761590955605690287884 15
435937041294487067737642712638212583791146308614599688138515495869934947775300910908956450628879874 9
49987918977330055395549969672231130329962237357438567780028847289642133583266974725843606615280371 262
74322172432529339257592492447411545058597603142539540190273271795348245344711818326753377256883131 93
5700208316178318579346955506250987414808507383726201448358464003533691270556211958906298935556277 917
8399088576495624037971430883909711102642297462931766896722340401278924999440030146467922020383 0421
61988077173466473516468109820565844567718498749974291629569527774417095631284596102690145423339536 4
43247908988272945116320992753074993793818937434674941245969372625788487798245921701280055758027 1
61475792168773028767724142783904590739127392942678843857719239680522994034053347073573334518367255 155
42658263961999930983673079503724486864611493730495761294157070706620328918110723891552753833666381 70
80430330556707275261677419463060608701555657445230760340530014805791061458163090131489886188
21079382747513047641248682780160190849678442189011101839255967801588844050853939763873601419123096 0
60086888484095875909397988754571250989027721154014401926221727965465649895992143756911429020241115
79248731265080755959728472872699682789149315247674585192281108652677009891927943533791353350 0
51468985793823713882735357271717893830142216485717141029006997282353232884846219281128940817097974 024
42190544369303801749299703208434010873321114536842359993209089515669085964915227667229639694224 24
23418831803521099116402806434847354437983200003628081997685584925617523929774165820418448109089 512
68364708462474117396127148810193294586738194325086126535883736855992025907815396082128790730606533 3
54490598834474701016808696373177373258291394021499232920692706783167596814766966752080774826807 1111
6784929358461984186261721109253221606852538815918559880627768721643818351604955385279364210903131036
07816243891471254331653700492954357313119180428419650321615780904252013312485605320368591448176 7170
54976690739276489514055215260609254132995649181716644372036661368578767332511082859038192154476652
86225246359219175013034284393319549122972792697635137486232078789202684450835204124961260947859764
76405051344616457799579741920939413873112672405794054104620755970157418271819156561183209665877 7408
14676916977483766418386081752524798077152469480717166443720366613685787673325110828590381 92154476652
15539979963895457334995046160296460474745985377240519452042446951551057601494221445985078562898 00520
01501442172360303944283424448708887381117569584856881015884730508202936051430137010450699026815 8198

945951504537485723487980716132368998892862049934134778445964826238868934495641306886939680162402256 9
6097259058369035914478304640969184582654758153494500776516075819566412400743361172866660578064097906
8506843839086550136603417156612177413653524666292403852731631584292475107182073761997425700526636287
3424852769709124146326043911646825719384474976527412791288332764753404587978756721972850802515877 65
4905655892204714960252104012016987644538174938761539621762099818669564349377520407867045502757495
4208283438265272799066400406308555601579354158018388708877140523005777747910133607583463708160374140
33581662171052377954778031082878931138989447958611693922813774524852313374165551870724300 3784468
387344461018587649662483023535290685620889367050129145033747129770829859205524051878310562165213224
6122880034754732559325711750916176594166482394729133523705438862724084837055281700296298090586508 04
7949546245822401357141584900673308445738818119674451429115214342793926996556225719864393919606775303 83
7468667222170355689084250348329476678415036783681989747562636076056173959495904537872887088011965418
7892476530782025541234215275554653752784851164219300210400696015444285500717437035400703359356505 398
6517857070065820998041957449512358764478124861041475565904500553778745425732766313738825020172648969
6161291010481279326374567313473871166387709984542067027002960111722637627272674521018118655910525 861
4533881970128991196384735029017400070680631462301308053975457760287247409909750691788261928319716472
1284958737918035495590854500617988132119821142982583753056350978991235940544860617902486260767194835 1
8442349058719075337294433959240918535280295720103839420962334277562087477231701175275687241012245302
8153879584509634857983910451921318634956903039125641139059641254019436292949491921353071715344840968
420710642282898118640267635071607969350470210023187109856135679106593066007897762553197898250066 31
515629833544391644492580570078010616280321022019504757986565811680933242300056066489673694876242 617
2410119907596206943468594040616410095711553454537680923271287235399907443591953983973721101334949165
6927142993028020314774560505446618457532305901707861564430381970179149567653940459435606060 50701739
3893132134625850381073195038674483556019491822805167730576850268843254958908956081419950752565189355
4022636610084816146203865446477107121879516306983362021407461243273856073300741729085341371467646 20
82332653007577845780127550787262783016182508700842818778453320787977894296485859758192842049964829620
4273540733853100546993954125461947347217039952702270357793858512968366440678898374656563613544478066
683846697651766149996185439896235023717671102740890919399583145146872361287138497242069080950442 2367
855102876286804251881635840261629851807211528332237580970905745356917840193816002439654325782504 5445
196940348840924340857899982771915123030947380554030823155608186071615834395651781063733110606 01526
4964148156840431946235604363017431750920771304908868560472735513309538084750144865176049756770360 783
31544722925343359638456302152171049873859253102039789581433366384155929925508944602780701796227 45210
5550671191316326262572993663598923830060969816001708344300203917177191630701637780038300371126418
24146014987041714805662603384977517560384954919157914179295368495970628632971745221494526436 40000411
6851275873879434866688375722889961342993866402364058944824529124282016580047816694184780636604784838
17618565155840174603728978215856590831489306511779198572317147647241893043153199908815499713774 2072
1011833196861968494405184713480510376244887581817272334427215708740008524939194933981030831 99952288
54262630851814551410496748964257681720314577419760554011665143719337637218818650652485445039 99376609
226777074793998014225859866214987191247013874698956763689618342407951357037334868399460740014838 1880
9109122785049874225639471028589036178924869455199904917116231709297408195172163625062371420825 261190
7131791876380037382099894155167362903105494029053725379895773700088178658870490439209102611 2781111 29
35519422778192147007436373735190083294733687322396549476329928275318351812624007118920345685546809
503026304251921987773744326372602899111090052198687553722810530557641476148246398586371897 72797761
9373087670426497044115114128900621261138853060373672295958161170213950300741417661128483328 860167367
167320358804701580247848646398079995767647079233100445662603370736158992261787526742601 3536 00729527
851144731299279145245062334020906397175923219795801119146699283960660909546200798237452122450360169 10
4115632219532972695445551518284094553953860438779628051431267662740064360143136939 88239098318817 6
0597852813466394695605558118458939807050113575168206806948943855291350882854473482815176097154238595
221327308613220412923307765655869270095700478430395568199401596422335959455022810565437799547086 2448
5590800370695015608019307214648972673163938925538159958617022059139702070900638588414953461627642 604
4283738912519184781985002549826303585100634103443763340733469903127035583080112435481586611646799047
0409954792381287897186780109815204978806050476668203662967250936939607313669337472036724300318966620
4997506525201754713578657346438364676379268501958461510696765363902447489954321997231906 11693296228 7
7894612265668306435189561638995745077221094512799188532444937648778904054422423347097958726521769377

763707960407327897024558295369616412632465754921044812238093363807649499671963486155406090914159497678087746191227708286844605325727180192324631999466767892603035979575811827900047139409217193998740760009997069581634479233605106166460077875593954578581356191179734847242756439544469001853703038899472199830805883405836720473010593371645853777333738261881997860740674509062922554889908344358447071868346283907929470801169686394851185081164381346028123477126650093708643498081061192511699839069112411278849225010740467001002498755498035240563673715644048348352246102196750403342268090196091918367697539184488898103076931360368464907518519208999079674849520642528150398821992945016312244419290790582175500212464156438780114736353486032318176976992568862342246361161413785584860345729184564597310145801442339103436619091211751414322400433814714649570280368174781573399255278767581363359753605595393840176867674304746201122791642138790162717823963320257354064984294315950054714114406637280225906499077297215318483539621059207509481870223099482546729701848378246112123226391435007317776200669540599603740376544102871924178666242088214648278239438113619744675884359564005433840353292121785957488140007374970832504832785755627877386652839778884309372421199545555354368165329371988636343681790751801961273509345804109446551725884968026121015346697273811614985607135933184212517820869764413662803131959266299800800499938017991534491839141860014917652009555057429439030632354051291327228528953318386128090002420185920683012455768851599860366700882144036179949757693010158168404896369505707194161819229869985461811066431624215434388537448343388807199476674230925541425220464613263330219254123326554225170039623700118772248801218997398674540232783101115581388876253275234665649792845609117583937247695692229581647917057753460630170075274314014131440347082620758055636513836577977785666987231433499772367553788033995966640307214375360223103069573323152377229874426905668961779628232189871890962510009667381607994856169911256164664511017559716379949944874517358658677084047168447477084990976997944707107264628013946275459233625336185844576023868690373404023429979586621889714158804557357565078504261021206974369709692144094113468994263087437856931812699404387145948959983500248233904352960214104347987103068353429770829833207751807152848088283370145995157366923408952220537876531109020004087766773452083041072554159390052576034609120814028417742037713070084243715867293605934806649256089060062140073524622034669434043776297371417295657738443594005473557783364787173858831995725380639809960264540532720562873151122978157160862957374684566542360082900430576533154127168989746228513385107670664275111061263511316160924243466569700709329952178094757112910514811469846816362099491211915737590934654666176085925249964296490499530049587607816360047102673940074073208436244200346144947032423951767853242453480993775280280525924048657628467141218672997407703083930742651740145003237220491517016718491281343898195950966521920190813700036727820825480844407705489494977352900740939641152159281309852432760682475952253308028248314091145503496480558833631436378086926850984022725493059305803387770701212000289803251790104923250778850250908326341435485168772200998294771996023052438855804487383316035866172342282606941786040706999902507500493233626929932140946580996465314285894029505075993837424671389269904330889698937171221522509329615283223140457224069622519312187694832867420029173664435680663582105532808047672615156230212428885543250193477471628043834193928479334861222057352401287750089411533192309765082818111896355024958467106845292644845557427712134745885612865531692904976784566259641824726692778344002660727974741444083706645758512376709200044837292085600229027797730864854147461858811157024028731490423374710220512254210599660703538395064823345926097124425018477808097717267392473194016550396078670123047395765332806585083484125260088946584673119157291745742241779493354979891900782062130861080386551622375821446483540605072432929291154590962931318592967693090324380500504705798029574716346564618685966633481647375637582530295286292373436789494660108835291361314666383280711983717491455000642982081727450615751153316431785068605599580414470505145134434661354349132377733690004775758504046993549528650821231068855209618997124393434111680987953962481708529338557609900270227949774128674424179704757178358915130101948311420886832494331481780233765095332429955285944368343551927231851780494655835099194309370225681931802468448831867708730934727803805915525023120406422360947098853016075837054367783741464411605342176004593264890125101094368883949375010807823551784950908235899540726474620437428881050839

5588426654869276504840328312182091922525491147266406362203079345506603988385724860051085280789875092
2954152053685929261987791689221592052220551985320147926147108591884318686574115963789931121609989761
1097963356544490273435909832894786788397496610903105796567898848146337793780972063390584010251514800
1880401230886131297554207215369649270580145527502161467808691366074981225162163435485100633443491035
3420533934694524974769550862656936276212610915726891851277303763839795715936693067949298648386633844
5448631276085510051818945811041464708454015362914582274572901534105798816935405528318673603825088352
3047392021460703493319106727265127179914138372267050300428895154913480292363239288242207957730865773
5544300787149940796063032576958992623920596601385554180341966989324367050612368387644621458919217914
9256545997264421416458801833242657385618786789598523634784235368090599271199955978540144835966062156
3295163890730042756825400192358883203001911349752328613006510978769294553113966889555834764370807033
5692464770566831708735818394804538875307517147833711950768797640772884648664979192318450231538105517
5873407189625387985376214371724368207104688142052492393655087044891355416556266667131541921763666664
4546134197065884404803958628401161360336854813455050935903141657252576055258698787030511931907553636
3454402515452339582336587160381608532281850071410354993068466278407506123683876446214589192177914293
0064973124185463655995675129585020308227841032613715113519839061241596463339823900878973879933716728
8720356789486961278469180298840007702404614168435709646116237948458002436915344623192970276370458105
4668618648920064818578844704613762942820674210703480363084411167800614815023913690139665758030939659
0441591251209695297358127309790325832927479399692824383149206259029330928601033239174038383427457113
5139845774783202448838602196807671633165349867343540139992159821116711151254425853760916734327669
9904005844497434759055642063076946934728356506499107776795892650798323481084018224489117074966510406
8759184748576972103674326815148420195697220747344822087928699608362935876316538904783900487427479351
5152841972707861432196582841706939049681577256828050122020751092363465517693151572323810501802258551
7518247822341364317881652068506613942262635344697289359780575579260783216673889806612519238800169534
3252226200528899561086132879950759573554749552474243083287019180305647708273670143445393759376315501
7509351851587985050694639052319582925230975140524595631400743663136533185988575773788784190261411
6895445650421603370647862526272470854341759779500692649026951120275026309543674058016472131592679453
7943694975261018446608580007831271913160212350913829704258702336416046448461284709186343151986536546
6090267618224747180122535599681323522386679597369258068959753037126480244820458137018203432869086637
1659833775710858330103687434656774110621814317655633962100638583167467480918871707737653558986220294
9987382744254792411634654679406036502474602456189525909349966735951227127320524011111391852302707
6142772309222947288101331929633997818550318293444263322064698010129517045570430063250156250431508365
7628326741230020495063971736784188910051425253052496886256738738477796628662164973624028836930961735
4373373136677583583570179441145747081249069802297768156882572257089498908991206605540252060801324
4825120813756743766198679642412986059431128300054088818812331334729747312126744443118675943209106
18417810265573533262138773747375534578279041511593132185182858877441772296620469138660209349694250
5364506621333173259619876628585942987942670938011410541539608363619248744187916930646357844480772
1565839931203794011832448020155671414285986372785984697070407554361823481587960114136623524548067203
4561475693978012289565852162256169671298369008539004383451036299591188765554858501038242000728257383
1915399075270712724127201395819955356637355128909069762510703048920279452591556500485530466782494374
1234171084511127271602210941931455401579975257597062357119650920779653514558284969464124712726272
2638568898076835163458821474438221218629646612627082674226305719504739238630575412116648302620097
2767100489826716132451910352878362071028544446701115230493323226930150870306338451974189270634069245
0913768037295753346806779350486647874587707376219253052874813619851138575784542929107665429283407134
3502250882818892915244527783987472190081314380720430207245467931758778466626656846528861507688403122
3212187332960724426894339139482344004879478181001567234526177025767088679077948843575976135181772233
3032469991166575966335615488804049732012975600925269823496022332949604235511787278552924080285877667
806337022880002218103354707338193738262267571929259335635709540615727780572441483111640428020017378
0453773569712267028220634931269054549816718499502811568901079688029001125658917905559234547229213785
2872328218121250345182059548096772277660713101778603438829573318206423587236993410098016172349638246
8683009133256534492346840686069390174735320150344061138347551904288775406077581916178115413003992574
3786443591301907846113155981228743338330985378397280207272868923202989397339481981775713924749169
4666678961234291093703387932512337739713880254983507066465501643565968531458265075610705724702998374
0904860172437641981422704314803738452936874360053639126726507615680781314200412241195949403929770163

428124078720808521666306459230567032070981617225290269602306308129260227978177435716138102919298062
50972902342140272119169380327132983200837285066796162815923582866865760999044159915872039418368224 73
090366949690717745701217771446726835111945799235764378946935653542086062973010467307198227660774393 0
230946886154828109515202159705255023615647835579419688175556091385778521922259647769941023057003836 6
747623550697882318965569824146502867863189211312240606180938604883264517308401695014208215732606429 2
228869111525025499360218005104193260969073847488323914024155853326011528507043066093224442482524144 1
746074448844534185528241403762264430116084929486645829985520542671517405454420268207034332941069337
272642970668910813538484014909456938246216847992970706873787740628928325451342260408837187219903835 5
526747220941231267659633815572502448461126959274697523814493216404446898951622400692837095862738068 8
833629163724004187595864552062449742675695830507984538153471523066411007256923629342673693671091313 512
704030545769699684553584714559181928377124625835214124458120274966078477181056994570433605081685095
894537463079518395003471725194844359949485446852681297359019069065359890335695603100644796826312045 7
319101154449655511226841732515227738630308135975821722905976613762764176616347690912507016199553759
643211557534396621688271084036751192999600088624632199875418094360053087413396269299820762966801548
809052335587186807616985560604341753925240496799964651760002692691655169875265250677395085130408053 88
445060926010964368810894202685615907389985099082501549463918220970064845536636689968589228638215971 1
076717976789750115428087230391288788699618678930719828675499197432627093410002685103259296326881852 4
148957912442276487218014529821854997714292943666893783643822296585911141794051849420735064528325988
459122593949274445571466776515114569290313952537580463375857964158116362187227661518982412205351322 4
224882561677424596765412540331784498838411137607350077200856972776264873652783389163614249642106308 8
349308903320848779428644049722268286185994466893437948513485590582820696461239464975741223628313973
381908745580331675172723866473603363229422165312776358477306315342973277492314973942733916813987 51
716501781032613174550637765770978869597675742214937163528845687419756069536003933733202065164936970 8
149229583433116310372740909862574705051470227230962962287676893693773871239020465167427955962537570
4229682701372159784961803623480433790639703585582422108823529872394968853577050451828950991937634747
320351938281114420925467393367822925849653220800152631802914086468882484513127730626798081559547648
285041727113926403153585040583548821872063249822563083781144749672788333322002877167003436960112633
377140681492485609922237095518547384457179054176873927917178516593198682887564181867768768410191491 8
079399611289070751588545524699476508217675677211646636427687199852241009652130504650883407275744896 61
976161460840590607410221444976223667994448377775461200772349048346804799553637492002228717271851356
066168229123239218005582781418581599044062405262546367710244383173398476328514706535000079783184915 8
240117595659005330270640384080707123341254904300589255268336910947101885680630974863894076607300957 65
914621565050197607344889921706745959901008710843618344332778765368258772308223344345432927526450350 8
275103688527876967792475290543516348471778306004342306802510416927098404294232615492831376121879256 1
598055496179934882230440432328763653188612557502640787057061635149791941954779774061554624267143364705 1
086815540321664452516231071226012308937104755068250860324046390867144369077324413439849284146788284 7
147933299621527265676834010400353802720252754184353253722834489706581156085710599336667745198566047 2
200841601908600007880595268182669131384173418199055901053852392002013540514610372929598408924440579 62
926120982981963175121453533210390556518543575501550633023039398321742416388765814182837549573824813 5
795353764156486001991020976533392075484934838869281611240904286051688972415625937579583189895007292 4
596046798855441909311232298527184462335677735993334804074037205328034706976370027157282023073057696
141487196642042557275049195036632300465139541814072510893070944397831211073327668759280377650717251 3
608755719376485636744855888825788766133918449302119188229270901950014684234363699275319954611567138 6
857045074414321633904078029992349058144656520449623351543572010554182060440262989131818118316521480 1
386554846503939174422768998320959947345326148515338090881844067372693578429319408508180054111250120 3
457445303331397380446948852770989980569600051914114142974729690339224154441598136812557513984940 3
015890891283903969068237159676228265437559056160300204527475615279459651918250071334871565451017527 7
234485087001066716220881341145595773312518768056485291054073928022211200328627810564024398003807019 05
697739652978779099880031620066898403721551962433429799281009560816160201491081915097937929036494991
206217653774823719553245806361501300980865044242529347835698514942884633840196508628261926818181331 4
535227102216178688563979618042552720940196904195278221897237115760460016399009088725718251046884283 4
618913181202969641334893176005480461049509279897517631114740226421967576451984887245324003253302518 2
483077491232525668475619542393488977846191575794422770855820578541006163487982613985550919249337 63

4419693864702415434328232827684568414269943837235402540939373073804634343828207196553140326715847016
9414833139149967410908125309984316312073108505016906329319101821105395585567039824196291695207657199
9272307176373081364238940148232626547055090384191663957458368747626735252558119907191836237358436871
1834305909221799544996922330034287082269330565446768367767558779608375718328106825559568543168045747
7689684479201244439748747005737254574087492178275642473325859338271836011855063727116823814246245899
0459076029692142808180577856018865526909992792215471270895924017947650784451441464551717272454847694
1607662478326065735413894461988583756749847678053698392957632660226472399204136521623736361466640323
1518541515481118581084433985515043673480709801630625247510197150466954597491414101130866168163686041
2103388074565199493245861160487111648627998380248125394189901363772731113383426677853259232453334485
3759664716208338537412263555308937443901951541575679942453238033408307216841994789968124268884616597
0608333973371771443649867987671879721261506319728855406319152610386496203137400551441345721044904471
6705194515692273936673249768573916031054311481689222515925766878095346204881819135120128416258147210
2780965979991944160344384182326231448081673218912379946597469447371994734475461447290698588542010162
5230641968059723722842337324511406540283755263039707842698045973210194934570025250559594698145422107
0283690676511133800827197043045192478092511785627291035262791697425806840262730758419021462844091789
8442561629867390910053700297885506978889231900288554409934577175007878492547917137854322631465566153...
702116916043172098426523231396606548893085383019683401924136867142069727633671975814778723614330172...
1255520558300247566655717115576955972073111366164764921241000743253672111718190269864734965813010136
7113732229307682146982330958626017552721672584247759443215834483201646771795381374496444065673813832...
6457517449057480046505684021118589827064600254984222932072749822069747698062260826601109618485569352
1180662292994038015726101038428296088987460656304679850229299020929193246177160616038480142250886912...
3734248586441731267184746634512149008085532127581189474825178594017584472291425938199664790339087510...
6666615416468654700720220225459753480984235013648485749478855474592133750963767821654279141354746700
6281276744222299118827998135726907378546909110155681042971730577884247638537402679741523709462463943
4605121639245148210507976032685986609400873423415642353566766637531227634801010770454890510324057965...
9667413083115792797480966954347260724741069201533920909334582777447339616500935752471124170750635032
0886771741219349514666763897126373728461160408770630158367600111533460421219020331806800324666379221...
7979268046636376195583620471957455881725511200080419911461363395552011300394239159753566741136702232
5519019341764557314694820222525295483332977455605137307742517709774446598076075945466065931031273487
1555326438908338370277382207431456894458643717541210660493377454249047001490395946574594377123789728
0058429396210552675039566321492376729814066350061202360793595072500261188229881702440932306066540097
2298243353765243779941591851493390417225014699742533537805043621410935772354900881869073902695140166
9721483626167233238511176389727375572281168991199803995729376032217281928453409332584441696304062383
0035426887429021182429856451245795207347984627346195104552425613736299433014983937291110870529969519
1835336505348442428480463104765220834208593651730965844961302858336548918675622029416138432111...
6325768598256296720182890678112418686878459731338714041872970071199861770129310462930969498935481392
6290262659333068210515885104358569494564993886957007611923572751102941969289941876554368842569280
6386143513031063844433431339619697973356152324266641706639024036236559357494425103835682493854306636
9565628381349757831909024277049909944514111124123020060434223288925974914382315759079844235177845411285...
7242594935333309857109787553868398175482521217518103082181765051555634524299484539045775626744065785
1759148916225117807053288117045974059759864583008005610322332538430075098113431012045083319042286546
8587799440122965616563890815952239403439226001019722176573621711635990257575290003386602274925942...
9355961294755185136993932496927866350652388267689411676389089823146190959683386123118586714957572705
7568639974542711303659181838471780818084175201006556621304762524946682059974639368806499907641342
58089308204090263232835178306322176172064336320110403440994091584051468783262604775680407126154842605
3468338802680940447930891937342309036384402549244811573907631680266467996790175567187064136332402887...
05087457165871395916429253614402597840290871343774441758956655811307376296889347527173711313779000318...
008272938712248798691412428008450272512254635272199201876242350802784479678433370236807361439928590487...
11111254193110135095931272766112488854689194535563621879329974884586783481448626980470399009162895289...
8836891137461673156170241004515775700160377837033957253926135405325011465741922480121824316985303536
3945988575041132622240054109705194916292574379281916876204926756847774860345138396946799094353055861
50042794448311206309059344324700093753671005604492655603472747780790783639504518762448392697983713533

52843484638881332304397927748022015296192355486000473427309507867806253076736433322824196132427055655
77945226606569501618926256269482380056950336491504711624285796896829089690508363758278283892895520363
22393096042046228781063765811178274276253234050556458534328739368288105558230201554434025666056611991
60833124527637674938321952334956317862584221341665748262887744717933870329083625039959451591113255505
05060040551933106785402214013928930774710317833410559482918371307692456979522064140227958467058688857
75784687252894700750851850045306875707839746663845073601953573707378535670036258558206673724333862238
35066357323527255029874737157104957827619393563501675928367423085338385735127612882617320094878087312
97810990413720875321879621767507234310420409380054307545430893475609999670503006260089395875398030047
81681712719509393419106833179242945159543851121090917226732183211816227957059544782253612682950095486
22172312363166507257218636453174107239765038264726120630862723390028195309088574096005768787459135373 14
61515897681028944673460860109973678031561291557189237969413783512316274075169456949086805441787599792
003655469263444618177373227167003444266776046094924812058797064658497352880792415315639324441257891
7793575259017863758256938063425044305458827958137410756022875844478109654276517670622237069724197918 0
21522914548339562562284607838662926681060526376677358438248733782586428850121233292472070960759600682
75605802893569806800904247794144805224614080192982744535914261300673153997422039808044537501195397 26
19448484495199114028190660032628026970954487165179021785465329254031123353363698668453024301596197349
22623522743230536042233756064177773162385607180504055918235089414216126043893732003497532632963317 68
39639740834087614896605346765002018018095017901524757381590514466840423320462519585964796028953155 90
77547204187516804221048827397952482737496742588221229084184267327354050629747395625186097845807013 22
766115173325159280125006610307845655227933906611955467698467069311445434165385872999911772553407583 6
267307205981923186415819683344077561735876011585410628859025148597066302566272781881934320443544 0594
26417472874446384970887538924365482695425677051955500503048573967192593691831322499523829039519079 50
074774192883412833931514360545655499274003400051475286011764913688197540583518130106428191854277239 7
98889115565442061329521483403974659533746931767971260593262227357889369439528143071554981256229183602
89709884483519959092724761429273847611342739427076465965860099675123050324376092528373453544759855871
92797451354560409284463123838891929763872245094869466433164585861170878840725956634464887728389814 48
06975933463489290084747933664114769169493830298448510466971167611803157567043074814086831913350
1510866268148110806626764448818763734119104630148643395133316943863486719123305119771222260512927206 6
28882836514602219447081969546319782481806230642959193438031910705672891130341666939068376651437334 0
0982917691778335023513137723780700801344937512480505007321973812385162506320810496669536517392868858 1
39774907719078242597941956168869722316752104506671379192086536793674998631486641017004445762995894532
1955986639580917941851056008686137283520554464203327542845960428433290425439861323590988961610491104
0370538129550774681638755768131629919025081570271907649077578436021697126735234381115446374 3779
86753486330097098407188529177293638378968670454493802201715742403590092277663302429744170450074589 94
47515007657427829025938963776349353159478320022542432672518423005068820208254972129214056337255815 81
12710474157018538931782939878712798274886567150463224664385553484887868766413556127610485824205891
86519723435333639572360994808168173974031190309304510004396180691234477085551492472966785089521986 109
01655248983577004961920373861655837363132652345982453195510560583790003979695177706979525402592983
67607431499008548564143327220034989033840428531242112340210049816242918589242224349575360808260306 76
2401546957880647296192668452751116009590537854831636712454848677882017252054639855865003582350020215
08516423663500787760715198561452562649515910550747278537981540632716819511175410228892983782225453
36756651902512162301691819839425989997594117695875215092796437661963360871925094446859022136218571 0
4072204996638635871299053055527028410467726202127446674545215457723708236444533570645225149756787051
71948005008025678240870818185487538392258881641762049036112904312811647832076693485228577606793484 6458
65766909520610087879064301681675339162115205681084467894159123374146368211388337577326140274626466 4
5150989210440638180830560804013701008907417193766295844421691227908932831982677233321279200768091661
0268387238432454420686797560394823655820673451550426730881815855803931916173275442955764365162015 1
66446685575937254097151878369217329993841578827660226709252914741823418663672193112269842362086854
52041883128891034878565882710587023709613027479913234822581376961217046227422496101673375478801
0634085514866228699569152716263350178980339827364448173458143299001343256788932258760716511011 5958
718714501057834762975288743392745818272686000452775717724788020989656438529423881738799918747129758 74
11654133239155651089087752576862866168071921090343265030146354274920447493124652014732319757703612 04
50117479398628215253549720267179089505195085909795621538912180564878996785730688499639032778132294 01

5207305052020960765468832525175264101355065145029802134353365875559161683674942894412148048659982256
1119879720866930212849950166479523817064956427351552584684628920014188138574351067080680393400649398
0278204084501559743111297779680235822043997189182740819520245230161759947970828058600328892886741062
4713286909469466843904201205774106699462157238511091569237592785586844939773606880019229349147175801
6316525474727285404111782500315404941649532114745075496654802310185518447420493368214986786738789249
5751470690246623904739663020066157810935792046362771350394697843435917414264377803119021954752269208
9547968878254856457071355666985023275461239882302786890131757321917167849301636501389552913159011742
2489607374946020294851279855924507737935690813338697470374157396558258857374134903582781434074625542
3551643968904607795838664996649944507475657969935149513597320013303372081166289167650986948072944032
9917612907230111140465891138242727263296851760942841879398026052395654066459330078965741576690867192
0929452539315527437352882800482345750914485335424049182058302117366937604978386342272174869277955040
1699280642854714019368556141102765828085704034232587586569465009202308498308267843272774869277955040
6611100435976476867865384955027225588542249861365736317396148099893608474097176495471543340623707326
5869752764592133742971412420705112256449499079144192444832839379138042063623800297002820623931807561
9226485397394933083251970873138458119857017100249136525227198684083184047846936429025111292751766993
9886203438789175172675722174421453857611002834874321905243332939928094883906152940641101879375394113
4303171916852635531673529853639867598665611968224723486403409681038558504606729601476131074024007422
6534024368037921068446153419708080881220768071106438905886342157852825822877436754701408310031138429
8534935282493585040362345923926688311448077546710591069406547237346587789419248732364370240202401751
7597046829540255079673098444317986354134645953902276574320351884311155975463523304523550552928506 36
3115240811767846454714132798134065927467320835452822766343059530547343093817503778723176464890964910
4425587969404117052776629702325695246367177584587882028465305305466181722096550981839675444011936
7027827210833550646667522809805862870333075787382108211292102103896165231228487920328037663587201 9
5820518561825014424096283660984633225284110157417361488234146028737749849149845904629889032849256083
9573991405910241206845576248034590472350936540898759284508157427478621941367169943229436893304953372
2967997800487191020347824739491797171976726253884952817305136008922705243085564874504121603355151493
1039641863808926595190343501979503662299371070610980328347286934541513596718150629766374430994932 69
5704953495914193527271514187376586945122363610791880115242166244941198004465572125975358046399910239
3279770838718475347998543603066177468548167430253509114142253635633838835435446860011086384802147927
6315847935107106227221232890003040085551034607680665195521686567189181334850680693770039068701376254
0552466849767101445993496184224689804170186023510196499013377865083234368694437821639202188508599658
1311126588948291773489736110174739245996397882377084643593256464055446635456450627816507891000060357
8958471411788114001470454565792408479338985904359558304841190708596839082696964630103312919 5
4054835663347075482559716224095724560789952275137431729204329442145717570211196649273295045892435115
4224989284182930968016270999901439100065160664547672330384646243951138306777012416199103093992894 0
3981160616742076033829175736056604400871622422948641209276800548400263546046021286431296861760486765
4587332459049004458514468976001291767593717828778157463447497443174722457141782495933226589599012326 31
6543180978557825775355049571779819902461257728494463535252170334339313436451383042589815147736236 79
1994181688328599187157217228765199479301424824878759641163109819127666028856647531096116466767189367
6347212909630086923962342070886293386312556128734065346790974199382881663850602787328004850544918201
9933787113082949339025444084544528311079067175578158009386753600386035576806398106423575109973318402
8068507709966620139938302157057490439491154383090561583001130141099139583290325869583767619707448919
1132631216409818434725600876621115969513338904625890504021039716887529763411565301637614017241741275
4156775733851563409689244304108416179749753717442561227676246976799301102619433763646846573576108
7691613043613748330479441225952612415148981699807681018481895780038325763387804353118569719951740840
4042917573207178938770895931557022742728265816149740801332063433840086890238893918333142004515961400
8986112156292436416947254313561701811792005595928543918333436945191945143171541350074200639050607167

658190582421620289769684409055245447001281886858143549545420556689838786244832075212174109487364741091578604099102585558639401309155175600279768653840756823639735834723364825533520524772296088193054082105328131130488092625504406239557590391944317141526527540808952986572630716340732243950379987148815126244606312813852410741143597940857665785251694389927465562180588692066232504493702921288512745927021376748342927777412718922155406268330082860090179218753829486291501624540724777886699889007090050569250698581837390135076394365418723293031183658168633364477932900708574510611399148091109971279294384939136701584241896910938623802725764521667814022361322904680977248778268641881228779934329767071638705695119901896323092458999902179573134517425936032513690208839586085467504777322929062846481720319507281491180095950512581344757157655467848443623776914819179984052906677154792911742669334885856708813947344848621438111159547544985280405742256709654256465428962467422993215792264672117881452064820869276936119217385236785674096279108523422956090632666710480288192263713909073768444809318444989699293106036870095308727241026213678870697269856973400228036350475523688004955869735903902069193128307071061791355676742158772575982291294318645745477205132894587827375775210307119425545175288583633445935800552400893120991882570655486639231448086237441459530559003065809529244042617675254206771033553684955246546435770101789796416313038674952396108289003225178158977822565761387167701321480211552727673111734163311892203845147369902316477647593880601607553748338232854825950629102393600165447353942982876682073712192752299907369744081134237202013142304505259316972775590586570031320224654790866575834512665456826305546193147547187391279239721184692177837620063262466814563224968776560280104550069682815286575425982348340092647101283284306430256976526446941350270621458319110491793194143493017703487026333460168710291836083741253275877631540223039562887769644074939503146491804133514202393881524875937371592838326600354064289535416985061992235303829218610486918255269025574988931977847720619919805017972262247637181445979641371384620798209083958821870480155631852081343051685237415842174781458011156305831176787708977094256915360787809113628435482402282839456580419522130311977227965959838840936358355901185742704482508788265356558199441663577849241763188148362164412222323634993894299078409216509966112353251201103142338355065493889092759611053693980830238610371304756676260555583092784943578519785639053474326745056049940977085918865106554530081982644552291740877465478991615116348753930674193023342758365049641568635388344929702698830212838507501173072253194741218860306060608665654771424490261810915095583030742760828948581103520110703030587092579512353389221572474247856716767883589737676177211751305869343355367649036743738817044405436987385939266494471535038152559632505530200490676462445852260521393121962997057825503270457798044858956070209902153444669537068428452178214452386671046984081075061673174785697189794278788716052834504290022122439834339069477674641314581807048184216581860366937640989433649381426799919717981523351084157064761650276045386329995224403389384486986948794924415099694764645468976921648614239235682753185096540881335323603150184543091811589795300960882380182295483646650730834615835739094675311255426108040965297251747622572350121765782432201282217368454018834091125636960586741735441883030291042295409312139163251665271628120142003938524629267702533556780132121100074687510492729215067733282463405433051464110169458015239281064871651193748496284714520592286022164309227073291131485505146248007131399477489383288426598056121218316115303041915518697722069640705170655830745429556802055860647919856113720832021397337158971801009341952992408631804186229969441590014844534898620679645053077361183991361144007305930420574967863953731729127783584821562914800308429891763640342582197758590099886164229150648818548939194649244424427073460653662824110438068746346424452489217270080026988968400977049906879061899434905998797123299393168654297787345036374108604116821223280321443018450796681914202371152463948220177093142729884554336272986473983606761973552064654414860620658273116315057177242088765562487226829266602037148511214624414964999515202973569634689579349118486138732356737272362867229569307022272826189362420914468343426468002660771995096819058242162028976968440905524544700128188685814354954542055668983878624483207521217410948736474109157860409910258555863940130915517560027976586538407568236397358347233648255335205247722960881930540

22165093651912429912989094967348869160997557084153856326697999287231066968120038735043570013902925124556125259003322197307426285577090519268291420781601592068468518949602680324164502321797845493500368
6853112007439987867535318804473478513954317739606005536111440635364216181260212638657309469059325590
63972194920380562876788287430919096154068691521231893628792498701245388290743083816074862738967877454
37823637094678891160341754045508916975405922739404926662391204064685581911120898761302214445697256862545295013041121719980910669115415746874858495998661990839837817126331068153364133204083616858950592
5150168276847524016347344155357829184992138591091124264880385657130438948294477742444060010996901563835237827612909596748151072258291813299828742705498735967744842208562067520671599421809118851
02146438813560796234541509213403430061299786868257960293849765918712706805152707141861960573903188851282433872683766410207671757126610927458223352290512994852828161353759254699066910180275940168162449823038608801382424988306996391762309396134854599451780067107822551932420508739012679549784401569898875079123165277472128947915317312845796906128837001975189510916690850679856760516587307479984144568492492127979028812424100884088477837315919361323440548263035674749277565413855998730321590540762951436378
8522298669818514967083486052744652644981763753211029230625903686358567994891492488089604126511073389
09663443593384412495832698268242760102098965945604773652398711801751865136733933944944267677542321917242795866773783806580166137292977718446605025011525812407307096301804069407496556849930872630421830015704120137922456787319402582131094546109926979422261858374651973232200837461954133230166816727978548254035845798586852096711956323580355674470587232686279282060364871793666055944117539018089837790280843442247968708856595271612788363404327608000580450948161376241757342033947798630322367382842571940605835474378389044641540657909999931856081243084635533967991722881263788799160033833788588454547198231679388933622837773206411466170259542085088518051290084316307150433107539545541990796465090231644288344740687197189354667356494568123543049612988845376092977089319942146304822076602353982432157616254876128425412106161133133648822454132448975455662653491942408224913402066175007211269430612496634133218785546802945912491721746410381867136215145716495073219848969572768728663431105360442715715428913668157127442833213339763343000508127395671874826350462103931432431997549195809096136562570024320166580709518873058917870679368295244048534202386527588450576292787146342219301500563804572513175940122856113554368542010818237158690133941061317832929792041099132745410547130717453300472338395654618716344570340185560471813815780898159470368494257868219263636344774282218605796424745794721701500258173333022732047386573215346948938772237005556940604750620414873316115686228526907416880934990174691276069271989385055648400243370326559752936860238742696015916450956508885490208984849887625009506966765359381231697790345117805540667349228798064769733991389207356808640542705632218673824785071644611430980954948809247373180071966058926964767438019702682241032886561113658492748669631268073768413367434448435491569475817285533594401006332275230578399387530325507216780079595410679816105128701235325189048159356172170323986252992391411785715560131667449541431362253134309093811713897244019512308210327892658536285024029590494094194153277686423597978387245805939513905679556048902242960073431779282618475193395556459371504687561997579743169015357214406687580868655452485004827357130853173124413579232099358194008478035530266617899903400164500470949101383340177229642963642334547788104605647714030170786181447811161624796255323924406800969011790922851352731479550364501613013003828067884533563391135173141024267843977071244855993926430774456327190903208839351852366033052369976131317334834526825772849605743916715539515852956333008194184226563499761732066839099307221655802854054589778169114890122196602406946082942278143215432436576101750205912960184367126183329145345884749623927097658169394040286387467768485985922138207034864983052250484391685775345481955113749471895212461814152399021533658511335253541411165644033399101481716305596064373428680323910031407094111326232632399839699537619227136973500148398588384971448168151714974590795901774492774451113062728420131635843764179209332924316744005546123114291616263976070663560386006560837140438303004134347463989548459451126970333775832905533641222535927524974345534833372325862570963384334203596640897285644317895876135181009427545203740088801072788481889559516723126211891250019208737338613365013734348864042522520017704620726688200704465618473896823556471471078268263004521490432410553635545255918421354196865113739788663097500814256431984740377591458789590609845475426078858763202314402576460524637248392024315470427111900318904204033381700098642286434180747777779834425559893089229069745701872020464818294167524913485599606198009898484748916628760419800602597001273656939362975409320859454667562340804615013545821550863207226603893401376730576253406555169815277785599299882419464266516768776119173622270209227833605250770480707590718034363357075638283659681399539076072706818136580

6575919866837510546115218083781191964755409670958249560178282456727368563121850209804703624641761986
8271774847822246349032781088546314151737181432979288325624993711562971573739011583631087044860251030
0496946914258386937065120377046630824216489443358000596868730214852492879538242286100073642036496791
4869424254773064472810425508729193419606670525645064069087900244040642473114135660990065146788809327
9138493846480654610178905627645635564452678797317660085645985904575945045293632732291403406240934385
1631402526002102085325002803141809837523389639583076237367334254811893427718926930339828412036495177
1760100346751920815833829363212820663131089145602014822523045528829442917400514389131182798098198484
3229029838696282514873944582039109406532801887540772094907478611791577001719038797280637623661744014
4045207022924523204540576280696579308502039812183784020672025012026675295531308349435347193634177273
4063602625796031365119785548566929712577804345867761008528960736931441334643773525
0159245211976597545908769502060561757819359107740362583576536008089376532813708436943902272298653222
1828843740013882581116297153345756740321498609755428688657987436900949705097986093770278357223388331
4539804939892101714335826189674003122527997303364571061607284968264026682347704558301545855748271713
7243584709948613725871302549402449573855889966053537090338925114540558124569294137888271651990000043
7610796725728059987482047989567855938858499483469651949308978149972763473305857071790270935682275 76
3063930497022966339552876337991307858593142078113351114320121020369871304216706260143575841179770790 4
5808380884980881666261853588355924200630530246434628992308203070806494107304156759771007752398558686
7594573174476709455684268903853112849498801814477456650509614898991517629924164287800047413850804520
3295305391840976899463199695591278676949319592733602074930918120556692462152740786651432352652070708
6787955864168604527735575020748767143337706011912940315857431076777795213590261308082898324883948 32
0949988456830767241759299430340209439932270827548357388507419917136940049879858619423446279608414447
3566520379282953170163351181530293127230254335629105545863957777802211653866611269335740729436145574
9056372007128254481135578340290160485176052432969813550274714705263542935264813662388695848981951679
0476124747446800847725887139455273671088784750842568825983963683066764766451330823429953840637149396
5512602596412691663955329422216277976078749552917485688421824863746324747798324942983235440257156760 7
9286742595284943389896764343657548230757547840335036965376873654980223987801192035440491288268359419
5397184364725540905314210556663207320463884838276837926105500380573953794021513641366249674935373241
0440434862382336249204935534482857905306545277265072203465929044320220717163242358313783512521095764152
7412446577626167543609470974335640076904143622180682993551510913855657373411948903218456220443877152
7004821101276120814078245264988636103832650848085252951495226355426406487577143594534042653382666 81006655
7716951714429565559054236819339387175320386411552242884740789638726559965035453160178284299590 62489
7569431465725329799565644275381025956667255876110303863545950868484208170230903776010731371062342933
7807454750823785605494979876902139063837989528600919904560260320637827290761535970387331101800844901121
4811927779674839102728820575597820535088346150021903483765463110568401425042106378331650970093472 5
9499426617045207232691017186806893159869500806239975869483897052416122301717289403904669984942721339 2
9568126161004650902845621267573941439279503195865023549110471685635783540426485721275402638812 87194
6209203813254648116170313586767106436587660551655133113317022718232156877362195848216856465284606970
6619054395401406510630973336513811963331635949030392164270853542280497980267149118956364251748913441 2
1426361554780892145283670822169402598711263211438852993196304804817892962988201123807490130529424 9
2948016114333023900806706572137816797198568613029030129939944512498469010019891936059827916973051 47
5943464960288332896966081505634505660937812923613349058578055094564210353090736019598446371216507319 8
2015642422013268456687741832312310418685156434120327170305730660785175385097069171707917125285511
7436278713016009522089202424050305756402153727369592667997478107072793723912355777093468284756010763
0127913119953917628186159430382077839824322617394676875089524023642462319045416738 6
2358360482837439278866547759485902892040201939593770656732119490991043352855179871403502030760557820
1914838828809464964820842417669924567583122624780703905576531412632602429224362037195329185547180915
9644318568520578823501030910761280604457044251479975896088808281259978623877435496590049296732208497
2443458243503689780365184909951214229401566917453416838309035284779643067608611599763678720495505795
6365166938345210212051724671890236358379083391190802068995968969901881223218552528693485736518886301
6045294102817973600689549524030606488944684834853537117060799430547192164875943131412697595251661025
2290957537550950933718544900072907676126346765291646465580371533060205534741620555668380872331011456
7060821971360199116696011772653512414405109362036010017584053344689876534900244758018499 02851129056
0362815437279676288312381657743751766245640457837049685690904281846741434107660754984114657421 53343

796282523773935177587703994255213181690173990186164214135439277933470876597369481710103318186376892
728376366023019205919792959179148224416394031804147790028857125177644841059315644675363309241579702
126264813042808389337706723982286543417317364814245629661807931369532509112875469498015503179945166
122841384464630874102798782095587734617666779332006361614129983611238785269844967622494946016222419
481882844175972508965043238838826776211538694490722314080038640966747955565960336586550083450157466
003715498121545591770828552690587827462680189548409854806477673225930833646432666789519813230343847
055425711893324488033710276608066426197680004014576819261412342142109083788260348803987158967469186
127595035419040689672781395132598842118325610948747352748664367133593683737190716713615344289027252
730570778056160659161544235891078464655473695634970737221781859123010944369231395220301011367407345
705952613329367437932120406159970890681203507862354127805416826582353742593856966435762710973540865
230333395749249771995346662569428121211926764888665256315169706607240021939626684282515447561496357
33365845237724099687357953227591900979741551721334845333578681422873993851902093678274021559991420455
644643838160009990650537188148493816086550357227064177438662975167896665549998788957217902623090845
480646518569309255696453172241089451645429767619694361393513384459604167285457399141508040959
446135343984501427618054220965984867109944082508151323925231360695106267337367922332214259952302229
640904766459615450559484204881311441317204646926704975974905993511692043902760515744667739687080324
804063437778416725021988849435409828211600072729150507598693656847220169410461894445826185511600415
494510628158872485140345190055563466615244737496076611357787483740038862938848861019502812807817927
50349584057529284529838909157649132473101056333147813464026504626291567537790921372478289700319632
689125133021524636102543583672268609282030777416870045904352635817494636724551789784931750675390464
416033638472405464980750039300245766107146606057194951091402482327352669122149601607089722072205462
810038730762296890621526297111428927346339214378575838167995709651297512128824707622937565721348906
361860141899595000293934330117463300332972907834026382527837960530000473559275468487189299720656136
337515374779219624955179692200855731479445742882259242287677732128859806537046540246199387296499359
356323021311084824249501800067571893986118972621842403307783178334458570361181609413976344651627256828
616878213013425589073818405734222752790944015079633506936308315858425959758344133931666799730480514
104205162135621754090487773302273969806564959009456956985365843208356206159345292542418929161730522
097935246571227066400541353921262095374160702598813126795666746170932371740523629631960893652984442
0743022804976641640382829257137163603061762596724995717615369585248664493172010960853457234236254503
854441441271638476726283333081895855936476006163524985906328874450325511377681813053346646699501547
4932420985686593504901062114129914177309980459978865399855599720886527297388216508774800198668603163
056123011444933193578407633418331385977273234527021265265772962648846204400532377509270264400915992
654248626771659965913245715413925400153811699661404144979220598528654631198814919187337551855095811
871019692417664292423893754945163159477245311019841450800876115562644078821720935112593426184468303
1073794000418382893605854407065172644916885787285452650728104911722412941522346848448989734965331556
939326855402116655944907515310970832462344595701968564326756803854451935868733514968195976960802012
537990084001054633523364189127960544687635703710651413568371551244836184919250949941411446246321784
676671191164877674448959946443158395848718188466274202784418999288032751244966696486793458941329860
303482928876260663713644580737134010172699240031409996328975932823997324878713822652547419034882177
981954557079637800427801458791944118907707143580110302662454293625150543461651519860793423856239066
551545908689970098727578338564769103346863889942896361916953313831063514443194692997895215042743027
450548912822404065571683738409173741484373181971188226411967029514001048449736868836048926288540745
71246015784688794778131708392027701850083959940135078751064535161654854503534678749015340275140901
464567541976045483308692169390248980675092299229407155069237778782666991230158990938013372850555299
059934716784235078673905803655389520181114771552751612837266568705503251456831582959065357006080657
269902272143379149237524221958255515527390476641551242308413093279355619405005324441453950610949163
703871530370152810088754080933294790986591783965408974119178143734113651271643824052441584288760757
149771141471427950829588702992792468332133705152675643942311350262877689034446663632184445921715758
924113199632987541312018325222678696789964132934113176366538896832051191636223996373640065062421869
982230644418513321973198591015636256982218714854708888378220216171014912432492165323865576985502574
254785968298124948068606644449351918037483665508175542257335268514038987865030070402899334381972301
961473408648283476126073019822268614411798984367558389159008846999131405413831939181656430884349788
915171742974864909638389673430651712173602754537578343113521721501826959291493227874737425724572136

2566263841389526262793021300099661963003232522013138218844822215385253127676763048551870068314684039
9268185487653840563848319210024722319166100913439507678555138370482142849151016989753390789756323399
2177890388047633681374846516892263571623071840641566324924108669239676012160108144560923213374291457
8448806124786377388264102086180249513057338836941585087823197098151586711709517388028679580151067880
4493390248068909905291953284466968288254552920787080905016614853675330813369070048013388285854616540
6413320250693835596317424365884064726157576009934784114084062998236648235748554353335905053612627428 2
0018784805295304476986322636627829632741637011531118234081786739876610728127325778515392113807681541 8
9444041763294630490061864780759891264283257299873528716127741833680517563794195244023212888549117741
5065311168183622698953190049592292508376260805003317433385637848674958223105863188940739807614496920
1791751393353298858853433644997913001657128680999951557636883579690344998947234260419431859912204658 2
7495644137636777021611431270014347716120164648321329271182571328791058413578619311893745953236310239
1270889013912909166527192377458686417036480120329532875161201291706095927090777356167401939117441244
7124601417849679728249366145899072550082434997089096809636416891569620894519256267193430471714563 04
4323998155688693543372623026149800352837166513591216931783823097964852220626854188473486939354384 325
2998753765119249233509919166689310689393430992917742912606797283043316638758401237022011217239456 1447
3365412763340270584541778577485248631649999170485476984320535312092927399866107526313197643376538 0295
2214163742360237221850677911388725805767775543742535744238997963358197140322779356139744707119461141
6517615151326743193251095192005876749184930277695596540981225396357711694671126060236069439457213 64
8076464990166378437484109735730098749733872155726595976033113712883158380306249032383304861952114 9826
2358667333635943608153309620435231806990586725316679671989775739671985056332039162769296127845043250
9302784936557570466366500504235380700021043379054365467916532116237013870815386679328752803991810 2228
7966754927414138146006565485087779489944785505088949148052882658768844456627293908196144006839830805
2403725695064114389933181166377016307513930445002156616091239778765007387433986121377676314373994 99160
3580029425394150936118287792564890199706361151183343773228780513781700685461893977078002754504705746
5744096115186501688721678158080161854864108089632223340991247492258081118532699879573620303636 01202
4863397052946712401923986886269983543109200456223327916998441698832120180955945505348853261754254953 63
8515063118962530921766529165824315900458349693937068765865424328194564764785373552515306898910 9796668
8781002569839068711319214254198866419286668754537724431761446354156657628714055353564287851801766 2196
4712668996243194827310093974171042590552437496873330365968721460188999669280258223005002 549474319578
8736177439811819412948004602950547368669231007938335527797500383591562081647286299516181411511824 304
8649558704764203871577064527587236770808180590408423523775775754008846877561116658139251193567319 094
0209614289011096489957150397207159050424785782964191398186464569806900388364679836798101237694632288
8360915854430104942614870503750034417768599956759933050225323476525468899552580628931781172 1823163
7950877845934372876421514463873034437300984280492495904833395497838050588846442656516711172 54089024251
6568074345760295714812822409478460558543210753456382184804837562589157517060137146819642 16897177 6054
9530199246953023901996148262601706284818796357991239697001594441468677531856458312725471 74394 4500821
6298283049375695975213397439120310652126169622928117872149914975372547212930687038508756 50615027 5226
4203373041216162349639788099435270804166933227218359322427911165736309254664967124992961439907 50009
7635710205091352172026748783818004596833106999955907782554648912836743394515824785122805 746103415113
4356381056637703548798094032652665484582486458290675564358632459180753460775801950589957580648 81377 0
4343413010596884392991793174174742867125143313255177595667391206113546736483115197935420861547222832
7585604017733891732111418623652027644917630825994309391671301603555506639664006376991774482383984045
2727764172011229122906118559512750665064983546029612962574754375906955550751018593507588378946923408
0884424210404435178451018449469776602243257275763372138266732831848510419137225727990030230195198814
5702121716572276518919027375580323980856028541791086963303805028305825415520792215467550998816607126
3796669062669622288040105219437555359934786322393330880742944031636339297431845742947964530484812662 7
4296055477893724165925475425727294818303585240798970601383390192188164731155785052643281067107830425
3827862550735751441780943794515208769644180392945533710695800288060095926155424042953849322366928 2
5186545787154155054376817216194941624398023669701764585168224184823494985612261220525940694686843355
8288000423604267164920519830030400639082160494841822317937800187897147326541391659694146628544607201

1663385275508319000328193491650079157574254157264221073192131424033453260766000540883242243039536428
5170101907195968996222168572420582700804488458661600852468547117664403374541716972573166435712993493
8871819929175946631813286486618479719449399756737688968894218741250518128652265436890284789523945318
9075978183784293738692711233245224020704855011368071301499127539063765677619504008247294577951614725 17
8334058293631438937472774496789651364811407002313274619898292467466885589843355545576042775737627609
1205804848027199984955395267234262699717512224621696901278008109844442553391517508360950391542809560 3
2294024501253444800135907132943742644076571567194141395791922713654100901617029910319986570525840029
7998434141590790940476127073830478178955109473090662575049936351422786265074676596040918727372254 77
9472975212156735685726097885664645754101381930184782123365156997722636151699971162308147353868744559
5345401386655999956253624369635168340979755054564050102772610837878518203493705332792138943408437 04
1728605351627431809719641851581334638617864058085289932616267479202822900779806148737877411733583027
6131207960256078488189046862289627843902049983703685368810277112776491648078939944439040925075904833
2890872832414244933523646308167981840710198476936630553895402873674490337398474907585950356060720035
8784504301688110214426498847297177449594105845200382301253166078708859906630371280412152890785515342
2131215471139478438631378025693727576221585767928915212630006697169938472634430828463068278319532124
7975797640301436814374615264691989502103421817625936597845326760250667448846712460966866173009 7066
2524501769227885205690673590084316041245928474805668946978182767794755884701037881345716318817494428
9989298716727868572546655367737283112444617673040875235065054391728319619803605638090014071189077122
1329209083766220493049647945786678841810035863738342709135352809394255181980793497854644802496427 3686
6843355216708089658368204954333674108689934436229704144434396109886127019621236168294238935804471970
2097389199208230232314642350567106499113603577319563884719181976988071005806703870572501530190615358
5559014641266966919236290595038548687337612191372174427144615555267452280574000580347074835030814855
1071539938916838373110927502058170195123131177677851844877687268042139392860342132389958185132776 66
9365598180645571158854038012159297126776804584819088091790995730806801022541165164298547985978 0120
0530325322575249974103543108838019843220023575674082733060839664255593659517583051615349713443826367
9902877179428908266042597494911477071422970252112586060390052960240254582648324832575575756121613 12795
8128121568538408559162780570929372466124367205888681576790769303505178957563934003111252929772535777
5442005699664022021383473566037691695441318906191606144682581316017660986167723206634974560465097 28
8265951275397273336869417550848721788938864061838464003716868968509185229834465775516415629814905434
6762684211445248399413123038525805518486801957088922618832523553436587364583479146903567872259825575
9755960216814522299491973792689788065727749960206183123619125984229997445917931919070044699554058436
0693267316708926126170133984267188893421249875569902430595059536766540275330755030949068933593376555
7321753833566489041072878037317426730961814074515620721259831472457483012110660947978796294904381332
4270611655269733179812220473427121226641381929147327894366091827878882764146146976422205029114448 41
8441381849237635277149146090765139364183608081450429276576158754217052493940928386373294735778423424 07793
8209531262353402755035057102890394313368148199515356610092447742704699911672373898955163904637498396 60
3242741993113944290390576059074155553325065548215179292254764254508718962213113516699330454312000074
7182980065063873898926256489054397396866794321270549392337213363591063334844185146953429821
8089686874083237540802626059432986205629181754412222900184021005925843557050011626334138911164722410
3293543067992468631553900279513923299722276621299513099409795053002725316589511915124333040407885249 7
1092537241747430138830317970184410857045135768151291536244294925037526161101183732100465189614678269
7244426178043464984407081819464885701556647291249400183231574748921227215054856761733105517328675555
5513727525722807015844443069091168420794485271927516752338486945201405843654124419006882995745900543 5
7430805615594652422881931272032923409033976547146518111314592519057258049351511243668916540226006
2755458017611743528143148949991614671893523814433684640427672953871675260813395098759620778572778924
9855982834823289177208117777338346633424188492279198052968309263756731847097048722337076247583698147
1774211480432136309414872547814926308746678478952455561005349890788889845921184102235978273731652128
0197485413419869770543953887497390145741221190480539008552547517769182706070979677126597248844196823
3810582599435298328267236317342426715578296460105068310046137927489065603076363259810279366112357062
2546093038459223095699447464899594352803595728122073590021484674870692849701878980716170871860113 1
7039698435437109684351766492479554420274224706377160003575229427613743288102773742437633846548182324
2545858652299370190888474776740026800100967319726684955864544676707987877175135398088398320732770178
0462499327861880076713309254338928428954733998046782679145981967469019830683989226343929035718573309

59662853884503112265863257014951784436813918535839204296433758389492384812217565502103554067105827 72
6887575137843597797904449145269179059270350881467187781681401490099155462169014256978035035958 7247391
49761619033480454991698043894828487160573309708072050466548034875571233312222486247330163998671 3795
12788679864381025542560425357927516241316245495527310236459199301134961429522185316982957104068 5148
03822139888376963907581428291168164390391655747003216588110006242071852457597469805276517097274 51530
08595624687639951401471457224685434012807858564304393947069971219759406459214429010712931914052 62653
34713416412866451075856458812395114011779550807321636781016037343376015731556349255659393736716 56193
55004588107322335593024482569699655583883053413186676168998085666828277132235687068122625484629 821031
31760771801239058725534724204741520016176660218058824619496648746064563878996315071244291538840 42324
50756032140477624258036526092205191482910127157745924142628710445597292699956501128606688546274 87571
87665066969779602823041381108469308787168483579092546278034944462612254586120459971996080031663 03479589
9289456325531254253443173998536945980583602867451850810533130476475285362874570977776754130771 42430
023253474040903030694218291168164390391655747003216588110006242071852457597469805276517097274 51530
946186599372850401116814957781425963612401478096837868885112514712762231791533148704487120579 37765503
04016642970150767388550477383288817873121246007452761235417666576881701011494289925734910123566 46763
96258065112713396498422882454502730566937089237538166314643437505650299631516797453409382533 02599
70256271802165624362551179869721624983230956587196829602546680650000467162224023966538241855 0730658
14460635905595549641982011396965284439301573931052062830891492180634287155353700590035047086 46096354
10978841060965660343653544490617007078995838150561453390650477052743156641760972151791948928 336481286
604608353071881948053442177304236004841191576164461309136759315286393239638705407205479884795 798861
6996218064420153535317900447652558227276706459363505728780427573485589876829668923357242880868 062463
2489791895841694457900295928632288829382800916032460635332012382231247377367840617940901041381 5330576
10283008848864941592255754380940630258626769891731846156393679957705382514934153048349312243 4806333
3168840269976702447327490618973633454332782082077440266707097817820783120857263444560947085548 5214265
848910184957350318664217482298567340632852564208874664253405339504502310275449342901916000084 41503984
95202153256001639276693664778095841265604290645256806170955852041871281481347434451539374647 54963442
20538926109954944289146367538957760558424841902583118172885021588583788068989305945215392813 7465408
877753610042351014931084607654329094608679691001884038416225009088837411244830646123775233176 545420
148684333531525807526673468582177646255479877158003780737152412484086925690704928900666781140 2797916
36537433395145161411107226456176282259912478849099284872988870529808306524599355173740824113367 75797
2723356826803378116357549983476669908238378145543916215512856385576818804480343213210293942162 408135
480262308822241912311925661205255016134989977537211303331839809636216624718633004541360038330 530579
38607902112932032887174995145272064295800542689317348183540808487347093887388619010213621765 0259
591135144212702101164809309303749416729159362456878600346602240033953552251424491516060429976 47942423
4983779611349866591027178292993124652842563039824406078979869734206039517007531875786306161264 714500
10032081159416960435788247184765644413513385091862089785746218053947270574914758885846342665 18241468
387985808048248820805502019288776349020535515856240018934643342686366341174030312234237172519 433745
140146909287147290589015061523895622846150314617602570657585862095154776828356254506431626437 6076
0714178578002758760804833259675887885318095959658148160080652129817919584398592529518445846927 246647
70760915168517463856471876221491932156230101271310528445074810700265024464178587243567799221 30399652
1838279141848265281625104614721116304790782242023486172635064391415298230055436257014763699588 784501
44051305748180118673087237596524652046435093431303966585562939251568103816946666203136394399 1734489
671398298156274119882627235109525122182170475068625339926753779784668099954202412499113435470 1022056
9268199940421665893392849856566734733257949343731185204896148039807491103310339468334077786 6936873
54754911077611269470298782112564765814915666919095529367320559572279295129820907151577300917 8847796
33743002145803603144778906200771554904764805424270513167766893532669437359097224008315940237 58388314
76354214044080642129957701653732659902048363057798545770670281905871864830613825275278600758 790702
43587438121184097439290836223989125389316731842659559695584958145573527311910746917982834574 376785
884144398093699453232606572599697588242192935879145436934910439042263781767557322498731475695 7013416
04149887297960683857414839097389448234081301095658316594628071917844798263328455336359745121 87626033
253578179240687001068616506202782368106342928741450813730609548908924744506622773308385268109 0209081
3243131511849533596968989039103170864371432842682490648126662406926704213367160470574265531 382424342
2085609793500650141268388679333024186546223074720253207178938050317199178734911226776218997 084782360

81116007980195606821356266409245951405941919888659605278045961898798897812460445075204391195108742 57
6253503374285344359966802548772112938561910122074296511208884295468554439251909040934599976232251 36
6844939593914310582200549776165119323633557844773870072751670887701833608973105333061674009407487608
48572870502540965220624984387807914212329203806188310481723210930017201914851255372112309151571 9016
1271374065368354640345909017516919077376312212625722658664549644959123749618371939551519665045731 490
3486981404538174573697589822751137745630786723562169766825239245298159046756218528584558914530148222
6584053595263909853937176619710158783873278238375584724428825363726210818665740744020130772139829403
4324598455034889846916089350460604604029720752430522405088484098134455999990758681315475222046561435
2331404526326965352792923797929373929190374869671964630718060724784747396550517097475096606260234290
3764684445058224722202132386388557145132347498203499660406621177772978606443414673662111064805331 61
4307054651648294660337643562092912703240902104351027595403970654729979272108961839673150847030362204
9840208056686992254659535837374960133141790035370052464560586947767468897309441369101074420202722 38
5751420522547081908963499142194431740411237485172889543080184574091031999461128055643344226228957934
0517365029531336028314329702493656220118807338796596866093865424493757197421024937405701 93694
513848256986001935215162568971401835116525496006360110312028199534967046160974620829177365729978 5645
713069651650016227778852738340725983559673970824046315259176738042297431785283156944449545036 9564011
0936985581798511502721191591763800482968531307898594703973120653780203227603441629732969801 2059066824
6901430294939952501589890477105080253835157115842805178152642640065745840279170365562834115519991 7788
4995168857898088140509686764825608157587168921378526143482401986700859594918363060974293809243990
0124089229267645452342219220757841378533424878833535474932468597801974307389167814296102044496628579
7198684931704497491550675521279111615838057069681965725948152986672133352580676789995964951 59606109
8889306228869661326312175575758648326279685711415335026472140795023598595852764882031363986562813744
7612938148477402022504450859121825095624779073889654945265020370785100531156965622029849937475086136
1904379629959903131190080431975245809400009145582174545266258052740399813169325700182768421722318051
4068200502309296617727018544600370433014323452514686646561052172069290603672027335396157770828488823
7288135889404534490226994538768784657248578192875582017026334879541782535545557549068250069979739505
9504613420534309269819057255312303387133029128503192281356963960128385345521159435691409774601 581987
7741238198895866727111427295831082708986392046218034573455624391718562497121740957925961923078180
4535552979777422106889367338855485881072236768784859382251540354099482252905928348478013601350091 51
5811453292144235396298755483068015588531659665633944781862223963658069731261285120024220474114337 5961
7769910633391463758052035815470306229527721703514821238530493265705584461131965622452385286064449330
9181674712723697272029500262669493867606739072357441326860714781306327962522521358345146175319707352
8090113543146777394534710240060895178155830757211821507995005588377445473192612925383049981 74304023
0702238909816076153110992888652872560217949886633964206182092131604317680117781542966367350787241918
3405805978019413641380162857929818320868424446974760959467546593513927192027080666700964469126425 78
9697600503752819096615375661855458062643522949216579083498437925743197158770646868620018182320480898
2564563340501384966557640297083448061878669689312163513966878313623511774979419930548228986619040 06
4715435959592275442777794396672633729746627797753573196084347249181190210119294392590380260264841 74
844478200516856843034664144125006122544118533069294998065721395351334078869245327059129149828074
1121071884134268787888298002107119318415476906323213303566470428019983416257261051670413116849386770
0275509498844108513693169564448607593170835467673690177738942973154551145922770111036084305577182412
1223403292822987443986446401919560923000143949945306004425799693849177239781614945113120420486863791
6752530634900665239580440289843539255578484580722003320292503465974481326140173373348415220872649858
3672364880564331283046930530487353905968489776941066248996816465510182556276908923306543747477325157
4823464207618269372020011128849083740841566637879049177157916261744725335692110279631363639619333830
3169096058563478651583641040952185421892539384536519000945682188235121967853491290747273345761908795
2770071453429642885777891979700517773331894256747705514167095014167092010796390010750520599027573 74
4136906107059254179657940736448940133684621259740377694362926710786480691656941449476496275547975269
9750611239206590555602998061827757923211986904515905942490767601449443302144753811078861683941 73626
8247379536204857866736619434401837539950788735770695697363489060962341520330327366441684091 55972675
0606818691954289729554967800742088808731999842293318016422639183011407959704912671956726619387623534
2306778374503739921556049731619654537918413623760136660987343740561564616345985238478285233197307913
7019825090583326929428640128896615562366533668086796762690219338587009470620408502701789450516817868

277031934278430701645193131391148579096169684416066209283732083338786764148839135298925848184530866997588441288965867024287556877312359003496164995760829237522689365557076354134082655772488902435754853975257909113420179830261153474517489394228238827710449742344359228203662147297399136740367101215970943082487534476980106697699031419407850208010006384516220354274895328569552580166987140127909455465844685317297663885922327228023922957255162170439537798680918870851195550148345006535420589588172819071594632777061363476009074316518417732001776274966861929830048478422225166252681241060317143631945672834892810958904495107654103618988534832669434021847931347638061335551520236021176365618271131545323152483185016002550353002350998118745684013978413245041292489951063561883988605939985186066266983743068215608935536408037221056922170621065402903346895715239006699699843981971994494884736379926526213791440855451262773768033692487909647451106309430481047440825975290276493019099618286720668008381247708280425348545154944826733517709915865139720744453559629062029789651482279964382284624100494925380966317158494746496877324294171486011775792546480922293925634847344849734476876789725518676844578041930104358838478749847191575466125277421065198340368876821770985647989749664179637585327608894833993789803869359050038859150041822476926213916322211511707329740757299950592161419534179545395648258069575581914101054740858366976388974854435670380887776225342372523648958652528686070011120737664441770345792390220504029211833635920768287468191635734436212258468551849117378281498933173294328687866734127709419506140678430955963646611830093772355931550084081882042990111235624950515865879877793320160602302539963958208885785246406838930603148815511881851063392801380688294775533868576870062873818717550962023016788908295772993703948123553225117730651413774790534389379645194776233680444566103772824237746406531747191228550875257012484553030349584254775511129310417412608969411450407359446504502916769772328830601951849785565324355223479476176792576544582052660516799447607227050526040253806930549886106509217465565888873017888384128959996591593227930453738081443789823577978122166391540010747412133609571689636670057484589520547926161961736661796689247300318163307768429123981774002938069046842005059303567210970470723585762676559715286685740580991153560586905724472242083498594270765145780433987934167957681379636208339039508329344955805953637604854622311362516792364381354247748419480435893321459881942136057929416741226159998361085501672440264935290269294762413425825638723031974350786160646167951012185410632203081716611241488676440338872975765844395220221961007373434541485089168713374426783595270561315212526207386933218305935658930621030492920953559451160121464198930371896439869349478415890922090093907042757821905935709430767237024389005345312096700350896192224298816087686432982939714882496041244651328082112489224188331444749268803634239182966671662822415678183967524379665967458116991492813690145022797801359769313865394554720684577771745913853617847926953710369580963772279061281236547915708886907667689081937493406103684106738600541100262787008747059471065864514391969805597069505565050150112339366658905717331336447642130706567573199190393881189382530752108828595161447506809468839245740011135470274789826186212743214976651883003196033345622355344214177368117005957663493676223290691848803237345192434912465665329718238417396919865871333134120710517073683417241123447237186724941510842704961554950379550152872824860538460676203928758686262616984382280560158757868309251223040159989753848922836009598075215949025807471577353748066900500153498535903697672297104371592109490542624633960855085082491832326584393402629382685637152596238947447178336999661802011547882499133236522159665340485701123525827708188621501301157793460027439516892752435518239624083915012437924668651002227673515415339817443806293681884318702195394674578806838733045026693482047409308509529400887069518186325483548249662607066502499564681941064704071227311004215441855491231634073401961809074987236703389974993439243991658806512836679056416957365216050282244885210757361133174294735635774830984929574307530604504128402971149873655233370252933328593415448831374005815017262424647751533561189773427424925968401940681333381441610749091640443078052777297700938276873668269280836283467725910032841281289357876118865330379943915710
9

917313087684148298147316743892415070812333046638666526885171522877349580865070902110271308011488070
5251086164181615255674817104786219400105612800134647100048960081618136339861314375953783573446347009
7380223970667333178847121742166237978339505084945840564804300591120184173005022419629811376432555086
2599366729792121239683362625433079201974575553798577860333993611397314797358588046748266319055503171
4751074082657545538452600524391171639479854543854640477445274442320820115805894011041094538330920475
5983130127803405876733811345178470423962088493582933994762948927492630761519594758086352305003906352
6975508973719632143202797580886958529773162198359442566894666349865566418285278405307103946038370553
19530705781433206165483720876373054215104981963982316223946155540162448192274889889915481816161014129
1114223327424807105120278497238490440264296807698034403160101365980856422960754069569557133508003
5595188841456265319836545998149354703533396578804949761273209004853311539086974714535368901571896984
15775481177794107604190191456527288840405188279508490082645360132429152288889379751999559985968288936
08911271091030973067570621865972967346398430619284295504292105466916091541828035178323684152181865
5679634071112430643855154156248975176397718812620984087360982992383637303027505941877062626357378481
36886836829333969889277444686969662949968966027691600369860596890471841716000197184123626519979301227
8741913927227986980227245952294015243220804048175398124008025763713606703033447765441140427436996205412
51450482700210515174118098254926097747214620553667790075520087998923234635592526127377276954567121544
499950148452685432597408757444502355950317588809467867062427950496812231738195319346995561510042009
2153608326032153069185727225992986006291535925473431806464770644415440963085983320989428117389795068274
955248706628273280536479247410723546119459442653797874986979596432214495014352061663273608338778957462
69008896094162137344263820026765753391034189247610563598881700835396449558521922407698130525217319
501323741964534289761658135407331302630302854708024851925520783232374548326796328907125627475246124
29296414094248502905947415575992751242508506996634735376429407383918207419091625416097284459162561447
27217059082938576767130454384335990826160862846719632875160286080403534720853621937494941366551991
085368923731274850742948144519654169382291793153091803428722389455877264858456583583686883541716
1056472125564386401518568769577988275174222434711115526234369721170763345046653272476633423782119587
9117615783397228747084512798865992451832791641445982814437270189462668035792333375174806410499318
21884550682118360856897563251181387504609424195574432639719843107892775247235667228839552735223606629
25987885985396328772844546836949236760648441227353682921913245547040744216682067461265670796430120
5535159904741708546163733620270386592273806423126218740626890903998167108615021958265006449778173573
31850521577151608476313142971006824824913916249289255612638476738621823345228302857013815543747650
78465690588394825313419981396295736250191985710580846679236491105929088055068338217718370944062699
8269197068244705320057945408351861050613616560945082423986504199873162383835746064545218873590261219
345316048583392126663772506208519054626237337328114258166438424988979859679541759200369130045370
196275360687183013764812127410561482448795986351085005487349029844775297575349396653530715055154410
93865374166652159183354997382566893264526882635962214023963409712244268267002533955826932243805998
324269084530478141991498931107101571694749555763521945874576862963018389905824201278786755796934723
85030744610603483914598794977948711391354629025486282224974389447438682661643606831812488598723268
107661623053800860292158338144328453224062354518799888174723812853596838289500620225246463573161
2663603643148563553160504991541441308872154144331745253732106513961197330694878139130968733186136
6191394025720049930809999534821942765587113235218778124571834239764689076306979340608138301890778
922762154846640882312642981642123941745697135666153982555543529498172239014522235919257419434642182
6647768886677250217123294152289784461629105662854200714538834167814891345667445069291290631913561846
9153315855813507647854132185945574577015648665261070023556816875810595967699899109720
641753128848749413907086642080806276944501319670197035551810041719242513642744852197815370357770961
79780365547150100641818050275407358363121007584869042036045306374304388761523828983557372852602790
0965811358399008202510641500684886027358154171039936269383226188823018995912710008790754166287483229
44522850239184867913815692056863543217041633586370353243353038603138245178894425200804853014804232640
0652099629600966417769376130820806870201308834720999916455805747029726506424850931007184698953106
9174322837847571536750984970434834820599312232247519854853455451980842281450746416932517141716603293
266776227819083489722751003080897525205030246649351264612784908741118303858779985696639063450520022
2393456268657992044238614847572428901051143288838145837001486782283330164072761140369413621157371699
85605841735445608033688890692524353381053971593150452592094012886826761957851131323839763615662128
648264097235668922065045948831798586410105643818698894289927631481613114119783948514389640201616144

0282221756482734200646153016170156730451861843770521270625734225871506289598548861762052296168865652
1583778424714947986648770706732481799084249744141303077278122172757039389853107662656948276197633287
4465960455935018021232313616829469105165367996131496561881423079790028197020753072041377449667453062
5110784565792463800382738568602245895406143936149940484842869698064832620184643807433365583986880899
8847151429570101451205496684289667486610186687651297313962832144102683416586035991143893180762564434
4609274750282537324722496358231441866189835337323691680869502922048143623720481234723921748226844291
0666117849435316725923850410537793110490815729580082189041099653740522940426201984820594719565468 4300
7682203732839906146792078580314306210807428878267456522279510975301854924501080667143565220340 9852097
1712751588390486131129466673396409924349264716231223460568160044054572146286566256168928697468 8326
3390433176286370809582850819096974176212074055085110018153302081718136717685434872720891726067431810
3868754909964287428041065285744478313994874979824434353732018837509779534604400528119785269275442489
6325741629879428825506076395165103877117664673766387526975088379398903288357402112295635432087437414
1624949105157682271174177119322329453919740921365908600004762024328188811611636449287155599082998145
4358403717095653052756790061038582200502574224292569773732947648339790217827686204487144336131445923
2060147349748148691320774245973643394920073088574820672241184792682405330459869275458907072131545 8415
4047723222208392080363241272176311628918816433940306869317135582680006351732097450524378305263342 9820
5158572929372906514408225209314003833442860094371776965082929411897120873729354559629329353499 1399224536641
8810063103071050568773445423096458954361418630876241494447419866698347713405300957797872079 29146238
9626390884639299690639310163833638321319539379304280498487596279495977933648358480037841610186331741
4981336434273879803025779702183008865036418106221571029282021587181445626643551912892390746449 49104 27
3596927665231146956658445818114476377375235012859312659257607505506519843356775432091750690 113587867
1619549365520291390943771733201558080722733191001779316519681542280883705765075988375970 21754117738
0871379853389628968302629816285055301107557758978534823854812508105014962885887182322180 48882601 8274
6164487902917284184000293674697066526008628284388308813356335802435105776352859335154 44380101 02 4994
2174781896553188470873781552236440714542829548321311895375408182160748659797776227910230823 3648 43617
2336029171933992616788286898005789119115696128611373040422299624444564452288117974715366 35791461 5049
4381217969991171955339291546798588569082819110314718381129967865651453059679813190800422 18270900569
3044807483008345642515023609363752556381935174270343629268402347764138990390433623518515 3502630930095
9664448987760417437860381723420607907143711944475395228435351677888570909340590984262800 7555002487258
3065055751125935389631517182606584600128932682974216179130811687821272103921757969833700 705560047926
5997432364952544759584834847404096171475662882753250917375915460325080960011443270411987 31596 9680079
6893604075977501803740941714699818850466289826305034062817302823197991255309183628077992 51133 0655577
7258958054972193754768545034660721841155705309674742472328699207294954176864890518966 8750773 6218 4297
9151157585159066981546344699735695187655323006152781995740879843326878658438770370483 8866796 9634 471044
8810366625529586283434384804626781221476425166019766422702036294508798790923068997762 622002 33634117 11
8201539905017307149399611155484037152282685779983851093840045282636757612605639843 91352940 87 394557427
7648499241414685824291385791775623295402016824786676002102151382666411013004217800 6234432 179 19523686
1776961388050065925090700252900155749948752601372319426963769173901232800260299278 0683106 3027 349010
1100314060770900457926798279313684945163089509182904830296500190892833649883721 4396180 7932 0874232876
8068649099204449507359860528957211699519040408338984083563066438541288900714754 386656 7070 8501846 953
7900325716436271571357347225105929586511098406845689715658254557256333555947567 77747 61802 01568378523
6403149785380983156625318580942578515081390461546524807271983290957729580553008 8353 76439 354657145 7913
5927674664325448384545582730913986167177799594349945503174771436597966345645423 4578 23495 302203 37939
2308962273442866717466798655682537325145377430092237210027531693856659776107823 5980 1886307 5075604083
6771274187145896698345702753848703603042584280212778831071351132772777499204648 30373 4048642 790 22836
7600773900836561591221962842690363666029854323536317994514703396349461386817617 7630565 460918 856551
8443445168903346245846829124143175358657498912475955608960402921553939774464288 772795162 23061 2464779
3088265423574801071780915277401654871046826553570309898131083104309486675394151 161316 7665828 140370086

6865547643833977042757125044749221422969080622787608544404858813995712747561301902637515073878311885
1441005371031418316347433675117492320531768258391503423545058602263056868702779243229691617545953446
5744739971171211920778991581218062470944155069472735140123455020462906778337817621419189014690880707
9818100673485939679726693487696523832201910820873136965799813776062143345608391307991994916188426151
7662011516330133652402686506039217254034345528675180825801046290582083646881107351833729751961456122
2537135677118878425199402354396224247751737334535338915072628823213219610084406315154799859952158888
4711599783751001062597243954442905647837242827112768861308729717667450387784470605049817926326301121
7932857796302134172905845069384231278574470982819903689333323225245658784864990947175032022167285214
9882775465830655901595843516185552054897826736135796279574933052201119161084208023699613890043993283
5066345181640768389987922413230641588672500479862191487276091383655534567175784547455676041291332244
6095514831253536112871452107430436355478982136977200927930137261704890828039309998355740828980055613
0326328048784070258801908538773031881871273074613664205091087496124190176913449970774575283633978323
1561103564821247090359536816103938413756883377157827938294135462391627057605602271779835321108182513
4419902677286104968703506073372089355308640081426599165087223597269462486048366658911303958507510474
4947269939264526460552266518924976453080721310935209623157608488311382603647916682468008414223588777
4104611556932270788793268743612837957576853818407467782098467726736724469111891654655430484797104215
2397792892067638966358968356641805894421475396226231753265579725268146631176978012993711098245340 5
2292709330039962951371443831951350573178056663961041225778725280132528384833082997288182507498851 8
3147176842056425852484751295545343386763129364440772405000721199683801405226959470919911895518840 53
7007723165819521053047311491449137585879146562388719464798452652846802689647131727151501040126023 0712
1222879224688811363287433466723782055811179420272164792256537042244877632969701292769131500425202259
8108009997988559023568904320231478538738724020947448385339441778367943233256130938177205017343097 9624
5324951033905554193731155678109600573604690408793506212987488240528497439396469465846777436237028014
1430030440166179011657976026979051136930909194130040439925056649562424638020340508948029000263762 20
0578640552156227401553880280034689350975431276218613147605256989989509498932383472879082539984234639
7579120047011707656479780085662647461192149320181526856763348584378214733117677538206424894580965820
0413472695716237083720779356597620052278022518225296320032536927653198445965838103755079283425227858
2591745643500626084344163112970195573574748518566661090082130176889791760811867452945298815710766 1286
0214031901965386239562915131391540587849630720194218580207054805464477547695567872089601597179073 61
7241406114197727650252901385992813531379709256036120684938888215192251813217998093335978213175107048
3806021942235771089945917470518247086038195757880028161556161166585331628472637285890244233722087246
2933270283531749284100443472630417990086941692873722278380802882663780139205329286772928042365659125
6451489454928932008400609684183178390593490237620500022433912426191828329180108477069574155867364119
2801283496458206833115754626233916641813906198852245734368306773983610104642316464133153532352273752
6644533820300409551032192889761040287882571654340181783887410155412948949210975077746095597715247869
1288112448292842571974688204885654909561576851286105935267802656143938335859217867356600596530099 853
6546206425434637910629888573898318354369679219264775780379709915199776315697301911388169677700383 136
1743164453785457007319360521970015173541007668696013054372758816070499877893650174222466174381932691
9286242930106198425416346048533061312244441005381190421705295021020394928499328022918752088000318240
8337953638642237831826693545467637857030606403992213306997783815919525955558887819155273311550013170 8
7949016677272842630344949747135658767281439426549205263006190800059109449562185452175790484418298466
9384194955881207470010606370835634410332054500161516572407201252986049588947079721834259701643847 5984
8658280980869054936772614620215398636294163771793047627218313659680968500291803114110797043113811 864
1849141586975849866022454649276513102872758629386704002130005951781635866440778748117806522568506 5307
1350734245894460268346208345803133921453542233875444863038607087278502777777508677629920222402710 743
2459134004028928720902658654473562108425271922286631314180112202868765811959272712835132421508816450
1912089966021973736772018154670989399273542199116232645064990200284852366347161921013531799 46
0181932739607212488739815750393461817002282230279563933543167655527885029351632734594726579510401 1499
7344823774206587573034636025051615613927268982066048321085542322084072024003091375158254174604210574
5596153391984589002739715743753802242641556777875930793480442455004144500804634412405440897265939 9663
3007997972592129223109419470783842041830759662553943244560963589541422566136737969285423717364749526
3375569771626021992267349139427236091784676196587205759703367639965915756363993425058788943766830142
891641757385033888100786002182503880088175428204767842725956719604840605712100796561649269169969392

085577449942497756211414133332432379515093640966413408084475286141579712425085059258050810646205124 5
722168848201979344115322991552482048598975130100755988479695867649524880279423231241980738873590666 6
496491511396292448871057045643419978174371093169216647620690940566563479122208766735316201439579444 5
762166386846601655053394793158087747107647644547839610999272348900285145797974434466004405117604558
617700843387605314800885367357021156503610454879928403753217591482018829939970865269064553815917395 4
184083082629694214229603619265688038545999145315531953051939183455018588454934955788237465098560821 2
984240079518041763726731295371159953298713794809681866455468851443625835036662440852558593062441470 0
201882122046421925917234593398961925790163177529238655519765624923716284656853893447270690882441102 8
132584191077575505481596675431195699443978415163324068894070439266081121561861902530322071390646769 7
326836815066112376928002489683557371372311288425827689287823109567811899669697740948934872414661497 3
972794941266107636940295823628303482422371763319971843239990381281489393985868272446460540716992594 08
451818357188910943512641768469147790922528378375323956019474534909055101642338078119596634482620714 1
587238135420324049317575753370793509694613482852746513771468342056262632197172961999170866638304381
504998872606717267657248389966739493478186405997311990052966000353299413932648526328093008221083677 05
019712870766990024781300851305272966844060610164378230719136307049422049383419600953509993987135158 7
925219739428061513365643134081615688849017478248578366068004039275996290863093790311150672767194187
688246283075188795068973905489798893999883626631762238054216550097343843607589214242046806810224873 0
738778094787723527390164557430678984175586097805980115965586660818078353193002712597379133589578973
340087634431188614869768378811215108771266775723101833251113919851666507968096448532566384175831166
949388592161416674567555382378604424455712633396677214622446741586901295620545652768104736368260978
986496300568279073769198631560129161425693064781006989197020524817616260304963399712736676979574289
555663140148811846154458200759316788356682670899194556985802409067765427444507985684535490547587810 8
274277202197966003820209978659519699449293354316949164917055441294482092967609251479903225889004953 9
549957229930180621792621124071101622095785527048886053819284236085295701406576742213483133260263421 7
705437309535701089078073022135498263628528872655869515685634662594731325851956591384689244624458344 0
227704967066534993872354752090977064881709000110639125571874068247047136722570921712228672111917253 2
204691459692215612297524374555068820561736954940662733252604137044629086544615104425600763291794861
451069225887944374815778912508859111341722330315674973819208639722065135201582800869824687653737532 3
344666978377328101114526159995886665010821688195794566602903527583321493728643587459967647652582 08
605552447339121171297976729814037989097286506976492032209927924040206629297956083333483709734371103 8
313640741164847600506985201567447227783589996084255788427247353061170516020218645681801484237712858 7
661659119901186881409516094024580682619905112961121275457411927534638229394753499710748595614107959 4
010471612965889159165679925544442250779692987057697058479143040118254545356111724273371399994326721
271193231939913000689026268752300420447759813672847437901286717096517647825594600907601267703866659 5
545097404684858912713323990972933798353508097267094853161645430790871573529955075169242502752012
548206333930344782848948625326384370394044505434363120240495681948503862902856933719155479296289
884532505765976820764344629468981452443342055846528237505365957053624480381842993935441506813
507092075744497999051667690538020962176636560683578158732037923130276052862595077654529057952721042 0
413852461568605765628218747553890252057850800740693372942987737463057925699377504026856615866168040 7
615422161938897946385363383112595824418797097195492240747959539519422978658251390769548365478993 38
635688061829056793407651975802467421791025332169643776004412820779904903233081231761123799725721 46
928331218471752129580119166422739351443766983751998453064578736640777380699652918703123194595004164 5
141202258226972469814614881869049872728361923812720419779161557131344799687575703393540193254612 89
360256544031228921782203409543443883693546040954485024790180847629703419796230216029451611247691650
186005981164172743826730728952978409244016569577657255557295761302425335479095067619819197014241516
659357446039596736185325510610079244346653294418180701214764603012486296399753339473247459834339008685
160467240225216136263334375200406632342549930842141858826098209436157432408456471082544310188309077 26
258570251121046551644620072039153746473932152920633797909851707477303806053628389815119695460644201
711392956688531188328062643435313170171702051702648528131841251634392473017135494506051586713611010 58
050469783588061150722926127576939521408030298283970434147265709132950829112534317694308075486551329
222427649904174047481341076368194957397746437968651830444656683983184841948244176059863821383782353 1
373045228594983101017622494394053901864942713232427248943202272085052667040161106549298027268862995 0
320140783550068178326612442947337934708704036933947444883640146301430766548886919995190654063116647 6

78194049730100375198570541672527647023085919922739984234149334998390978640343907692242729337668926604934944169471569715697155470599042174219258825753425922999565986426768451410612384687880654457917717504027768236063300794375840743669481319018301570912676662079893317001720205395490686409451497740977410943734521094285968471727023406450671007142874586355868678236996339079944880743760883533682031211373244515710404186429258794496377751457723511758660881749387826753877641303779519060504064763026806474269068794144744826935195393809701784967319464528914389222386328010128343141180980750479702438991935727250527060540470472814746474462082224540747141694065206965915670010559177013648971322784201700335664320551146221793313232221711932737347771577788658376969907911711317283537401935719863182444704761544818457702532441248455096812811666497090604322747183366809811678365157004468018768170956889324184407362310602566274757479697637477420906354840294917347274873081201950527096375836074625445479352954234171227268198133990264190257637361165697385508744130241104699836432221001267286821809498907623686892633844171858526283212965994121375179698905209247519987894668451205783735845043148908445688161384011588039698729200985408629694713225986323131712732194289141354943233591229372423096601546164996985579627632755545779844544032230490493465934666330885736959602328575018026850212000712714207655565772640798302072919480651293602177611360939113593935308121773628431894417396213554536050024613878796067604080531692002503944182571342161361444011915228953065300777905397200339662894179465964177798355925342228777465640003439352647735178351593330557199471598121250380175595457774595152071390843361001153790490150789070567088184457411450883051229970020996961130971218937285330835942880990492700476790612809697182928394128918960521323361070521443176014995019958953589718336907828313863423685795679656341929304976981520997452887831214597368460756747663060603968791743542138503489532629582054474876924133036611832868911277831754602912791295567600741873322554290824475967304300804589663453378198753699486335053377224451590105416566588706989890137538880243275851514118584543536529874915727188653564082683272251101470090989862716772241528776898622677036983969959782816557940888253718376453740221818841187448133828794719644549375700207234696253360159221820280817972604912829242072671298004240237600224648500870889885723158842214694147491027891554652123059424861027260471269813241493814990176735748960941182680504771730131645894682345409075051861661015410701795541759759828042293862715551077779967459122466163847477146011178964858677219186104417530731906039973098127182191067244244194814711527834303920653681221424655022693020650758812764754343855734742632815492770562660066546711848765207267335654731669662177069833645838526830207663377786474895873125521947859905262932469843638246857387232842778983992573759790403828529778739784663809624174775914830964299155281451658037282442319575209277526932305975246124505643557559140534014297675082443474662439858347619389964144402514113002426886937195215783044869344384073455908669094644813823535320421925347916988980143917862337601755214594654274049397747606361662420182061888475586487380583089379600294714755490167949590428156184287251585293982983632066919799567389478800199587980348283125992532975991050207350736266025424310513286294034955417639910947632944854290444810268915259589521457320122663991033216800068635048620073570348589610617902776607181677068179366318045383237919125956198438857609518028943215796521205657611093989915496884031377549393540856953449900800901270384017262618018204667379599446829543697194232579220647470975895251070020377417194263893464967571542725040469387188353727736344487694459508498634887761683550588497002886857269280542129175858131914228303873173827552868306454333148688092151697237876013755279811045966988291777962410970709137037883930450645012459886785895788637091048216529811520583322780817799156082638433465541976031504888419198600848406264900399587816932942706196229745312713653269444151463629447342752416087344026297978790021463626962620120375533883677549691572082415463924241350575749432317953795546323852419930894060511341098518780558840927355911671617017018711815535058236609856919392270006468228835899448657522671480046316567521942019661830703452163928231825112778342832895414242172602600810147875803021229475153440371272332167086572434181080477957841442651899778860114020271232948237623382689209641111630110322175485080943353504340776283415930030005339134180446423526240038073951095137011158509534732482237438035089352893521658778097600325712128354195312255082041269876035418900771365110860471819410808843111074253639171389881359753193233465157986477504684575869893919631188241144154303570400082640126777953841106665096407904987522171722641564904349596267065632899045783760113685132624683460213484557564626077735218836021662216821683273267524660090778680934843879198156466120824848678050644296958460210293045372463348760259962242700993707358710745996846332376105029378740807906736268831432224002632541834946497494337876425107862940494355970889509680133762402481909558113491329791361379976702061469804350149070673315063787624020662390702718327904802851482956061154740695997256194929668302102119305025060226826388089654062984556197933891284197588

1135820072574356168576772366593657751853637407687010082486727325827055947358759499152718354765991563
0391378571672985440402751717388265873849917746309881277693341113336407022766678610507168850450924177
6817981250769109509091441158337030312240850672305458124543727125939201379371298950497688964104658348
5074819454011857702359172469012966511482150074342299282812227997873756699200712962845547283485924965
7682030788467950647021947309801494977443099497791041466285009066700904219602986633110301100111327823
0180588984555896639346087537285190743600228423038962151716195641806562700099995601348969285573932626
5725163640020407998471412336038886746834087295761469762659715075739705145216740388083854201531984940
9356208349418084482165184390718645485793541198865262240327191774938527578225038226933593759400549456
3248668816693702136705498488652755267963759254295311527394338996216801552906883513703290248458258150
7127636005483568596542941065170419630389669909871302275845345904165117509748613302711360435317037 1718
4168098731418195850315648707721494788546280030926277518456650340939212044212184230421013314028569 15028
9035417750909065422814083329824732595632824701886813189069773412240098245720477810212442578679 416217
9708245494715165738807912130559425579829365910185929521458891989736410574890894887746130638684259775
6423252686346343967798383912119284815590463257268442980690380184893738786626776016467556930214317529
0992680576999578973176289166009235193383288594182243444483733745857062267504976827361392921256590388
5590939178962936286792195698563220771145167687115170363840099018030189287795686529739177054160961453
0699375915962278339171127981012984300277489093587484953981173533815768461395084185638045832986728493
6286928780127547239866389241352820450578035388536484709732754492568009427559651602910345775124513594
1932152689509014544595241691613698064708110231533338827660229870284100724886605287601573054043021857842
9642231635313350340680402784441475200826890037863547841565363689567311688395043354563178010166155103
6438448861498271395607957291390579008646612982441522813477864422645334901586371744886261087774283733
7586277208196830696383410588110789389572363677830118699970089743365600624571898381501767465512480 89349
4477000837345306194820022438725236049547571173946629136182524460576260094886690256581329311350674437
7932320250380205435018529949807781052599853044887550724931255065100388571908248547838993266028603373
9356873644557467569660119191601438807725928766439451357947607144709888932776053074116311530139110492
3282193110588097367047432344058908828564002679759435346427421078414906154092962403387910815657 89909
4988157126902366200432802152895896157752220606214579882840491823383657351254312693680379872686847552
3692007777213811796492400361590034941153406833557382487391313539191477581585276254133758998420
3618879551380731734622447712358536727905300835514369174708516100914754482240692245575761039860966 35
6788289455003336043337208465567251648964623272786309895094558630983011437878258832122990671329181939
7652202572455840081402413093213342095068969419077870202615357580218210512329081352831950915725992555
0067198796875146302345713152953633128866961298875104615085935633839150702622440450567951168869 46051 10
9466276723760723750352845553235047447489980158288045753005046091690215964523830382629210630322558517
3215675280891358488354854871241265327424716575220157935604338198464883776605527026552678470835601024
3568608832661178809770313606137790936091130872340772096181023901578532920354712719691056747947044146
4331213143956096751662611289444050366388193402864024805380728609879198298048731334990048459325455073
4654155537801466230578392118511432796651242618604478370665694246510969746211106625572671764171963200
0060679461544447383009587849363123252352851899857198091600068141919186168389015181248046434719012840
0070704117380054024841599367102849240755930396348724030137533991511180444162615222008808978968333
3245220004404297312612251617655694246030207535958932402085289249243527073965651388572918316494763448
8104460321439087648326551617298762199973683709458549393433833258846205940157298446429786277862823390
6904492327659329244991240461338339690152475856010196827612373263446896932592034892191306713471710997
7033736289595978350953252359827128964919871104481561301082918581273929221101792694864281968328624692
9210846438939269288165507695822898733723861578329428488075621932309574063083808914428364138116929229
0832336756904067876535761639617542664593802436019747717095403483650856887128855292429351867994655 6844
5132086630207489953169878887983908989545654679232060581922617788376612225335473386370967729593023 99
1180556622809264950044987752773346252173326389389449414654038212432180736158453611404602155525975366

```
7202239198881050284681073770437723293121597536227505157238764529436348387913528988787905821550111104 2
335845183102276113343358047897141497715251577898482989361644918289793177903264672287494186754532691 8
364087982826661910595314252866309267070173240595070622738667057928733882718495187976068532280614808 8
186465932879942779746124529540952671249690441046131073909192112396904683518446969930471686617424050 5
192229767529852966848673483681162576635255457522440318439224623325561652255141042359232732018883363 60
890136050472633749620582370618484689529986526746636302195871613693678672483535610994037381399698900 4
401535889590391032827945728151139758776274744531842930536613637951698172746052088546469232041759114 3
500135039507534848961569249509002075560557543841702290146078181869942923939119438241100257051821388 23
009803545858781164004553031127717926661474377483736623746020207353840348230968767789362305616789380 6
731785663137163538704715874017289052724912520827168930437398229658837763226587058276813473532994786 3
858277438165477936098563156779341800020147580442455937717830360395631349135763401350303048407399275 78
659534391275516020753086523653852745037888597124866716500398774192653741450111604950710018649301535 8
257730695530268422875848340321047793053485220954907925735639770275737320455079910158184367980977313 0
351644989780684770092412465890171794998639015234524231446407331462397045474526050430187421299438212 5
706449150735469297093206380290377591188187390157910382794684926286064935577057927535699984474878454 4
845580050634603101943277182722031250809775029951919768802465896398114153371350684361011310996298251 1
092340076886268173814285820414110607895701149506894308627965260288795353976116845807304352996562712 2
092196855252288491697344907527823409037934366431756882408319294978354314731449399148449444634289657 1
796553328504009467593996359499073386426656621874810726340406290494679202538485915398026608630 2
820683710681925836375616167488300314934301281052995998880618862997236896416520456806077951010091774 6
573081454691925813273129673025592058716565880420339131539744159172871948757014262221471927891106602 7
507612894427322429136699466592706572510419631023859817549079869943892438890089393463412138281362194
710798111450300002043390157920439312553995192226090889997125630923302714291250144039818705004202596 0
878087081358686490177244473695219446703490650223840969305182593996803942103858321601640076947789192
421214895435937440099937854372856130362894314167480946744737971676182086415931683561842400484542707
621856066439597880214216431665190096072817460641376846552889186181960348825852074845556619089693189
055115991626908725698821763884637459550647377391415806674076650097783414934842161844038388310937601
539388963106315487134597252904837036883415016860751585799025411531953560707703814971227446960968940
953052600980817492700921933267421388744897485958260981627811239347933827705619740666378971362110111
506206417832730909438642104434100156848794624469675999352470493084992705731112482397592335989864528 2
770415482776290851650379608150578350982302758742157795977900459206088004354874347352982814703676657
845392604783416704594176917489468082537728362160668568691135194523833890891891110912798514587414076 5
028525906720911115279925914212843554397507587498522071750099462414469284632662625182660249829211 58
694801076576660549161908778645275866719423683443431434193812335002128977713142500052922497167310777 1
001716851188822998892084729467521062242455347160799120832204726268215968866450816735096164393399244 7
751146236967030823020942646253694319134703407424587463524088261910336251980464113711612273934979644 7
014543812884270106771962329369178482286291205618498280909007323886154644578857828584097096047744 11
265916188951776629166252716872171876251651363170897961764648430969865502038327499036901395661677245 3
347773594284105079107689335238793554981878467812308125958197651326534225639402346170289036070816642 41
775014401517898713394071954068036840143777538230232741686512733144671792770675415259373709342131975 9
704093148149199488987228678226746965193156630529145957136183974955247480194339279493656351149799084 3
542657131020197144345952011778759458037750787462512809490505221398981478459503317862578489997750091 83 6
758799219228260550380285078100178544158073124700713812448131990189182879151729748063153541840272497 3
198667605030469892140012207784918654167062889461437398736216971466570430354657042071325868663 36
279283212156148810545073264464398983800241706849219912974800942850816694356429578343441236526169848
230319409486334102160069882846231098338701418928078207374832054307495416428963248139509455899337044 9
182618664265415081900582072358841898289980191510351770638816439664279347043773648026957853684103 79
866531939037687658266085036871810357283664433650577172278666728793090809972219221888136713093480501
052572641573813406411781621040678714302389166811566310728760536945753778478065058502995269117032333
037167196011235234440747526812892835586828684026582531601094449517018066181081906441054524622 66
219974605510837665828858129814451526749038022695678115007030523761609716107516474358393341040779805 5
772294775991343313672159786610550734227837116776183199485930275857261397058658470986866795339609462
918867805571711368786100785500743881857106412692050966991492894227497425410850292688489395447482286
```

30674993942734032486837449145564735830378788605984756918370045423301286627585792709510678969053335973429361420032882467148804583531441386307976265974048930871107688512323078607934245122017154558018416921640760895771328756451994313802327207500512875563762033975330060539152081158801054294431785558939382463564583104692415828661178849010046341407609499821446541460392837510247350589440734919649540370272245345722012446368700313848339842309183240249015795186593378599282947015985795919685983868598194490544129809748419955368963696935471172061818135855259262329953991693759170924286136864766772062305444148858262537053208245437448229030278027065751038001710881783281861459865583258719554645852534195776978217729361207133364124028163497001437315008855271291758919608289738898020217608210487754500133361319171319796208564149148137963406300563544654430478144609304545428797888157247969800565600799594388364820283160338946679044511365530731423327605787108207926760439535354664225367818502682922888474386393039583740097385870142651086344620470517584760549904572725750315085058585681974684292904170671661413318850438469807937356017771048117683664107793967943620268481911773148240899769255512882631452780126438139137569639489110846479984997677906299545449515731787008400712480283337111702642723015442391360206618534063071130828152074975371931468400971062250371959216386582428228940836492726282455960515525235059120784146707605675873673361894111742761679155530171994367568479285326732625708045593297864425666944433892546899711225377298404289632547282555336798903507722310316422225593686622402932897079007749621309571349629896492938787597564476663310010012085384091365778048694771004653842197085755324495654302279840491862319679062531769931473584526802758058867692232477796532393559917992633825686079312971530659535948783277738880713698149746405066251751431410646697548078882506210803693030149523602476687171701778304594493982135466681201891558951093237791066397432171715286101290073133841301133111413566848007910399622453150596207162070402855157026143358103007802778165023775172009678562401690788151274926741016063023195786733309667419533263179154869008772125173072357898092522530225632265419023399141038030993435384582070680334442716519277258234253690592492756444995993189337302024156133921728688211168862565852709987290601656571088073894875836169521706209563268246090399618378897872018925638707023785356834418100121493461723067675286359901128088910089647377924597956882500739850238077509591298155530662948953573727641323856684678177814057638772348760403251294676471534135510631406958562634633689731065163807435534312610069096745376501440519811834923531887194079939909553892895779875047727780327084660936154885596579969702595623820284617474416482653240287241282894947810298822837746185616358232334667185868834958184219194388322041292566218213616277944900410223801402421658236604363913231229544346569498272220679082328807024151375324162312274775730325612470426854433853224701279778599783518199543021487594778160761885270838202084834997471275528202579649966553819395937982896137784430079530185400369661661157628681207957971615535606620273576267247120993482629114587258566100302026165460068481438714608372797925961459932392110170397337698734954390474516654195473774411618809167007285249734723534800924594736914410223283436343840137207710013469640670753098240855303576886974837617594401545336522149188838320213279173749522508591040941264626158726646161917696317331416633744493848851485951358679018700795817364758504070965634445300631086513401545686454609857793768228464230331017327992712144695553139665054817276401972062709512195388132535096325094744398891087859016996922581980648732516163148722533208681152503877839030921096397314087014632627748736199925160369035164018132286384101577891383454820869539491144102238344309384402258904452354916067705752641775191736090440204483367881891568970277893660449983167555333460997434539066796812379833136398445862204918269859801850628083650581758563282867239702164703792611573456804046590607888585936157260733764794736283941894928357605048336449401134986549120821004813255068375628197025686969954918762197639720450279004466756395119376131560064544864855250749799420850028954444995335745046836622766872082483164135994803070601611822309151752595284900289952934287376173510267424188159371599489096979222577214401390912722461788838743680805196751530

314791141433573207365859049304277337984471291954446454304400959751830974182337618633781152912801517 8
663600901647697454495895732314679554993893375139495643626014549592646737347216021885313265446860088 5
372077221345275103105952625307111023553788569164995911692208388877078051735438398567867015096378088 6
864757576698253235404424542840088268747777032853382619976254258199293420591217977808277870511858452 30
897298563876865112750727634100359941466012227948957495592036099460378048483855259599108171236282744 2
040178498021103176278778875030363022636160976667580106039555479978745699815797334233997424447588453 1
393345366459175525813475504634426716109489081799689592267064440216917680510050970108447352631235416 442
486477743878776517385347897501402520406932991135325561481360435329683129289912953561529040275913176 7
734127770463653085221325754863079335478829963834069993715154224943808824064733263123350461388215947 9
516917592545098480997911089233113778953966407436983457300976511706075241436288346609180038490635692 685
329640055165359978579130606640474557190848662504146276335720442508766033207641200276837147202583957 7
572548308176352281706577594153270832662553910968973058504562259369849985756227021582652652806200252 5
441648981919609558212078989726717646133800839566487722019320463671881723954704930209279866104118469
570486847004963864125953067376660389421761893875423752240187458159722844252290797737229255101808867
399083988549214491386356253886378916159188842405129981936517136925916957791998849494149771151943157 5
582630705994857586353474963975685597038626780540072008974502527051939796981252968831191645904520975 6
305283373094860238329627213225690073767533911168294719812770574264375221375825032008736375320464500
057389179346592355770836220414528501390794640674266736018275379854478140769212560856944810541661956 7
526475074523902545170531094066263668247457573460704065275753577743201023913341138135775033225611439 0
097609946952139817702841461088413608565918393299393135808195270769270927607721717779098763854634460
240516904994977475774828730093397894061447364957198504829905134842867070991509330445796639195355891 4
780109443450982701736772409790494805928838465752690821406840936752248884246612560526950978010126057 7
282287359937984997704573176117586941984674742904846012319967446286636332600764110702934889723601 26
214984596841603187424531584352054891859045356941964460633379884931531169547583611157497667740703154 4
805789781705045731922588154943114390257934504998955003727042361826456804158997001369770936461843182 9
666069073135476364512034680880448544639479886105468098469670131795494286681388276584446505851797712 368
142674585475571382290726346031643813750146542919453599343608206282790725318865751797342457466122502
744301909224296277693665315871680944424278624984074946655635268045027684357821136273406943561642515 3
792322466646582371079321459044461110922107039760456510101286970593556579799723039393868396179918986 9
179915938694708632424160101610309887378543956773148297234896594767142721341284004376210660200550566 2
023339519575586451283024177142823715776932876797845227571042372817246844546162244727036661864054672 4
925014467120456437835726921447053879865342470443650654919003697853203007372006141837257113091116802 172
286721258959485746004353295033114695796564900624462269539125112528680378263752083462010919392052994 0
673531650175837826327641004899446489071950821478410208366960964415554899731185444595313232788198843 519
364669075619174436378489462636276503027206860925188097236088479380162529503922552102831859119520952 1
603087970922306318243049051361251785266783098964840762618711211938564523325880156358456632568153659 7
414006351665385278330279933276183187316429284343660511566325528087740547669670001881483129926949075 6
061744994556048405651692066062944190747116474057556195518345387406404646465233320351037784467588
566669901728752098224471156450455567243110919573466626779195035111912484914064555435257725496566392
678522191762385475381971083198322848394692229495373749313772755425809647949858910268635945357613551 4
660375139149684512296745547842471779306074999924871503917371133105984182400457742575917866712195051 7
428961196738988890457931260778363161297825694113684507888434449786640449589578177977537633175038686 5
692143284656590763877035085454922881774592134644790364096719378848001565990852627617509773940164374 06
423215378834130050262017131975912486458792300685002533813548915131998471093397988436544604827289154 9
710760373174451260971717062154915230935580784562839217995093664990440960915699421493871124291466312 0
546913228606353526874460184976497513460031398206640598310398201440600948468707691618383086
238789106146824197283200526150385110681990583517695632365696941423626101521553817250369391988202927 1
368544401062943672645120333442448082952850778650223588668479872923363345262758265461364764488099277 63
316550213007449452989543396175266117490169121300581095542449122673852851223483659839780848397524645
101205882674283063996089076557344363448014904882983571898164096451011928985398760882296922643542082
445684126875233258977838524384542275920567609260795846727938663976401333752981741038008397136698750
179251889900509616272530290638512244815629473486671807921657743950233541879714637034135950706887960 3
101501646447422392328672923717256538429014706590688672940694842542715905205340933901432108131385458 2

5970434858093148550633941870543754820191702268823175109013294136171877005809113345796164220301259802
2043107382966864907037504453868442844087909458071383896209544920759961553665571306844046952712994878
3097134507882757248448998973497039622465110498777004179371263198665504248264504271322916123093246164
1353303916768945148358569270216603190656899930352729669425409329858504714285408386710889273443108736
3457002691119243249637729822940094402075202466200664473839172044257548340145394080354627340990964069
0720756229073746841311985865787988559142521470803517849673589369335350075446723246313125258268704637 5
6343964309594790703928961259764866705690718158460299360428566416523782713760774562109031798086571 7
3199934312817969129429620454569520124331544608359576565078324467211258500776099680 97142990621463722 50
1837031998516922005081391020537898391562299250064687356797954066278295022475926656176868862 93211625
6560500854294559032917370098208588173537468783148206447026861274976090634339415664902642 621944959
9972627879887420686537748400124027120252747334436166813022720475458702706409864346757364 14944556040
1804690256490853251627167095127900529342170274303288614132330961696475560201838621599787 2360962122
7570421099687370007122579942951869630041295418877962872409327846179804864790769989057773388 60558382 4
9114228316374869287853814814314622428398496233785577350860511010509261293230994924741892 67513161881 8
6207532574038684806964904698279991221611386656638203260191359684025164924790644398333003180789 52369
6165184056141033843497961761781821004743519090706548064032050691171664322099215471240461 69424710431522
2205104885366382704720835222072281792307243786214629860668300393443184031897119593875880 71981150483 9
7750862492845205376609935703511384695978429595440648575150506208497122812336063954148045231 830375903
9230419312935804702532239639115127937593601514087053514606298995194074575535486686198226932 46955382 1
0219472021248849044426365539530539453300405061335996832016669879477207952586691433189909211865226058
49608814962683654672349840481438109431985740368377466370936580637339294587946644516826114833 34598089
4813230142344973630008177003029015416740179574436162018422070535776542316284564276440937 55689715537
8728695987224574072558628891627542274001393924890937205554561607842415395618839990558247974029707414
7881435727079149221043557745460934900537368159682819680197715906457966057549425418144532 99247981996 6
5711964551885655656812151334909804658037684761826601373717568372741306170464438082924391 653713296382
6165307231060682333889999510255307327441209426994137920101046632087497154105874458261159959 06877240
9428697875946269429880547786077645800522940757030844063931255565855331231165546386 5541030 07
2699344773118169078662801195532248099561712788155639351901256077112884694732263635778302 596564232697
3847446967479381719183498441201028423519219594794118886452077960231343063217174354069248862375881350
7950130195421256853896066931254279329444504604143997511059306830758986663120897091786738 144310626732
8572913524733573733413996498162256056419741787984415602133012302960042148911478485752643 86251814296 1
9136898382227494155863957874082378090383414087929764451188880233720098864012236174176430506370381076
9278341189017294013294623856393085315322442227688185717288938851686500759048441570973663 65393941138 1
4143300708417983117234397432128982693088281790845460835227616883923670670181937790638751 868898267898
5926122487707255370774680909173812195568154246781758249863590845775282210873370947100534470431626453
2650987541870482604085049432894496711511556404503759922183153150621894176807974004026780591 637352949 2
3058021891056162459013019130734230922011877489054176347523478720509575083012458009760894490 20142283 2
1517081786382971517875410232779377854072683639518037224129822462360116672620238536295804478 44894
5589355612941667877115144459521507280112452897838988924398057926113395117163372724763220356 134137665
5493609107607754288639415375091128355095389101756992734810580205247498860656941406880598 15714553496
1021164041299207119157382239649687900615771795957511675590848192969043559344490293 69364861900886118 4
8364084454709229880820186964350102416589280672893461892128839980588675472338226628746 74898276007213 9
6917908199418072364514382989430173553032251147891184679935694376925654535114088246264417878 17540585 2
0462752091194551715071532232229003178378639299745079504136224307388747429588420386537963 00887828 4974
9101536714044872706052239832746544345293565280769604772215030240603296082014428740664630 90243695582
2017514819021276959194854806500251596627667211562658270387072145811474469794369850676576882 350350632
7908552602752457845213931189712195956022277801052170563139634446112700212586723573709009024 6740694 9
2576669653047973714276265705359574487908761992997975491547095674568662669331781602694924563 7861334
5040128648268256468741442573431384860068426879672708278086840324762594167884057145941148046291648 8
1285200102610561984527854316767343019418002873928094615194818546866310990795150448813365012546 12214 1
3542791228391303694726258978002733168003031221980792227233930323396990249323813630178121360113136831
0388677339001225252410376288392001146874739783019681516234484584417084711540563067122150252400600956
684502099654423577855478800658445306335458651974851727110468692370821037423834451621446359 1642683264

697674851191611783934195142162634345728069393696716440345231117981331367472558103023300335647091931311841549327154187214539550089621519955776480953224728170568464283525646275430576200001721499028964513843833117116585841776451061858250821047245469809855260571543347364077660598021844622302303576492869835354772286691167255092926423495539112977808670676400914176047251101842295428946972427193139578251117713736466128861099385951544636685556640632424096551152712448740813863395840329068301663750515937889447003244679444205006232380567807586140516179491567587848547537980361332045886720131374684222777393383010159278738266403520793857372733187872211600697886849105687673735468207179344205169186912073390602304035953144024842755249884966863905615181291333068919805095831954745243476885587738310414344539933978264379056007932657790326877238933597040892906379862359132688276912048235322685331113094927661454578911147071816782987954873977966493933404999580445111442637870550061812019428565801151160496597514709363084824478407151016568819151547630853977194900534176719944035467919940544603932951225808709103202246970808460520359026902225743269417733783302077850881645997749984355147030418993109278863628555954986664521514608322432702939865354711714611209786446464443256180023520004864169280901251955587967777012014158540324292727796513998214262542785336548240803764657199183395660336509497078523288974588149334773099486397383182019324468791919227598192870591900590754067357118523481186834643477850541578282244907985840870058612052844716335877723304380827091373968895056845555323261822111023731271368814205839838547470490005360938657617777304878307855972175388186172720052568368736721769154412628775495032392221164872217861522620911242592377467055016125582354615442087445842149647760255957274029184716453158885258827884286855889496508427039429899354427475578488409253424367596461265438109202556299820515414801937470048035796677458741008841318118725823857854484888166827711303320509140899210276338669469093512487922710050927746141495291462610517648599069299238175990244161879497023567518095144258549957651792962105674460022141336737548079860359695946856983797726728528759250604441846435778240362377667029872832370604527442555068151809555886835891475330457011692173081904136821961979244460246499343848452732374619113711082446433501694801550253284184628733492266134568944139002470663355470788141605485679668656432666981303178362164646588208300503644548129228849664461641012502375768228189455118458059573209093946206486775093802437912896018008279404748174220663327914712842095137010561943271762928685790909493218055689790662862891543775486731986937384664195856188997708279900169246494251903473777039886301492011183528372327919965077715581634061761969840722213403499493842916855650515174383773539203225611599297568796557485637139984984189981872386344774773551501353507919108180826989536012524782547043214097784693232943801472551258437608609981460219969863615484735762920809967022312307799997668926143970948131311761267250290202512501176953858317579332482374758716019598930968995429457152380277892235685048764182413910232691491655744448451246083051457874175073871176744173551101307645537897887522221852058613301075912720128415816099012918291185666715739299809291799149161000466033392295307076086756566635965341818542256059884318442721810942318109063106059677332784039505960669772102777361411827206434298538442465877284718517883633752819325425830054996146148373504141918616198629106706387911951762050556410548148530782364372016539996520950423890411674976436029024341956762244691512408945656803232777781828040235371727161613847452643425617030040388071688888421475957281324163069397719048693210177484934395220889779774606413216065912354287306634985649513842991511937541568268804640344074003191648752422812899084960491088752481897662954439946375362198306085302662976729772823708841064030679839570455079864255624132956970690312606937272270497385849063548119466535336602643135554561276200224543650140928707010384502494229968246989481931106380584896383496684231546895857394178100478027431024343617063937322667141404764505532068523452777271337995989719129335109533521837623781448282123898331839226954036294419447793933862947820865391602768654742216021917454162503479486573087486963621460507966572115585682647125640198326872218016273702740854612351531487465319393626134387278498409826789961969122026697546120334748717152720342378485320199773941089399183239935343608383194483504017170160500197194079556929169441248268460290554602480556637492567690235856549630563054990947383386820022532727510147653820157189689176438617603846691471270502090101667931435388981799261389495820568431525427173339845849677842195422849693775382141987539836280415138532709515070677831761849268674924518788782565238807875593398741899239410646608428419517680029189967446045623476841746284361490524099614609132990782507767172994873799300497060952190521959187881565502948624994826388639682384950368650057999208125916980957625006322052742977572235324464631299191602952782838780863743794848781600798669184494495203389091181996840014360193709330547221904717861619952110929112791700197667202021639669184222413495352640541503434784937300948107500099230370915388201550820033012007604036740047502823812172353310546983710100965755488116614281

74197142241111465396494088106871353167996748974279372263513581038998825451605536725102136416929003217312234300723995569142252464301262735666817822859016903399525588647085175428439793432749253399892584039232079835993252157435842523037436661389938632000800645747508807093799846274709665708094369299361070027345381568480143981800226497961654983924247214953855166490613388699479448764070625660160551717891131051578981248406744154043863432180804960375369336965075024967546596535171500859975076400044559542637011962683350423969409324732540732174653657712189786335453445628417039103781824263672441578184384945382562034978117494710465895082321408204782053999221708309637924719143570526892737882963017205984163967659793992468451202167315155759406108501108401501493958481324314326483617063835229338983573286296250064539653232340901663455349761453977754354551018002272987816661057242312430623503991266927255939838704468224405690217527208905973140317194993937576065170443081784358468902322640906702558256315652710399198787449960056696531169420178903331930791287640450006274559035541438501499121604626040796637004292941415210785179395124892931131087230436875493332119971694115822422532345269916514842708074964982432091087091302719220736052823988903337764824402482164367448928389327178724630129521377758406567665034225484479527343892962263527106924829572233723726052121486755901243751068863616862068481075325255190808700823937566799300005256400410568687321345774201100430212747969404626772079602886807545332844611639636702961676361061209564091590392267597725612770823369101797932402760094779050493905949990355097623285524569201492338038955511453694537989642439077531543866107961725493579716448034461266623538041455736764262514459057192580222293064030490431773991107745994805184843416903012472105284001145301170159266417603100466879843400676366135754159383107394902338459599785664900633100192580761796592748902173088186545124915610084564921917339841849364007892425340052885127409782607281844993623344396773834164303286170746555744709588710661228597984383289888277860899498259344457062552084669336207364561329951753464990966070993431256359749029656718468035188787644371927343228496757534874380605868393873208710712344196033089306023521935023796475301514159372862118229525906701857585590486981033613106195370441077208602330006693493559822989972003894205071241309633012473989889865016134460416369764129918551398564133480244010903820420980509818816370765602535426852064250474895868089179466861177193250332482302410598058447663804552137893230572350200971557476025937720767760874681482134522531630208788823985355668408462018877633389382394005969382347555811966044053660170851554314092863357092159448116017532965843133347177302711059709051781158901708660299016051247945070243012323067026122791714151068200226813991975072583213035616679491261005420128645322980067268900094820970758541021988488529545960197473063613284296985538522652381613089665914509124886813125953536296057660319750429504118843939724705360578947986283171400396848076421190941427568120273245423319593111952395056290622611009964398948381646448745866830754857785328740819937572748521974137180942967777411722239364135603322119093344075567878381130451998451486289800060848386942062185271928018778042486680809295122973729446317094600385950152453386835579685905846517230067044888968406108630406121351552038742139284496202225754658582086698640604986554258859081455309948434938427338421786450513985427397429095857008561462561834952700228141732536765397946912752974701317000638354159654463424496835265059485344744721078056107810829649426478810025979318775639239043291785327634203752297565752743408295084547947015245260899313885783123911751269225566757288513340439769625403931174933713994495293568010603796944595685975249877267348079073267618245233552121621496802344929252488655145733756557659455570923953342890311952609170248720311152195003731484313740709266544929252042319900734594639311996535056803733428291486584110919944394627232845367711284847362246063313602859105963523719387163459869636443906854053223193152413546932487576730463381703029447983522602051814944458504961203626092337527162355133526234320721943009358815033935989749335526957845727831403967039691001707341493253102206326301692523701801202442268849290981955511719561208381550144858365716651026908664817132381901486099246991315460820019927050473056887614189329810831023526482810812485450220875721283441134379484999727920258354341720442598469327409141782143949280179974565987369828742826826748442121371546823275112853416523703165307043258283372112371376069591399375495362322221974659619332529074042487602512819524267939101756371975343004476178250431153315067582562735343476252539142515275704787843767885241871969346241992700802576108392749762263654900186532064549951558029083985132726273219672847830253853221904792786893895387803686993184466031043634327215469989881118644367011402842620261504738823589974728149334325706547463702245188728906125503027379190264039657741760689897698345664647052047163592144830709958430647172750763371802071459744565251418850503716377381890296968528440925819317441005557608985029232701060089825722381739454369852965470690018738404646564373071319590114076131742822038833136029708526

14512349073071476245340502454237636666855757401206040598869556301144154431416969860740322078860031326
79055391786967416355524025076765308563224273784714977840646093468129918719028927759763070159840887781745192690147691030030234979458510001808866062162868011109151161040983273088089954337551187183173786558776487358854490366879074169182538313306362338205820154885449827883821742437580549238159729646019631058215160709190326318730933813850113091332770930342511221097215505610491387050426219058260247611607505971037943664771529353491718612506611636738330499564878779365697911293195878277740601075333724327900097324072560703888008711373096596344570960411750894447911916217455169709648447620461741281543536128422601551301589525862806957063924443547802390516861338549861926656762276053823498556919224896901164598660961567955415492157581283540920253931708073781465078832035266513457070655368569928112426567084480177864614974045492118431227621845224410884715075701908834950224375698850494954241057266246094583209596257287203099477654010737398558069966198019534500506514501054924018861089134173060759238439461246569861605609094541772907364009391321956768364333299619997965424348026569406836986706175874131676505060271358825744337072164588195226133970545205234412809231730650519899545028585383522873787286982912758078242098261060609015092521109089899648929786294301411140067628774992078172379474620916899838961894863077306027491026703883943352474243421236712177024610116024061118571687050824404916238943435042702508136491551551043272507479410247374781922652059335513625530181219642499248821299260177408010373884212547447216029695002774268507752175866917993801362584224173985607669916543275706123045286072807076318947058954406184421315713800339848740868094144618458542303449514485210539912489324554866593053355845782771442377433853392082412776321659675383262506430638806955691562026222594694142900799876983402009146999897956832667090414139806863332515652569687878992257432796139643702654314463933379900081985753079788176133743609483928302233803279732038653646242498055114022469226478931592944918040382983690348886386975644148556085605371772287371482282368642781136572039474969069372399156100368280751411784737825622127759959167758140385329868885136552323139384317422585356280701915440077630163476477812613243802484218656698552740076410511901095452898382563484070414584700085002204502214769722042873822020644511165580202158137220466347663335175617990500897780885600932081454176443903235241939731547218920920202474270405857913543792600976807636298756614782644338926121213456594456801042227616524183764018673453391488079001363603528432174780401683146726188067936493046865878364631148328269804920222787625545401785091774972294676569293545166609864194814583644478909603830458285369895608006656661903603100453047200242263938815720686746900619298730822484900168163489212554337544475803887361586588592485975587427838599929541709912251153402542222969223436617781929653313496747175423079784301938184940432185907508056494831896792205834341478905102576174909975820012347244102850794123184376014714213782951102187358993917081686805598813354699137945378313771141354919712434658109311278332662175922275893699347704984352519210735157573702005457345130587313221438462060277594845348937132376629907711584466071375488963551670512601746640760728095967213272942245361042667928094697735836383498228114766916959695573270282791209979260863817017654210151115811092683287485950646262362992592496943781822291692343254853616041525142063692521477459809419890296814776443905376111917463026074170414492241949800170623866816946607989247516971850009428744446241005863803558630650636095727097734930709648907706301907923346170197456656248712884986738267854626894512094622905482754233025173213282785316517586934530140834632832008101152525413119367954561261216103197428723210387954825391743140987706525700603767619683830478692910032674834420668406673515088586091662976572277996926003868273349641875833211808091686185171345911568508931494044819961072502337896779899188725826866705377574335124650813798628134835066431324662709924523654698351740704265136988241433083552631945475425324845884093993784631867338133661031005811464551518773050141308156660180611322683529639711462401049831484646043952006163735425846982605212545385915053646201416593061524137743303724644103975985001568921027909378678603185176724046959915723405290038167607242477016978152148615194742705461422112569127225381590502039583051468585003602420054909455026826185095875426210973991322095025548958289172432981343154069257570588015736545351789274152712894131431826947690258885478659919698883599043885422784469473118751974887712150191519190888581326376998121590808969182335059079714874613336803970635003695553942259073453693326389696163284254104794629483793746381324612408253069026914650215171965524256788471272297526679202803841296615750218468110886939392526879439503326728700875818861049300859750196928095204823757138945438844241621756626373518033643277401734125942745582026754402788121409286880

412030002270631808259318676648149044725799446704594238281114878611771247129940234353193476389778948446256041745478020529553081502542980157090392382147214574805502696844438096038016443726195300440399081842658748795326360174773439634741333029275581820008464886098183291707862124370345940428150160414770581856441322231887793921639889612736992400225160535128293723655737319319389834173384397083090535333085718369311365037008190565450432984220993814900454454004445486182214055060528716831341426278964416953339802968795175366678475509672567390773471816933997599011898911139624652061607188564911926642694016069590616104437801498666998284332465257608825710108991959111180835030365079742812307093591984025480169446262059236363510719901481567744000121307254102560560159316840532816997339070715372088090132654154084084869464851337622748284991612747470532225050579183371978299802173759142434471486634380845991013925549446596674237301191215691048473306980508171297273138066292296784931657070031083055678897912329813037510317467834011208135309146607336751187621872232478968370382795012395449535967588537384466414372024700158852017613125481437213347926567224469179562586330030699458362891038126489523388807638438188092118073979276831412308688096917176552671971341129027158146752944276252132950153401153369038922138410133788327539359202090519420939389794128938076595308702735507730422191143004325668462407228903402171151273168905339467851873822784853748402442127739268280565172603764458440068284667373544829524969162142390338893181170113891315019222854107381814522925921851485752524761838480775838298666675567628883100860152635893536537127421777289515445831293470980084717828967273412903707607159708978389934279262557381889563194050727520873259994565450858685535826313370849340641393491374919790772182285311600949827036679278923587884316084918389533623615765768334222895927023939315386811958152996066452081886184389697200189336785489494221651236101718473760460603829616442733117846680699852153721647753632066072061232929323962407800474280678081245429927950935051439760211929113037802170395082950990699443939119550034072157414801775922997193282716303971339380034926791334165089113946424796613209472508981957237504911331291669842737114895866286493361339703036342716301283210070841520262768996306818969526666944051567153293786905003410020428417394295414458444754124677185087103019497637306464566186349406566159555902220388135713617682347784631098854748031416075183130544034281127035071959901420305881492321488412935694492964317031319035779377391901656912577178069851218011520363784718232894953445695553676473404058533960666301990692261062324086257493667440265813971924801331606041280430940301178183953384569734778755496505226719892039806465088098765959904026835458290840488743590121024857071905695585518229964816324570157034748335071625617787840260248079543643142774097520462960077059523215937839265605847631831645394608691291390297075293245752831122063165307794559970933880191017954784682220298137949673390087099348352365922152153174780914929635957680364164531574198404829497042403902353402659221352598530803324317858577406056900395174966247831146154707470110152410927099019069460555923968305614807726337390849969618880204526136957223089073682258369725346732930434907962315838626126344420486314103499740365508692160500462809793942595737077522421277559627928584177313693090364594363844939823392918548955367647354089553676473460939897919778070386620927463185198508519268526245382587024957380501099381632387082022856447402189035037020515868245711255202315308780901619437053871768164482939449118658976812285822826305889861029605283254438604524159887904370477377371368500213539382265354902342171089783710051759702068682570272957129199904777313320890674211212008218398677787056562942469382004237813491306663028758752092564773273167463696647462570612003467516409034789631481868213459806127296825654557674985919728673797538106742173858019486939322932978545561007900054725330516708185495508295733164530389667280573878280582588338674396272132016122364271902399903185069789484729803770517200091305626967205284316649649046310313400903781521749243406714466245262478844479703267080920479660215253622977798413035291824916677547333861097577175208925700627095398419276724681021271464465163618290368785573635570011836724154289540477483432843483747564304527445068084429432880032635348132660476017214226940034962856397025725628371882800895476023866630654974704534984376640385989464581448053190519925945149823361365390441231674071296632618704241988407865603468894409807356511171908841621313370382847527158014023337408746055090197337083604720090201650957637603786372188320038639008631870025211594327839328479264705620690613940364883094552023943726001155644876784475408356160399848851377292303432535009793967383698300912779979497170462853141004353493338226748496581775235312719601590617282846213714010305334392027034513019491070344617456577179219132414373649693969048965732825756691327645310834456871594499005092022404757994214851413924545725326078272641713056995813723489622724101513581463393598573917624057344643189580687674488008030490104731595819072414641729115982896970259377776750067320383367599920898141119898575770545690088066735105347037213293489055455497205353010557324696610538066099500702754752262676383165272527915483314536193916719 55

10302515194171781239124482761114532218866777176734174064523789930150371042110232238847769373142553
98388888374132460169484945229905107108955879321620563220851565300740790505924944597435103491904411
79156636661772461772342505771351604426630343126850187643029646518215386189201400711494551862692993124
58342678653320477807539009009460140438067643829114783029646518215386189201400711494551862692993124
75729922061152524782916654575695663832511478672525609757827406443804607071542461773873376801930679
19130310725974710951488737474969530349152024257187438661942079828728831057102817793640271463453279
34941390242082297386601437354856090802957134692265331182640676525443498999967808629347038553615700
037969986461806802893285729725290142829019704050111709396040179940077279073005957406453592391666388
144597641874262092009378297705225640556980299485145116253966586742236309839350814877821146971451562642
97232451972563939261864816182977438532410643492525416862643522704313723804662598556860369392975077
00699210927268702532645880616669225572963522442048586718120450934271666318905192624501498460068805
235094444346582740225829051295030497996140160077255644874614627097868897325672101929230098245476543
086437400100757236660924137231868007744762686294526391724897026767456792734869542632962933077993830
06951742589565375330331982513746697462587161719676711578595912938946419005195942311625542655890360
426117249989093064050953199964610717090570516938281588424916142363960011138453259171654027649547661
565903127914951736029421362818726456267767827689602966497224763374234588532554325573191763216539
350146857623916565043387680021015694435823608634845735010253174638070912058156818796838583945104236
95876705666860684295232559420292102688752911543692228947178983633062840897730393136992191147148
62894383234242910144325546512377642992120827639297377159406413974052923053096407841631655435542402
70608217818229921582821565401767669766120809144860005952996543762363612026305988951007723575565165
2197145167836743551290845111362955595308758670837780019199618165683100455702282789191613024036183
641223162709045429397967412823508433094522021606266842689197180814965483076922146809272437721832739
09295506752479298131156698231593310969899391225434479357745606611269046429656407405419288226478
797974455922998213530029315531271761722186319708982235311610694068488155752894816208209181667248131
890810462376990701322935328445008140894151893110987965407214618275748055858024353816151391877200444
85065798479202545694411477966399229790530227713023499786440344218249616320189181180432647831115159
81114816472645477617634978713086688756952976260303166684121463430262440794686978285987418395031713
2900135590877225577053738582054889531696018068632325879151070588932716859422071560763890785976077
0843037319077176693143523913518622363114271160854458040847500024799875823005870882654914622078726
80636374537598067416479448481497389238754724593438752792867664432726547955615698313724625104542888
00733489513025043675009241929034354360108566920829426185038839729213803592097852633481338499592723
1325821086087156279941191224265330135185468301475883617271648821814983502877885575396597889061068
286181619529827640257722566057247446559340325286581601438615853736161141630755729703799944665114114
40794271475377811114255133972834961152875330388644328160811651900971960558504031993456527982222343
72991817130543227748759230282529480265505717408619925237062968149812420426377685988195395996847506
120317039135076007001949208483688796385142405835386068968037754420706427165194190884481774909538052
81044824739134996228296437835613093745017451459267596253518444582440734844891009395575602297
60624484561256004382122790034279070393187107523808563892113640393583540105820737115659807256303077
23585173136193707794908570998470133668500208412981911223595969682225964544588321433935711948347789
834588974007603069304945402135727476634284866657596679093779764676816301584881677076460389706583275927
30814092995503945837813851437720113532362892642037783484121359508214072712089533733168788100003420
05761803890377556602679235794264582504634285826408242640704136419947470546617833543879107633145242
05378089999550347417398247970869632306577880661508788209327159575654482616883205894028796721171980
28362406561189079991380288320943433112469126814143218206702353906652549656361761513240156795689103
88079558258253240308003740724316505915305906941655229142226553465708270971438428418562650322627
962958902475416534576128240935354062701440070911482095483336493865766285662502730215442597984538294
191301499938168390326408029823488774147168249587848181800555206691015613702532736823984640258918801
0337650092064772911586849770591281425393646332561614938138098819347295892191223730295843415750510217
8760027425700671130721327155850271832631656732297416556993878693238288866660475346398736332850561625
487738543568830057262497407546851550544779206649515963229258022932641792613907725474952264621797
46086820993904542067512223201937656293829788088618030619528493132084967387233260394748697307938567
07594093947867425498820242041547067198240680737708036607512546417371045935695061956103385552732209
1093912275466544307152856539660824821860992013074825707698853489469154319716569722476616161766706519

887344489662458628625152225384290156087408995434217695175410335336340373475038196456961620860504864141458492224453185569392104609170949935123097068124336744992981962743873931320970401669537573173363924066206733660664354609510478033905319817058757712382737802524734990313350046170788777247607203422573148159707116345400918423175198274427763779402582949213724616142441229924935894614340088093891446700320346521898372728974239743797153198307296124285026965615884663944840556777668605819505314833709768685171863066764959789067888706831505251088021562700170042329256525775533715174488071703307144098044652421313222560288183751881800277819930167409986441062719982527455695968006164230149058371294203283526086813465820686611208819994256081792395986575599068847944798957284718971157601126635233211871997466584038021813404363355789510224096323064467036239649691027401413840261860400252104070107826599150449718939463765389742339594447809412596446793881059540177061710349317906010285848179428692776802516063469818648761586233701525160236373181789799034379559511508651454025879726423309172504444194331712013746825419070853107215208512693984662801771110262424563807915882495667434928755342215336739800701841459890640693638164153469373358041348621683628124855196282073725171164471004843392677129047095449821771245997353794989592501819266700991923198151396872039240781317332882274061834899693819625882779434181259782237468233565569661670703773942455538602130433493640075319533984910869963335613308498200462205237942126112738646977343529840614120293357476985428378850731423029608600645839020326683297106647485928028070622289411984685237118462881782523378018175736273863563765251145897691425173868579768133455331235961089216796759921620047777736601739698157005189305563209347680575208239247093075056486535401470969258633244753524002573951420808860099352360949792716495297330731217627022775828058433099277819938043921726811176593651677463485868115239136038173399380436191337087041783245836870732687927591512421327930692493637159988393383659676333285508452144871173498729828022872224597211100337067885195708636469326747159061232018112861922070378166898254061531830765384398395669071980485139529940333118652808812321707773513270097093437728645069055248320188640544591213111790441279949017804606511934621383518200791941711869477902086478775088118722447764394728760053349119723576540686820193680867718864154368080090685123604632696099408509903969483621715707186981568085994319275507836176310635228624435898297577239716171244390007852725787442545538826516658015450494415164016924498904268345734562617340205714011793825315539669178650853286160227797761824482128672946924571491109116351044932017080072370744963492569121890493135736176727370757949888198708148032945608027014254702799289156950749771877632590121855832116833245638324635426413663047442069533906874618048742695730852892854977557044655307253726535449909165505854490509080736081330077728805917621134078127022965723580325892854977187766017279646966552244526699337699312900316835211790528327287495447512981340250629138109549668572838779529050442889661515542282376017449137351386273975389837428783645708577787917772282941837862509262813945141402838181071249937085703496006793856336969864767208107940134332689922920321079715285432230936783898249784699775589775841621288966611830971893956180558988873478822385862983606148281970638853766387023699719056887914200150058932781576679554126561532407775118370166626802333634507878413669795089586402858492504048493796171180328045570796932170830887117625919959435644139714952826220987117939305641225493818743814308084727553083891726934629489967863754936115462433363099267699468701922308814839034814606660954452470388376120831651420100763183375893493999620081480099860462830624324904927573357191805826053324935512474110735860302253057698204994600214598639757850583213910161582089381025366404328385622156050481874565965450748771370450523845886878709415841423636958253905004461026067279434510615024772280622353096331573612514266794341407466520635345465839222611463235578393993709702620191622134788562453983020165541146208476297265257464268723386362315135535356816636051903571895244664237586619804114480252199013025958934999070183464431270680599004325613871121045068905575067907676256305384262622513729065225654942654827321349988484160539500471824985656018473144907426919647452658113570265107162387733468833725899778643672591741155701655292484206315218737919592464483195251598063394603742186735024198039034291728820585693526786658158201199578760698252396949687467944308478657497450836696232960790189307962448748437251749229859571678125574842331673035806137328720790093363369257424578919538675866761134125824716030708943168749649125716727634443030554358794396727983916055021351483332405448295912401218783134030459294416212927504998450167094648495344091524718377825576002939794250148934214633867588345896387026112729774481845303444231471312440784982148467358627326536286255010397360301794632125535434232983465042878299269795250762117660416274969836359499603243495791210300046219201462117222885844133017429091519974561741221425481510127488263009469748351230072170223802516362654245731668291547444218100277761276072978920

368258611620666702210658425698166382179841203878410264487735566199629191877992434897623591813120793
212628580852454714318097767637213645840433599276725551587668729268189864664577371815671372712732706
174460768370232078792424382140383043987812547006213796556007441290647108353130420289291215214634522
675356683968136278189995449676285459172951332462312686271207307031686740866015957099166315864652238
745623449645159907497920994590343139214940679848307225457243051433054490785504685756035984513141863
848348650105751244928756718289912764572460725925529643384645472945207641184438179028342584640929845
812175616938935965975740429295571529150768297078020239789169692521995691674357951072811035844878034
761204586517986663278274801506364902275231920574521535572880248195865656593665659364630546997062951
550456869148032744860281726828102742488640261601893418136540881702751903632798220382703820898351245
577903536165344698442851334548386435254810831284998328941817291612083907139068486878921381010367965
443525120422214656143467220149894200214383442012409701466236343858272692961086252552511892129071483
045667955951686533053105451019783216574176268285593469187149247242547912327152909308767435328387505
316259536391996859258022570941304541646785723973762668711162282093474137383953097453856228395401248
780817487662218271515701356954665014154527686148821945134110598015908637689785102904745476574382151
811345102502465128213078825731151825135732082734220695504162056314244492977578541810479619444293639
466973981359336691118451577779926435064354447789078668856676365556751013302039206410994467486972380
890185771237393338518911517615291790335703578359795953286587807344656348810329886796716467813578
794806403342214337240478678694063066207205650277370025852374393455794943717575948222394848213090905
392311698646096483735068307724944504687929063370538796371011704409037520385264029625703004417508984
800778547400690983159410928873587189533947358288473512024777514415286414333015237602025716546639313
151695802886647315383404234437057963764866980429174989729433098371512156861929654197679087543590065
851911949101426075596946536884456188099014507149283559365332865878073446563488103298867967164678135
384388603936754812671638346326081522267743361034256279904458071024372908605563941984409751355348
368989443858562953654998625886048571349208874795326091097703034650129404022860283613342235054786831
753104804378870682882815000408270446006900724460766081328348616471638169678446051551761653711887635152
359278589755531581273539168302801968872202551866090543478414559163466458713173467524042320513784665
454848521082743372800134867245068256794125326197616598871523966888739252407998493528141270965420506
976947910909196918431017573132250706433908790786086732900306398353516515980007405308268960241703702
267069764470297602604698157613952922096431184121990812443297453496971403552504286138984205987355762
554353472123109964180757394304984758358977867534082777525594534690720294929013861369117193813173601
646476625497435511923723190337566403995542447783944298351828774264412327916624872611661531195565153
183793037448307105649190863769246430392533328182321026516298438247988940816153152609817835820542047
757423919079066108906172633363527535883672852532356699956862701830812167405280180002553689929336938
788116744077729916542788044467841356200936297554569151807671331296375531481657990907293104137962847
941779072489099883994658129887050988367268841726245600654681144806371742950845879829312533822895706
263555895195747921004744878009906747134908177116597027874004848558461844362188044559755361397818771
160012097380658520602274673984321980195069095331623045898291761625860113602182893530594673628712855
704204874035807380175224160105364492072700313587362744654707779526468446408067183202372794201434472
347416804982143532190661854299254690273890239469367565855047152011417500088637445728098633553566630
531447427801085599461612399029562865316383085174797943117702616621107506672591136795657312610790687
252713210208934204306864426562190889801026878816775863391987936068177968001507328645811186433222300
276060997923728491881652219262682090188547803894435603204243307256873460617370232452680436617589619
974446116911030486050955319970563600863393574676825493273026102929726522199970137847456683369278196
326841205387250396773387410090343310659498597575628944088749450232447104514117559283854319043656099
794417786945650590867372794051810358913300780225183670471294785839325894520337371154096525172614324
582136510949286496372790675667757029078810824521987372794038199950249285273634074361722349146013133
418826157775394499817582444937381706204990354471166776178806443476742446096592008034162784725694268290229252125238485853734791367159246949973608350100908415991613780484158077661793091
594718856909961023256006367965097758892954357381862985180328592847704137243664895146145050192069446
005545175316280653305226400770893310898674759236311424921196473798385954525778544792305750680638
126902090132506307939519222133860918595065125949662111567524770317261323632703634237172046692361752
296437171794845104623856042582227382057466941639979217871159413559646469298483067092014257797423800
529530764213118796303250033849846280636340724983128555939809627486824431957818590388875500902591367

550437589147202605836213776664200909010705930533871909593483383302999128166144933023488324286278094 5
03744594199622771925912129739618715920208473155346980822917945574110692956227707464682922506407698 84
4047590191734774414664989263523613471702165066560590473280364968243983685148167372969698615723107288 0
50308655420546534599832179968137693852187938647137415299348482991880895775760669754960742573489972 04
99944592474776506558813098857791532212534825426618429008335815330042553324053214212483604312819221
72957838444800154600298950501182389646309785607091078801026493793765650947187932258338844608895546 15
29462664001960838911309128568907495244109929559165847286298774924629350984305416318920651890102210 83
687385076860449558367008844971994147118076441320231950290439872981974058817957555924943246941547994 0
500957863464910789359582772786600741560112775892715697466918567208804615859568973929234646086518259 7
348098762780327357831228242397149180793499521896489149874891195446601846485979393343279816952173481 0
473007464831253068075519706380791066795589876645300045068524334584450141943807605732159724289332940 9
5357769028928132730950373344137444842404965267264912307457602800339029746726007190696517893056846085
704862419219218953295012064993829790452218907115449969937914434089777511702614082584014441535739125 7
5267782112047795676433288753969726244731879032952501475268642593745614354879964023555025562412865641
888711495639054779427658572255441127991197852527863555532605490775416372818787942166756748578561416230
4336382407650378801560198809577238048798808776686188531550715521356754582486210745365289042299172215
641243296548596040476034010896079300019701881603401870668767349301168067625375834101944939599597517 9
9294529013819672487642943530349403840454890206908114517035801938191590053992538549904281164432095981
0429725600989398297814280181586790336160128834089182426560769657275473619967224533077566356514988924
03328017852909671591215592207966059200516647151130840157471985733155950814931327443860967772896845 62
8669479813650313566094452938658576216472384194726675264827978036461664616546009384320096386651058793
835840874963614197063437324407440617596395040543023084204531168926186905477455123081429286713145478 7
24936722627140194085807208956072954026603573000165918462286944297073535927977913514154572366805613
51033735941605798693509445930764592536430129491834110698106824329544721198860420889770448772257 978
5222935514414359586661710840576191296480249689572963361908106473429291580124923341595072790860874734
7297779283102211703600785545769509313647869059130192897508136820693747662440979739060858195703694795
6217339092242356878719544888707655404175386489497400125053265954906591158579312704816593456876477106
13834516572863565673081448107224136252029771512301152163664361755541537773135312607367384511161060841
513583749649991446718312384781726612841728329101190230569342678897476884530592788790534903105077614255 4
29962938754841751963997912067148771056699673222538913932234015644585015038358971663177969650337608 72
6334162566515216857077375304022479440398754569309976118475950363021128888392344959267003790422257102
80651298054546955321149537582766497186706631970791692260507908921887407617766347269293430000285444 51
729601647189015641206994309986169366851898334010548662959489946569035360554157029370869503683225813 2
391740011335789122838643818662142763911201676285091701580132394079716586787519933592260677409731015 1
1742268902396322001301650111700896408566091596419960202060399859516759933616460872516105373830608301
38547011469178633538023410741772497712969498238933474609781466633088906239419319707976848862203944 89
665218553085537196766750962598805177974726179303269128318267486205780926525652903428019088270590140 5
46637470013833025889361601715015488868935910569440449721201139245648474988583600601119303093340504 2
4819094808467898760038276692181719086769660182978203559920718977804292361836630667945726051889605382
34728759328230119515865551801782831335658309861731652593503080582396255908286572921386547219577931 2
243359147323752403786908253602164629481320488457398705136724638246550998731146037622534977822103010 1
454298751138224482136735267372107046667779742422508385846488729490960052186147747715199660246552260 6
962127062085416878723951599782686120792232886495599471129439200099760064261445296688181911180944889 9
1955881471379931481529329582516076010711662495155659372028930345139657761202158250017927165821147493
98982521641313983316292159116787863114903503708046729113467566457367760316704471187228697777493720 49
8411552534228368936469984659364928163890293333295967050233106318654547159461989211746960440747930112 14
16649749227696638793413605543628013565218859169191601265017034734921985779121559965545078845902800 24
1979283614688616696872940673265794485514120293316476780504320302389291620949699409462681126884776 19
4192988872830204532548355595645610497507526105244255809386273420635892776828442281229581464734736707 3
62255972829437263774877683992863753229345393663445058464032220274973415024875765991021519544839141489 4
5282238326814839805199707655926889831544715769511831555106187300676856804057123078324335686278130316
986843408552081541485255677208828951940358785837202454478168634680841197036059946920420502238950509 99
05474622524732812866755075605926657023819679854436973843036347551353010142231318596659835501729091 22

6635606839627244308404652362865284232381423499086766091580811529985673188270001310938215761913296593
1457997811998234263059141198338941256260215214745561859783940816914555513000175131020838255631736162
1545903512910618001537992129948114092826421522700318799959583363724143264712437805197524452586298660
9139765363120503485519687014261642737394425538704104093401178690284831732444669643839274132175880980
0049498861502455683932136997226039513159990952783701454801273952792565706687603394301212911978505936
5743941306541784682021900338105036071495296968271934082499540824216163965539009310071446388762031349
1340775627536250087774807984394581194951882007132052594033442134001243688746110151026668901067251610 5
8423245264855139823924004357219566830736578634636284713947718760514239870777535371198854467657293463
5846988478103391466784785470873151504290874245923946132610892919503710118254923184207210437942006267
2244936264350959550337583817195105443116873279605675329829916131257543568252456465008122805226368146 1
1597532119917066036435974480828856219322673936865606254293822490006199560775333065246484430727348720 8
8797672929698147974853263746082115171227217434446121372624336324011843291889863711234758307389741059 1
6818188166895595371823995831015895586515510793911051537159282391972723851893459502417779055151515157 2
6497570872428767944097314530204095907691738750096132648370745595341535133134490003875280310133785644
1911474235022452362008865534205354859562097776348591737878062122554402259252738484307765179996373820
0909619759380174752578307961563233626646373083895738467111670927506441574763282421096818667021420776
3268375526071611089889449464732775343901491089616361841443514027645812228960506038155465315732310 53
9157359227589113492569009356071497768870073195410228342717487574907091871630476238872340969630 253 40
7824746509725002722414526033508279170509524408937573333123198203021541635078677826550932171378171231
3716114212318124408063309815360743763950265425574723877774795160370760254863489482815303352021941346
4669637514327152080029100281629344182641082759195249518173693711365151453769757463035503968815577039
3898348715498840132387691533308680839615203879265983426272431774927682696354131068465624484349311 65
0896687484469347103405348229548440424944548029015008712091167917658606527872481257977534747880316689
0035108745856817454970704973599571113398345543688948216100537152614530056399124244539962580313280228 5
7815556753370517961433593548312607199002585111086675429079991703537006064368388657603432842134674936
3284798134599509524459413666885886363580165643967245987521380576361390214842158335086871769121384
5760044375608187454847305378791606854309384081019497206105993813537730874309360802543746183604369148
7411272220989011476728779493951733778941126562046748007712938053458407655616163671403281364506738
9621785209078922246439167690643804019848760595335757297848080536465417964743416320801881522629972 7917
5361841901057253910235261006661285793861284828721151144331217481644040056183601867286327932539692823
2426014000725989950970512646811985138070288116929747036513882781618018114314687279332331454315102504
8027376973590692969716888934545315353695151898759689752891547506970959483918709199985676 2139
07843263703165798784080761347005954231533063135629197234763111237017263743716988364569939180204023601
7065484741044827168499151197376191570800687543188967229491535191956193124787701048836490034307122021
6932716448737890748697716889055669789349557482865994538815934237990106692455338066035693093474851627 1
9970008372666259460916369119186683408760398160534571873044883787844599456410661946492827902987144851
3829662277069771018679726274836234505524986843346217582535199489358389264000821604257858150101486886
9194413913566869688314459862119923349722584933741055324623722317841154042257919783776260572497686118
0888568657979800450074365599616849105084972307353455998115616755131668419573344751329229644291365157
9966325538347945076610328968486987628172465341810383001694345217588085496074784900756730397047497994
1243512618421371502775616682736902616881508892134935073218269322806605182315763054160723036713898748
3779334050158419807105831531091345171439567188012503059886950631165440619618064708073612376753683950
2282439875286051051135181419197379564963638681274531652810635038346828357384405514229348007993668 21
0804493606862164600676634483031922495893155954045816949724136805749636567933079196367152569807186187
4731197459621434447874538755676792302136148597286303389418237450498917697400714288423976680628317290
1190857403250338545222254033759894400897932960374120546035438418200763258809362869789359558085094975
6506799535033458370680057239672681329462140754013128286368823074537737549664487670662370586745285605
2628768437103120859783070774298894233196659933706724709689525249854458209814394921207939343655 68025
6945592260893704789808830592173549902808187239423500831667258629898918865217070485809184394517 41645
6009022597384114577622103057688690026001961997563649835723153371164800780419034073499583084414551229
4281776021521049358131823966276787157371499828414080368819714227388302087450157398585239725106689569
9623496283549272734508268912027147525704270506111845946471828532054542723345159408698624119906327074
6182667956011701275258350864738849204602785712850234784675339031318786296649552070645507689584383918

676790667856947756213003776080245180827345252143205516063315040369941548606130675953365971680023592160894781404681853314710609546833691621848060558707134105519361745103710159921613619950970865013778053897593988366937873954822789683336138162870025109398023076921640166695991773931224839082644791280998578064053243409137186218853230491028911371610830868087486723811585974218156537092967446551849292875004065158071920282805299224441627354322382564952517424671577482689612161852563998936569440558351032818566413199762669743911116778094871793696573690376143192238347665546373088851777088449683238058186610490890214787876907311571261412345364963900842699447080255146246808750232039897736546071330432841705373441390389323484250318168830269138738845884794203555798916511772879206893386316082730043214700842467578529416604696403753841191237643274386841834063376908656334246919941284466004865121778043276898551840227751809879011069676458190953231321128789090149768694698126335988541954556717571267667883715576312670276152597776992561643721030108499313035437189899247009395752824298729692241071111800967264157307261862392713715782806114143002017111426482584889272072257212203277130099829499920651648922455188366142159595891282375248360621720444950248575680996915586234438165167146316331004873922932262219431953654657677812098448625420396083011277674528313406762294368841354220939496071825983260013557530996909691637349665597450373805541787929382030075698430766954393118727067442208400259867930924143662902527305487332037499318170389925641616744964935692663251534815077105717773031882845221753638550909028760052076444558521723466926597049347060455327562960679060663231026724420595594961738355220588146676672261526539099092237347001406561397500633569448750968771520848323518979421232525929201006950720089427540980852726809450590641548218430923520598808493156199837134863086316081775309934313568261193911313197553117878918315599227750248461596706466405284880921192853256995495732701336628910201852595715102098731818469439442800526951909043300223556760197787053090664264844723116541768542221191674862350821405552416336328561937291560265630374851963379831249371251258466854494417939893950361305635853814653497429452024282804730393196391966568679509781211945037151849483955346185707607532969762606421264450505727237623975657440000518850643825547281839696475595398026300097329759096766811250192035086291396430856380268780542422691206708466695667793808428879759063433851918316580733128864107796980277388121632164618197229799382327922503439232071557065563693152921882935546190236507169076166585341141620278688145206333435182444586927795980466279701274348819999836714161593004043854909347712321623085392496688136772265723665048007428664506672319728135654420431444220544279556928924816510679063605431422762298205106039468390735987427441902527429805800573784246772140507586893277828396493302953931675436558263682933422786709690862018808955961688981948336656830850488836176460312166876308659491983675260971893798508642961542780131573884226269097205749301238347693325994685064488503584886598349300601794354979546847118805815318348388546700900003623901656750909851974382608917629705175162644632104682240045354860639954363499876179329543924509327532211671851255361710720256583335435269680085114896326771841394897101579682237123237779981704531349779534409755378224041468783103118732043028771398412667428430830099637565151994256408235324995905280936608017708154802977468046987301844281722489133669076425734983149530154917971860386488641641366820315278415073016083287609047855070410279237624004973267965044437574361492630923299395310213857334039465464313803932847546016682894570398520289897056199050058566331681025424796122497994161923808081718695617683355930834329171118611940661424593023416858928325402281122375144878383804375333860052459837715219085748474091115965605510126102135739087758744852230707278675102694059525840044387691660888998376929711576754644448660686822582163159384153340218454302004456787861653538779006847946816115464147742567664344666786165714321395605805106852492930203486311876979067573730613086580498053996249830885162974483234209187093248655168599888218076342633618038030937080685652170308357042659784939627742574272212686526087553770342429721316247186562181502127210015338143526404467438408179213985247176560645327166630534430665878850200216200832224339754743884656908084822976109598909361512918143246941901775429720168260292606936286775939441303998070048624722710746163817080272609324491040716169431127216932459613466992507264935833273692236372077527056795318596698939329745192340658032028977873471602317472274608055716294488616376900498028768185146200620327101913512335735036490080845554425576826710914815596736553858259438221951358015637711971679367020622456863485469059877803465897456584478524088192688309227770546049463556158414097453554239477673427530500839360542653385834292408867914093681442328587900922305291527999159050090651924744141709100470084998899030333856339419441048643921902022566287739846266971296286873757221618817127821235633490471549947075886804059463249806132450533895452386272552456440081303686346872684137921937957850774989027824315838273235033109670899933015706641952401640802598961958294666114657087896993126661948681911951688997857138260889720165022115

50000459167591831413795284292401626307590692810259150476906872272115545836998472442221101278030084214
9489854275989287579626814082842462027344811563776158858566660628707597984538985883853973591506784666 7
7440735588962117649026388773046312288150082414864035251260700750590426855154587308199448361425175009
691988121471776942905584636475578159388806623954201893749084032275102063489232830934102823789408898 9
531725707278146698392546761505090113385627347482282653535901008952817868923853130548337080121637570 61
868763785954802148810067207114027051724468180760760129942225757837851526237298044375489744603759100 2
93991487767534772424112309983814691187957254447449604200948758632104967562261710980657037545797784 52
875550670681333011940370283684631830050668073506247064752243618574382697783983354212993579342251352 0
349235503103226324224496188486719775355803165214710619958410406221190835966907897963380812136398928 4
37310743158230935296944235518933301343708243264385327899082614976500246622336013479717464642080798 36
95465133294537755766227252446205078516996720198999934010133350720758077104038265583648170001882520 48
73862694727977686819841244971301436744461773836037845058867731041552138029551148291438196811165301 62
35598910147759480759411771570116508692185411471004985307661544503263624343420505278561286837117886 98
624228435711227804753731921404895699108618357580954983846566013084746527651201522516404600863882540
89210102461583536189885561567954555943415393774502910066121558706401665298168145491504870910453636 02
336962679007420452451078747671110858160388335049073019845459657543115162722501328826413273272004586 4
75040035973588411508587123667673305367165176672359234741082205566207389714589673612442860131844809 90
09966194185512659140312448268215050509682117223862123115265230071316586542736092478521373501526308 73
6450420970868697527750651953873468375231617031793846470783338125470305674216067595718124481724154900
677955395033949743440465110140266883233981384072995257794268605098038571003884722484823787587024923 39
21727160672323783759668280655877905336621109097334341997978443348526532435072253360398719460987064 30
1678895409084338318327380095549056808509279132189616199663626200962269637110459230585985793321394571
8121849704692397468471194090316286548267278160233466412458675531537220698657075888456159200392763767
61595539143811410641071280366418273848570213166476675524075009491137683189196875945945767797550505 43
5913958847686279204416073899438660597048755736032076189399290784732570121893863512434504025611153110
61598160410629347212590338103396221076910139206418442727572563624499218841929692845048865571680814 7
66789660936346427193755563009580265716674060696704507005597645239753093536692672134765632231052170 97
972678820117567338442025858585630995827722376701242619501402141241517017062574823919937573153980565 1
84474781497885071568714142353624433273021226810854708277975682257009325701405883696364043602801865 7
35583944918479033626350155432400879444552885841494511564094270922406588405815702746999062826291235403
8024579975749327018239144762628671972800674151064615784260870892100884433757773968560088108136928575 6
508018435930996339224874822734697948078406477677573450810648131138541202668180146216960634863869351
8228330213466133693279168595030799246914783325130820766917015785797169589363615879272996719400552366
9049880576522883405430829572329495666610618163109789226399407301157002457628374773576308183798161638
1190797360315872121983138196203449769816206423985075228327733732583324372882160597886109873551913778
5588140372986508381751163266745475940834529696709262816998084488901044363992947117335504917218530444 1
36127553727404380196616706233382973802404906298258609513935027945898750528072067318378647730248913057
905481562736693420315336593480354522123759323198928884374803732410786208605622462175566914066496560 3
25627015246939470098514711600994089128351947568274832967222721766829934944453406790736552541717488658
39010393192874133368705592913800272731477548697755411984013889604528855416461552959673062532883041 1
8555011888298415869115770818537363775060713350543268713538259132850979585614230274360395764960678598
0108937955654308759455383732132350076030071091789640837991887086196320867905115889137412620483854
1157848822039592359433270452459141147350780545504083190338470648101424868919725638116863193321649 7
6298148495396460838748495439694585152579518091162665452367335174501636083335297396558922109211569119
78316729184425792575750040307490446422709066432242851324544080760503099725830217902579218216908237
7943883480865654516580322492664139624733960511524759583935636607443829936773118971243814345071442 63
4906683367462679651448757604216490046578069637352466304448651394648231966040726712639274057
148060020288141194791417421096450631305594241005639877803076831545495507157300582014792515959046024 6
61151783936785601612532406275242043711925061379648508919819095877705809924908420856001038860239431 55
70640926229658546194059909869647650740890903710204107377322833000194344131318989829114607687934794 75
63792607660871483258647930931942606503765031220789605973721194264580743339322946456217152992869877 57
2374421745243510593701522448697613048297776512559901751454759543095729786727400721774104428762419527
0575130818896669903902513418046187913969391751999610688039579457814086994355792939886140071417249894

605120231528627326923240980270344039357213519034584028877988233814226609893318496674788926698874274514772695849546580773589496632043196842272930540234206993659346551759464401935696808875736513565527378275584560031781257854725797390168904462967021262486121787662647754837065335028591870593939017780464717594243236240820479795837934985204559679902540515940494547744176083957797368673769058556479182794887136785629764703619719277273687581742206193421215839221471693970525072478316583686909212060602977814422914515406119505865265141288187110649187969794316047996512884740365408399967083236314679272132449362698570740704713058434826329318845440510586619403755873250257313566527539290215236220741530144657593072272508888326706370582148843196156531041890179125548457874634053818750561404423146119784513097056545369126825579444878042588829500769682321898626238111107236874499214246915040404003133218668230591543595496718901355903146969137362240776144310312264860376479193926771749354077493544125571025588711228300980455394033370487328235883976509095216726564747258020548090638929790959497565422906184606794762581291478097037557429248089816541744855096727088920335142693606922157665634684417142514188413007346342117709133884972815682697556483084091332633671724772500819038332573981873743660364293416815416085287464587255653507621150730885068803456968232915354060692724241361263834962430252227348220766743282446251953438168663298333351231892641537539026849800209230692246918808661653425971339964465636551784185225207740037755013360935861416370449092199891298081723596202953847731659451593254575127621826588269650230832748779522126416176015483960583493193711720936819908328893259160168315237898417509554214565735424477030798689073885423655119403552999768615121850369118536307584274939246582474395494638849523052293904319208347004085959062477586926739003838567519596357188381477340819136711013050785355589760197241318688735176938896092097081199608639323761143676107035675754509856677260928307830333384947896302610290498277851673496566301792404581418889016861103714921870755368544212423002436184629978447099023892090870282586775997932900240687404128958781800155126238506698352076101525810292220691850589787807610317527440586023149271114853812525022120659086590170820258780104224742644030508149340088935536994215541242794024288330163356008454806114251973600079468367674392891368037545225825203564337492304061188134789864950179120395213927258935188503227479728304555377830648563922152548181723320806725276659643162054180594480709373373627385191779318699473024590811653170846477848979352443357203806146987009232700653873512378118935558807382761431387104773213306585242921769520098652215583227076240478674720185383578343770223504250814245712843756571479689310363418089373299722019185677528387336013394417726381562202849015960447422841069146815777696019357384853607566769164334614029548215017627807940263527540169883435434195316982597257270323767232710438850214575511607810784903672734137568505475128333182816818068792348394235790283046532570099969893392931956859422151042753161377858913302628339572492462565209238098216310022009938517838786291284756647499738084609884915717838740446198137448390687340591284238313992268355061730953776007733441551971134477801136367129304953999001983705760561472107745833441235752258798197320154370849321291768820404507449491093051786970567117701141196561695487469569955457946649024598805818205226782805544006305517162019170645496657728091231282730371179344888819275752509942568057204575160030883512548354531821670031202098881756656024815643339856121033521803122203872454617649277197607561639899225558844824712539754441995752875533816075238289074957943408478409819050499547223491767808998973555428169119866029105054819004664313715557369930816878792785736669646282662826784498376781127049916168005880869983507642972740292518890639492218318736923256039915873045117744833426791251685385329533405574069546616393992231256623733289130824466051596075681840157729049551578853288990615412343787136169948046390831109473541619184425279561023915503630620246485820967519520568630960898927424160266486843101853570320910934148995913294012970080979209338728524506836770635227193756121070552995770056852165473956309340145995070906849439995261099503701149114858536840356944392440181375872950668250507129554209918508765071118123830525509427097018083673324604796439163711962481671287816880866722863275435721997602279585502124996042934207237855661035521342883540324720884075362274770767226570322161459853084807014327113323279558962703353311330191389309874263357207481014426060412219499873952098240140940011739601500116374649373933374250093596798354959808168174942641161758900297445379024645100115732568115366055049218229071820849066730683734741354067686468803628105346636603639637050840060032259444236806840701090414791408074372432537342045294549232054854094472777330867728957284448351993062406233601489465701029537269319239800791374246128988062094042510718216870035418271782469358852896936838712339508180711265203231606942139435044164221180458659728470198747058770491745236140621664718268422765043098722130253292080624410136519019242976831941842247953230488722276199998130593434354096340873414940098361137992901040490216519800169653451050932011865288950013498554815797648691997535540019962392182

303715570495661140749069890614688031627456504309292968564920493914266325600490954672462450603867027
790695778255886429872457915515182788954929379136186433549495607479606299394332898956071018500741297 40
27902759832985344536654737382544305064921875584905472536631374537077658929988754531817075803194469
761261208012192364849601319145375873842088703277363104544410317520421681702024924246854339384818323 9
720117336727143759140357324974219076628395187262094877364228903555507099568064481385391213949440769 9
817657122176587769774137476930080385309241470529267309724342517389794333082207091984442963496836827 3
4555999275654308492145005895949858388972194908048002110310107746945750302798672045602966829024330049
01958656561523385066010860852470512592510723689039144260448041910688569171156055467655775413133784 607
34357281913557921521252441634267287935145798726236894724492454329593759322560382412430804044 1
5170323303546805930255459808401814193883913099131378039516586688564053442504597686306846470385293042
281253712888124063837191968253155064045865821220270334452510024555697185204494271266175570977907 65
22016314099243456249658234632741999669659190963031507721622959737941330449069123546628999767870639 64
4275439705307201015678667651462562096649852069770710283319925355222101790715425549106689098901571435
225232088249354832640582544332289933881814332460776202119276353556055401812466401180644907486567924 9
3045753030540635899858103643050669178836044633760072521059307210192292241895322382915890934128228750
510980146316239573671434576541180542227401565044111311050586337680869378330384195969453861337 328924 7
160253222936973132249379972928005992675567247975790953496341417002038663419614554418285905495280534 0
99977630879343399759789105686041339012606507861256082320438045454486313325106637239101332430231854 66
52650929391180543257416902759540719084176288677352999864254605851448070324558508800237492198538899 82
5763569046940203693882744668184013998417739780380225492914925919574927731783795745010729161896999328 8
31375419807672649180013064389905496374484069145828464261242049616787491900287079969791885341262481 08
79505542074352644179471940034479651091027244817919032055597758773842727381310581436032811962348249 6
87706680871902988723417289527936418706406948463980108956431553233422900314798101668431369479196147 20
60973085801361507055166513339144332448586946901204924773255718230581885747169914636031877933013847 59
7598611209617071626757731572976133468570481409796915486001251242060298788553533764784651123129289985
20040610540065834250345163468563030940509789836252773934123544691012936261701989971139567103939774531
0096305490105448365702167199187220346357563638244111617093788893854939355612500904793637171542240806
1034381188603305755612486733268456069417527934049702599775887132841401490259029022006789764961699662 0372407499
88832741071097653063416812979334593947425760724413494314516216704737673582685317196054281712 85887
083015931604884149219032436305805033072196601828090094043271791799057699323544388105032406749186069 1
57840628987344737670923422578224494543482808126567717321258040238692893611255653082045065306630456 14
90103696865856206381088416211250724374687218924202630265334764851006488240187531107752387990519147 5
130170115551259365095027766386564075993267772695534780632704930394893489795980911778925653744326543 9
37422785062716532617100491301276476585887813004808407118441426902906254200678976496169662037240749 9
9018383924962308016797133423606997570874906266593733463493311747238915673444128676762850073286742 89
347429843541691406948981446418541344524728510222660007961380960752701040772759689266151674443606065 7
9171195189270912069311570060784613104860095902797295446549937233831975300392998202306775086147937831 0
3409232516679305885809444971780201240607532058426904533209973279358065620778914455124440482552693035
330351335901481444516470171740967805413422095299180906029126071568339276616892674456115532092800093 3
55234193476248116875078333750521448641230046935917245516840949343564359202084008663972887452644764 1
681222431979005740376752041131459356082948636502886651386671870984019126520871938214610496291771605
6112413262298472918197350192326469347606735917920473460197024681202422725499060500390521739830882
39346961329546058235596168859614438551773255682572004086406671572614287421585629365346667650305 3994
372643377711755248633346546866174710294742570747114524084069357459330581066479105872577087039412159 5
834972080143473202166732002952917783114654794156763235809015934498393436031139470196023680643647 1015
52654833032290324984887401748716255541784323459838353317368567411948216488283219830392937820890066
68641656356329990025891242536746598742757842445313505681163336650379813616103342876919977909395637
58746991782052951266065435874802495310546226709828692382531734870934885092202989794216770759304665 11
581133383047832465845377999596422451105520528225985513579954331407804687382883091817168865047 4647342
0060161594458175927648783175100615405715334080277700173393486915972544838499570311690890502347 90
8269611277438910111362489365112435221732941277060381302634181652357514835007798682610689357810 83578
68111581663802542863945488244743152171446531882122656033716864388552794908372296727150585998390 20073
704352045196213006682618638971124575397983167483580286609024375336597037955301742860170982285254346 6

2822605029782281922967974950706846947014111471827123771794543954247545582317637072002093952480305502
4861529194258380746445612475663032936211194064385351261733638375785199093033778956350709984872647563
28815771858778297353705484562184061665163805211633553595615713655488304023083904849905346402275363503
5322131262857984748871208797423919712980651571562251453573762296496995785161894725860193040018788841
4077064277898211155038934041729907842887791396031909094756427746282346450496185895718672708930509
9175082590601623035608541002615065995829409418822317117668561034310940901519556035941976295271519144
6546570114623756476034016640733553010784066121706880487767658296034459262386475855747734128559099455
1261972615033569804674294654465241074799894679978464892745481446292704610242772449547512854272051075
1372587973508902883565123730751402162131170264048251331975042822098951384552813626884177326700882525
43172435798898927526987160398846333076402741020725428607539146320563341900517801695481641794931734
7812244472518611941008835131375554904941816728642441721880263881656275398573336004110595994336004451
0938252578802766481442579095548425763256667686186591270514838980441597550202022308442316481714578201
0230876136894006217159186369664756680115058939506917948617938811069135739111929119476457163842439850
6767609270156013876254738775551308216149178633110756769969326393636019984305639886793035036311014
12592618232432920230504873973555103880618396303383920224450218778063418013902925080016547655960039
8806917718524407509635151958193085485349436375269314283472601286321556955893475903752173339569860535
5818133404046834587120394074492363546970801353967029689205641532705761785074369410421620028138597409
9445803948437171223780859161062547291289418501433733201139419237769927284987921679570485720848262117
89537450995901605731914510330159219504779986163997351363430181207063619062564218296135571474141496825
1977845129239951809480276157790515052996245646768594108003196552477778443101848753683776930481208594
9347514957536351908366451038348117838304020087249543294359801839909611614425208081024615483437721590
0747465469707895682559178102153564440601396893159844515933212019827378778074605279848063188533935925
532640568047587139273617127439049444206400176045960097267419950111801944219947030763868082018915210
6960803731147927545046918070866330683047171276993884349752036150838640309236770796659624074665303
205885579543535898594777542084658544662131174173213638193761117452671984537710736595659498165968759
5352725655305225867778743619699554527008888504312759094143302364295198309281704636367105770040823556
2634578787478523448239984643780739800388210355197146779367439484397950422615553063937295775976121
9082829477616090698661599524434740720825487274741637905412032043763264976757883944911594852617550814
8236435201449006844903755254950452871127759024402682896602399138122666872871694391424233569071672197
4852985490650286693853139002780062228105465949199637531740871738526222585319584765741013650016888354
8356236992354401223596936006051229704848086706559828336254616249888405808056838747680592416721489925
4667697070427975947074432992401921713587689294557714824440048370028276093844436672655795920533328638
2118343620277464608717645601823692049975214261411194950914136593993958849687325390545616898309629
6277455693159627111291389068753371428581683265582291531673412889027734335549344788683553410612823002
1846623652602520308299055735996294121284036158487698284476721665060508430933235779163412598672524107
4112855560887417648349820714209069639040528539182261622899826666986485775949380590488575368152351745149
6461426965879562019976643810050615041800687076584704534771470059633072335779079437670642119611920158
2425444418641308896296689603339150013243279609922778353395891945266902421463659868461586
5059340714840086040303385263822463815891581183633596643738185621040582013281656985403167355638163901
9680564587348039675160571644904016838278201603100326068032668390468558981291340311753680129125576
0097036049925914526513977577298346853058553693635182475233378044007504755143509075612721952284629
0672210621607461237715537118688504003714786281788426461390580536475028946907239289094722635625622125
7205691977369329031393413587569782287912428335072502728559563234780250407896120197892164132387436929
1691397743472714978009969646729789539147270487581227501458990446238905869642949272303541293535238761
8921156458876442971363897816413221384394558034626557913144029141250116885199892287079988203332745885
0878739620195842849169998809625666397846140216095059729972870961243304576253129268156432918037383948
1915146495291988536197668964987775347004098933379727159401939180303124409381216360642720597499374
30095796162204706746117408573410974428749024072224071920084911858181518124726338523114088091933869
52473755179697915334836986207788473417923759000206964547789804654420961655824545657572601098292794621
20160358645909800121461108129748652676649377548555016380809461344038747044068074173071111491203955953
64763786368725212586641996518155272682610249104716189727921996372881405772954371894830012920612558
0088095864824350311584272504471441799240885831604436354263131199883815034474732739773265725829183
24868253221336201914847369762675550760047847475071302631527914424684583105426179273255957895950216

364980568016721702398636422151384913678946966518959963698189528929209109158145580415830296387791786935412183004099868888870765056067578452348837144892995803139722692500263442393337293778361219989460046080519291815736507140605213243665711174865186510958665531766993318173830344832523723928090606769052368514645582723843589209066695738354627801124291041420564745807139444790481665880981587834729983910310275228746947404696773821161510972471275609184821603213271154482879902209158099544671719102398577577600759370662369931528510617800162228001306895034828245805988974280780978633732375367387515639962500202688917156087205681980381321592713346498607978324698826325052172467732321585052767727690807395180206332392022289351307434245298760593702510692638789504893955632192116661135515559813426905757540944016636894260092675520406533365539514595944303364729869725224613028739834973048301961869455565752979106778727754721134723081066651220266183702365900835311812752978241047417681200547328540882448388546683741423365059125994228687922984350772627145754704620061652091400348912926039955543195783268320040354268712828068254965253831583532577307987414298463873930588432411167585453287548999719550230033832132664235652711070175079374880683078560332541460194332096770637493574135395330037478883990090070253142962598041526455897799394876475410724850931927603294897917174136213784198103506849616403938713561098187853350649482250675345626451525297740329892753756169181748537555073371637048051131082092768493599453069558121008825831454181705533962376276768523646265893677337334280315587857812808211157430619791951532712435645345476888116318086833937758278931522464199549300169784479090007976647619878336146456619219575754528302389984112801986210384988301577437087384108280801443728766819032370967428941970930424336445816131807477228213377537599246894948856887259048714181460237646959950801386604347059435174986009052318312201394591848890753040173686996125439466721399672314030349362286270110183021106675111156974413093694485088430863920946963800556700634047876561037082409804867884265850559964776275293345172174819545507384938113300423859464446301683723440199071880086077474584694023324552057248971165150373546124839533550370716635546955833592208900331481109310503562524157515546073932444462024389516294507183976761698709746973277318500836328590626338132577347176797086008286365784771014243655708737137294057536068519961990142326153591912187818324038266010409932276803870251828268990501392874943375476282680559264438064463585291569837975102408599405715559620169061180606385304794627810116368837115018556420832409881625698054524196110805010759134257423116274388612649920868926439355212150847906167359649534179203357299319229870090457311999116978422688536651053937230734148336277659461082027507201354847990537719775211020802148813910728443483895833745239607913126446165738853182117046593665343126495903472419700891057207310514031003142001607836834277549263847812555726811479079790178690706587063479451144162525321346591354161159377354271127487844264010320913869535451910162267754683709086779176383299513414680468895693528680453620097557985880107544175928524296410275443941749831975845436916715453758318798583064671534276462601661707365201502412509413289171747243537279364230528420491538431367186862902957069549824223482653556886677643797575821735368172417852613962129235281465101903304020295980863199433281222029898589174133129412548255309686872331162921846782131002620265685969863338986031149068251518406535826202849203691108013004510658289976889339862230200298730202666823959834337214834359411418600944102423948059712951621528595803182583624588407389192471713075627136269742883335952005433740229716897756514385002397963122083229688685441518076875750485099198641600858798184328260708036579415375088922473330912890232978391570165470989902590963377562583277115219769901272065427673634314435963386698379990914273142987712102809813540389905181965902575287171101772553598101988971805969066534622525599610871060290385068261037369519036598094590387568023489581209837818456063284751012262558117613591139727878696566364760387833095845869521297413602123039262307275831620171532709809176029470213889754474404764535418138440232395192710500836541126144987477629576646131529273040826246467017087921767316215902352103397158954705802422838270279714940186022288724774495150192048406390897784706393683763842470276918437140113263995349055391609284364993786220814923084851585691045365720342141118382724291429596098440307151328839084613953670714121052722050610253405101940294074975957452717492953907938586063863227169758831091315775480834273084500345820943756785117623829181332285007239565267328818090238219283414941449565542842602213790588610200418833919173178632327226069678634981468979548112924564919562757485899108511676602352010867035720624104191113989650805631077625446789940282116489206299309939504142691936328525056590712236826429134597500011438126624469319402922612431966460082178386024422263402909882607071413101340225182292518114507453249611798278090809090405986688739465434533741529283527320684520374228670618018577441930845756845900830486689521818505460258364007276520648231602447922945765035027161024023604827609189292591418654431079730615857216897581301459977941667168583567014562797481377628779120199707

337600915488505485437349191072444887826850797672742474988775037169509964568506621052359813315597357709655906404999570137621979292143842319021934015133733714638856997560257526096919920441679698230878351338934097212741361796713331802161065533514784012278050005605899625441087429177105963861488871216534202742021940010898234916321433410966455236456415744254716128061499486226281979471209953326569288357570768742314825654762139665761587018860883087352063421381805508093587106264331097921834012391015587324449789928640434008568645332440355206342945708350867459782220190720434918209816527415475561920532871637706698839126538932588300907859330973252798030071390325461116679061242603624634317464074292851212258409058847153194384313310747680446329529101441178853360814172418307882287955388926542866648434674012601752783005323779504717394619894984126586178838997327667730925977236372511240936935715309934453343631595721100047780613195625664941902610029205275667024981564837479664097200936148742842180671772944466864229698906010450055271820474193533036594764842861974188173599121091811051783171735572336204876797734979795164258297228610893435015799839631133567144207775122452215944588812353931831789842776790776195747512520272576345924105999269154185950946053770947153664423368160345377494478203803147994524854190241582254730780105109221383043888730097415958976243928516827241735402495335256497888361744765198146214873797373350201389963174984048031417473110535768108772820544027530157949921224822818831585990321764218085761179589830507631045793941516754013599164596088966112035636072409926071387687035353083602313716182758794943707880262354513499947100575161658408314018141609641484855569557304840323932205248542084091772149915796670505394049709130942603584424107356659675150594129765057268149531775654706723150313046360845483584572144624672088377626519460492307291085755171808704011926298599674373996670398429978562924491578367945605019382322891997842022914384619287710339811795327919640087064849999273641610192982828364419870228318235369601337295269640013420550427157165630034780171924642065185460756811038794804264588691923654859303362606440276948220974068354234243978014853171920260633603021898499877395705143192427941574283714669171775653622153838039556125883362532556198988138394131519059407836144156978797339022026664366760566126034177238527338171700746543287622673577991734420640145957598565058119852043609907487862010630095050398949713531747581834943611833585257563921246465585146177331430099874708293493646305014631674574921491274422582208884946092094232114334628251716078318242748223680631197587626810722779638741191448120760796135398449987832458778085584707914035840432277933215701389593658177353967847577538591986059077025714985199792918862071755406650441436740619597569024610752451363496607249358249381528623686592641392363275844595423510530266033702306645558408623065624456971108791978300610297648846110574242652954741764866252078704004909901790467103598496470060348646476171102949367262514970098727032847990599934789281851306023690074930957379371813869516821395468129591464986234149183262075502638768248950956748676320264693455175510292818249839119646790918239352418715525228632683189420876997759678736117498348588990308982463118544784224101131019114582133065280581124123005358964903636926524369193640694048651607563283689485719246133717198958925336526525704820267206476980220983714151087480827271214552656540049463226137117556522557855785438620484397274512811246989303953851327557208738586136332845154980999121622176081942298329537528843084974815265989509596031707675498664537413763046783260728838516515898281909583662442498412397675438199564138877339025561910043407092540587331227195150043907332570074022910892710639857026423394507230166256217803265052508088792039039832902390563040093083018130172614570730839500184286192901257381244218064366115969970222769336793770489676516002294892551841716903012990721201296501333506270071422766354795741111999219819664698709566640066533242100394714517812910000170032454064536894501473949749005690062242571460680569254946226479467048866362893504625320978470128681090305962783791319601090907816037257598889091566804943193195890596973623783181042947253396100728725746329776748022624482511578553027500586014154179875372211315288767244349544889393712681183235765079757375591862260095475879390068550537922635207130175199884858141373912082390955291049480886320773452634495606973773156538854783575430682309858090330634518463435242119359009917725193272329122989298239984803314307134208898676864919813766482764551564850978318312571966685940965467399146866738031428772560547672156667644589756821784995803697938800350918275358548373510238035096603225525565991455444173691944962615692433112650812479498775233971600989640432045163241566124325014503430346056756360644354019814710729774780115502323050576586429235572979795055513976023219507014587792644414739212118715572312611788108567349674367757908697004868600761048553967400939668266925229948537691346709983406583106232213642074997103676648809066368182809868783654476565052399611687466435038854549657939336678299942212390575467578961132002146388875142770428485161410379085362672854329928261900912400426930018423089741947233718827707653645996344376750730597244809046843733502536860175083172039515236001787907322728885

363703330044440927812905934536866314147010465934188347468928262998823630130601376692698821779885172112454145733784882303824671916659511051746324312790315608741488607081554831102132540133568685405588343101887088938761393732502340880796593820148048303164481123176201540243450258972177670052598768575291107994887617033468123201993231132192874341846612599870181746561179146118689268370252016529911989888749488292420616964965430894423463417530646262066320412705247904652222594748526298821801665103773915209569257176760513915129079083306308913138467076780713608298991899449053998432749402438897106017627516486543243504174682174047720535790729788190300647621795656051593785317469975436785042996228068593836058350652163718143758120359463898013538578900875386377999442527513971642857645585388150959986542599611112126352521853573540893838299407147671947955656533381033560920091651358790643737564521490021087374521940701660790742117146389209287184760160902492319104226715102906017895676424383409591983591142408642645711070748530076249802206736383779844598841477515071622932192031026050055150907697894319437834821122313171979698730832874683832939868019319165370266382003482464988882800995308021917638041975942730434237050498168626631446331819924499513504093358213264862216626143045638015541670299756701810799145983714301340032034976529521643857783420248049746048135655627876700141167645327657091594698785747109517077561758971954701469140528987623863446660752169184051529203706434167143445810148812459041088366769369630161221404303079623341879278070741455430961219509880330732327122514307467437929490847000011181578721760472562843687444029999034907235233647795614826072754304750733835794169520854118581414211633631884436139304608644048120305008737447043035198125879550512101543796185401817683516395531429788921079933506442189220638279260170808596615134092310144550959805004970933341826034628222661365245786243689332874818080831166321408860189627933379691796702389260039510884923222262487914699524694482213222071622818763375411744017716440826359777491004984411315866465552169347946993853458952764802986158402264099994210004334206449394164465158608227497279056804659105802319981404181666468971070381589917825990524437941647676653136370381649556880784171970666908881871118929635540970894493350838086720874087385891678280578464638730133563290056081755657051868983518288538558189418761846431885541883532205586551491960840135051091304296438673726701769209462568404821695594224381628363176054907299398382901877071378648219596279582737284384930210765170111412097127189513677811336345522511943256406092909203989203031142869311029961628971574165135312265097656638725415021810918945769606338265402520174627484331378693668353589272889441372227122323730318997627121875635903035524059334406804067165885499108922339510318528204003163077793139388788124263739945766173505780454864709713365612269105452680323531709465782235132634711977541566448001216476189153839445438270741357110988027382524358192945270638724302898388786237397269948101909956476339872677974381882407864696372061345750204043865140481454084863722948089187495333684538332918569261160013609052698074850778808097199220790549384644912998116044510512482015768134230369758457593135250764991726718978359204624413558353968020439011298808204879270599398452081514555277160455545091663910861465981010943648552995834158940501132217591891882740788585450573370754171980793576571347642256640078552027123596498418148780918524754051782988938587194509009264562032214567936003200980360365891403659282029705970423893401417849408983405889420828137541084532719476594015784918087988412786712883469730444534633001134078424469746761005216325231469607461797235227518889113661084728250447333876988089978824961745714326538959319891938094537356200697795660073292073759878773953340122426282463811760465552954932776015165444339879779657596421303648495380297336297340540953656602715666209562420401897254100269030887306885967583236348486080031367493378804698810817924348705535858612604433511134155068347210280388630798842486479599344269107098070530828950651392898724560947408991150499159326607612639813350418642126839879243828106390190244271673507646240245768175241297734377047211534086168041782949967650885806251274725995065595324988186611187221699214720959560654755219761455049109990697568423675215339297355971225275715087665659664502719171782052938851093694472103279129972997894959537217965414822046848471079713315292422565810659649076885751212831515757008915688390775159233949705557155439612034287580611588670398308678334081013481683943703392193419742431633446877167554010287905951855546970244107483690998853159223576836050185875678557363745857714841063401334897590877749058335453977057795352135901682664773827250855655781354887635988320027857706316422404683951616571696536311177116454122497180862165255308450802635618913260435962920096403238435643721295159475370293415557820560910346503148279361410656603454420828537160023113668131901964310287304920850041743703833744628104646209427769637855809342757875987841833340399660193542026714882612819488625539504381515336088819835281749473542529613052058898947452978919276536230214649271640863202923565929941917524547761440843060223793185676064830394341621875362704147497613963842986391528708311459385176684853692245247913397018567966189810070204722123318045419?2

30799439221508991833972212946648542669182478057987878265388133877917479992986271645433393042460911 28
474141007611054200897125853667283631486089863434645693410241748675676488649993201676076913951174561 6
3032737449804460907809064030467634944431558869897372150230602240876896280899677720082954009728621969
369799908562863478189206873431243425191257166586084885331322342618432605583673517357559462474494240 91
8913520207424891718440223182667020736467686101862473648492758014735888129615711645307777309918790010
29888628714193020467922752516710370807163943912374316792866821934446224765726040705459985968287895948
18122960996644984189543550512697462222284045582160178156384893241562942941023547244744065298275956508
5230803988104176753109539450829566866700980596803972387830710888730991670810839900986667030216146571224
09071470146803194411882916774100108676071463670346097016587747938619865572514916032126199719973380349
0164842267544912596739312397990074831055385068661830482906443355681392530449017556754977224586553701
31148854521455752765003400128947427422375583403216774265860294150285405959573417873490709801590858826
5302204657806921368634418238335855058044069078904876946952301682426899530301950384904574094772378584
1308094244812638676254526179071856678494915944757525890432985971556253916870664050033886911470252752 8
430195925772339924977456535991737370482345525690174683861816050968503252368717229255820454717814319 91
85807494916821191010614101754667530762028915463213429187226015691453233924467835360929239295633932
47736426588541429930289457142967643673232226292360240155503056432028370518644027032070094133089307 40
7897145934113546630626358728571889770055691796392094089754094967577669166831282615198053868575951638
8745693396126973669872220449857426520785733934500552182495973648387278103946120544515637979612030291
6594765746993415432710140747457728926544229966008021914307516320121147122336288689110031419826976208
1161023720046209913211643260706919868028640972266780902380740359335421449915746197968355714813677142
0102843682700410344318799421436138119770538705702515776750087453539287747201965450409062159447237705 6
51061967599990856948777593914911594201505099136774196405319122353927497551027522621259329031592920 20
6322743156316398835598947694912780282598450835836799862035335202008546055921678655283576498156695323
15858857238729888822191559448037870908916485672990721373860536043712143961691038569517616028475070707
41220885574454803861554929996011109008952930561509283466502880398315529188908659028176649338550360 21
1301004261404612185620272908635851705705207750060330829518090619335033657336926887231145986400466223
73484736298028779881021471019245854937487774531159628979254055017807474919647784067465527903931955 65
8138966925439286116812702608701649247175794760904071383841871022921733518989407640808097143188308922
0126727746887603228486924873013499735546344984108290399024614311248528848255524681487627399427149890
898964065884653827748820154989400559486508510846567976193302486083380072550353705752672616262712 08
9574838570781071679039632140611478575892731652066513874141839901415240806942716415312484146575073 67
16210143728566150672804848209459012141153970570484622153904550532054514086490834816933675066285207 08
504476168704764247062925198421823405671193175977385071213843566161200541291487091099966813318550345 67
552502739480560945533332426165004974273699236895955712032345816444506163980944636812010841892621331 4
66567215994708198176866591488326823185460165541728843543416704493091663748465689767634231201898326438
3910341875841362419674579946492022219798345930565636927568493597767109310304141130731235395642486385
0145550075794360426654494747022596898510266337438301815326070463610412035069829100774024752336575842
434925980678196106766125498936694745793203834801189180462993440204860547400539729198870648908353273
12882942134188943426144115598374198430965061807992482485995574739758659791783500162512479117682056612
4568789795467228944116120724622182150361118719603867594046340815340520931954899452801363923920455820
705023281591771107908638599432662526833708351622186327906963513461000189272878972239673342112248855 25
379496233480501745645714169688636010053871749288214974692856254740324906591107947746995501662902714
298465088391795743901191544231663387279050548931573371400843033387711793984550288105152253878585 88
5276786724546822526013941421263800251511052536202028508833681167117913143351827458269079362143382 87
36367147855025460463817135131076739357650065187225796621355848452519981400465049664429369446
2643253534227048108735843865153165747836934948173618439389101920993392079359173023513361343361743
3788943324363676621020575206404986003394762611773065979007173384350861190466728309191914054876182490
35409603611171758738428295310712978874130067815729007187202852534737368305268388208851900065288899206
71141417561482180485903016126993630220042457303654506308344452127181404811064265502183349180 8728134

```
317000593894546477780717800755411594479566368752313028096856384976467416423979403809780240068223930
4397514877618551014680749244431304936842402797966380697010721859444694667569526315883828526261340027
8056513954164726797847201873928734317431956342714686128687031868026805130778331133364970514243458619
4339937603831348919536165221985717340060262681642333152627532561526996660464742821000163078713356756
4176057061036539724403431994640755239144597000424882780700901824785204769730606818272868950111230402
2596546463916882653440624513894380086858263099263707383047836303898086010994899412575125614015344618
4423708749095624413019599875638910465209667545877660086590395215269307249475934637655249995739813687
0468238357822213502275156277174392239955413454901430780656888714513281337076148502576852323638293314
7428059668809646209984224762074394269002794291723758974789327985624247296590853215947205332369490434
0279662663074027313164322304712428965781608109046022568044881972470679934948937439151507550517355788
3674663011336512806280676387389443510734047785428449458103240215302688926709289273432162228866530807
9172552536648253192224860467190401188149796691897238390489921449906378342247258297448757138716393766
0383531958221258389950053175670095529364855078884042900036232460798510809447041187766965698527002242
6542148408230742496591289909650888536308725432732151415989181628756781130705162568510558151267135934
4832178026783508960472580054261710332895188363891032447371674832059178733650962829745596943624009255
6528166566428133690259307587404400286724317337677792486726102625684036880816938609418304354216051232
9431137753391065117317425791903877442755577466603040662009904063042605149202987043184601327389509099
8152703064336944690410044571202235451171011328756403959370242331710298393490082305141709289754049256
0701174416574343254990117806917646759647846879791515572781516247306058334526364851289816778469808818
9113210039395551118696836023267657819460839277758877356094075598291775428086114543301390045524655512
4291004911372885966068671895355711890373330064908975683351650049482437502013368515728499636967464259
1495360373941154960982344314351093202218097093597803259759598950811043501360621642003040542535251
8200915587623321754421758808594192994016600363439101534009403986138161418529659189582746862217600
4007540224052349144874115414450603504256362329696036597208236492553521461247652077137457479512200232533 0
7577273544066672546063855660020024685704460037275403923296087432532813924489275962636999746081980307
6121586944368125434647600582345170986588687578964346022705480070837900413305141721926594157615687911
5019134029748585051714860817315609739898187117889639975438593851489127128569262078693528607609610014
5004686282143308100288003423799080316038850406082976294182308278380860352272498102367705906046463477
3095240249025118717986424339190253045895732039085850787195225693201259024230765182777135271884472532778
6663725152809340520811671011269698677937225985653349519432693201259024230765182777135271884472532778
0205511448358644478230115471184418352293251149325726988617491226032840207277788433002018243512889526
2643485040180117669218940030138462303925595731289815372438169530773158947855646025489012335984452603
0584211078366417704984380422727756181463614970820529789404869419642105195952976344279449380087623752
7458736540436032392568120396815397806203184411751734063549646449468864312900565992397103980260555027
19134441217649315767012502328215868291339170943472186019906149971673693014471722324481136577364784334
2259996962798552982234585134519821814256759243498863133157643554985220161874700944848462457290141545
5918948887077304374958672079248383857434010982500628916607997109441836998747844395670792923886262416
0244369027154652760022493934903690547167448296577083073924210015283272337960335692399033882446560129
8007919176430314723397642035147442508813987203110473304468399440629881969793719757735323541936499
9703329803095057301944905176813411654435932990515291198614709570335374526557874245185688896013513040
4654670207075880994609033018356953660132791718794449541015603436922864802222470447675869609032096842
2563613405634836829717434394913450350154562712113070691281968263867332213184044414977037384509444 61
7548305453689936068205803889877247411952389292421637467845624985427985031449329533158554300276667154
0262962651669580914607881017471430699654817441998658473290401655356658576263080502414955884775334985236
4672238934163656532479436451005902522586321364641258499846796161843552340352324721110521226636091573
6027130213294482089766141037807091936558026221817849571220758511904228780008745928677362633322009690
4378031370895252076667175277182998614393655118371669223725419466798082166668111039566043933750372
7554514848068166043674678943264045371156658637505315120812713275492053068222000152569609032209 6842
8383385888272261667756834554604203873216650357630854083599997383442031879253515109883833853900329 09
6587405487398852072968379972293660129231230716020550097339309360503459039551443505307798616792471614
4327074762450851301978973869927099332578952464554750676366826464527152552254333880535483627391626239
2529667664587548946734475772733560138382737290053938966565922305985710484827743980497205838211155382
0098920966136946893177199111474717037337482698105962706129131399606008821877214852557889824960571511 9
```

740995507139928669201545658383431014260308085868849327192298415895092643571831409247104705184512875 8
699884109287359028743120393437627985164110324412262926310011096914955445030945335769214098033156765
480642125772776756252536621018085063681829579287160839823402147203536259820636455200852312805800326 7
168668344815110463737048499734839907210272119035800884324222116433444580022597779528179717226997323 7
238911215489383823913828729899317311947617390611504478279287691102376475502522571732194818147370630 1
308841788981959816299954108339024441069270673735959569971195359309384961102865740765063676944908930 18
558649870372897272343345722492789153260922324770228772629642491769808030278236217239379885400503625 7
155488753610089011456864982824376478150512482820550492067614725271465218966300496885795997677522593 97
406030511028985803962621881971282170519263220895174681586477249400663476252399854173196026161036924
195715977610197169490239932872743974658804365659364968801685286397751552247599976494185950268040500 64
096984351130737971104411979180057466549307802152125298100873140064694735659064689241814839126360000
736247105564819825893808874576453627742993768135876541917973572296127000892968471369649368367896352 5
182303893131039926337585965257961649644990890955243550865890255302785990775532590127306002355311241 37
228833954640486577783316157682986151786509241374742372088701308805439522592788530323943092165956490 98
407706095942612962824796778811135333262952874797540987883556678790042919451576744148678404482363922
335095660072754793914016971072318582441279892338820023779406397575365725162501335163672644359159774 7
506119257130162300909373451004745276180163807096773700943768059667142294135896008247553832459748039 3
206079604490501769207058512367261984589568309379680625434025095746216595188797550577965491955049492 8
671233251337556738716057356380028942990248851218801240568679236189247556048248749553282638731464641 6
420598853851477433433172591297317119740004264987222438106142110324979924136371337547432406296672518 15
657913864370256202430396479048900504429852624465756623621882085409494236850573272737762283655293864 2
131946178526062604999062547968847458530441305937394727730775350781935576273441069215589407275736285
969449663889092158513270610171061497976205385708528120957527632949857567677194759352152421676878017
343705567423740243650963517997153302057143111464013586402829024515173261076716920225250063376243107 4
161787476243110180290133180972231123824004466527025579113433386482338478240836415091426303214665473 6
617596256169665943312066598512767046145043556505676323192723803445140253542128091685364058606595686 5
009005429840500460935485306206267704765845643230355579621397040128414507115132958915455166928658278 3
894033915299223882332902533888572605849243307420504807749658189661060918105854541979324802037965682 8
039992614596920546380587713649031487744048091127428165482419714572111024497423161561692475479084730 7
516632660981952379567646387678825343150882208179566747714706801635105964756831889832497120460920855 6
997137574144650469347830224327100384214768793221424857135646283401032424128263276542081608944807 0
169154954190788990858389973870707015416653384195835071697145193419203744574382051040777729732736 08
393241637456285892241373653863674549504954305663770843450836517700464664638153286744482262904960184 68
605036883440776084482390700256776213235721377269139239309592523794220566769837043920670890348267373 4
452833285765991776101256955535019262055179939802151703124143114530230698589870103035894212885231515 0
644414206528495349362202442156032850094454546287414074018450857333734350776305942612250192525532512 9
918634214765821403830797527387376105273026392418224264154215090646009883184415256430726001468614601
161949130240366938247501714189422592020080670774549157595384542378138860870217866424786028682455382 57
060700785282733222651056334456649087436158229522645069096083169561726052652534915020704138021903400 5
170481956079258642819156197702496700421100095380047388039200472454678730906292796860054268202283886 6
035287764184783978631665070737509669088361316739147668310680483001761136059412558390261849754766696 2
172853403585921901345327583172600671055941435933213580593431965154632317833809081857823319571
680232256364543546573965389158512696172683566529545299336653616507398029873401838864612441636517464
669893892482737826454263142720386501175530970761558733454310267608916815162421264870580775063592788 2
007357177805690881607984334565970610092424036098487182625417202152788307191579766742885145058773813 3
761448400083912643956891713569322761335281604797325611648002436478133941949391998144463450338977230
307901722197876114152675849137827671364048145222417009763802459275416672698590014203411158804515183 7
936470769448992165195823326382816833363255130234263159644400844577342648919320741277155095643226103 8
689103857009585219216284418489882732686554704236672575075298431229087305419839504408942024141667808 2
196809827976707749289849712423880933541449510829425629732782669230041011618064786854216339301274558 9

22324247867491607615769541168830213454217159658460908480197196449872285422924913322695771889106521192
68235209734285293627988609211167917076294858474961499783598343087200700477102185689744126172310910355
86226249946839024978024823106107738908043031729059847704522430321003304957569659557590980889718773551
32748296333986457187784691064035564489612527351448682310530027781884310676814363488368688151979355919
48058645183785865973102712078058781768283476422045804174854652725579259321275422093550670915217460074
18634501047954448472804322875904278532798925864532242985233863325720785494344100713049181600750095719
81783809560002874758275571459591214237982410344201199042980008348466798477917366763391675591223307736
0449981783300027146207947153962607424019051778269682879307334273726355455968205132147577968851655215
78563821506100375744210687869817590879723105471878597945093416353173097134275573684804654936846085589
32795193878054835535183845795512788897107538526481259181997522714467314889783066814412948090438764754
1720328836793153948731927842820614083782111123855185925737202642344646616985206333845340600852668716
9998825468531836845011643354224246766319745613460084963085605743373590303320585846047421171983158007
28930013561157562071742314893304796447468064963812931642923523581139029669949468001450688385729504 98
80031742947556236767437649942436129590188781636342231949340725849731738971847387427493550985506472 656
968441265206780502192042876107362888938588530858073245685564388165788462809886618203203578230333800993
0591300723334132345096025973746052004357099860029814550957846283200151357359254602735159675644163653
01122647127864032448240077379969176465060238733966563679340398356807221965404885119322488205427 9
8091297110077004500227754612066171669155913980976565982271696317371323802330818946438128134866452495
9954457359960273403749531981034137354585996149549836091761262853953078738457075946329303714882251930
3817175115438350026708995826545263811037252544877439260135460084963856057435735901733715374153545723055090
52087967799440782253080775816995602711127758544868440276052939451428880029095380284854110122615778 41
49158140774999841496292240198891308317859666915388229009946947450247844902571367356972639728300328
60634546819859014808677414089210890401057657503110419221614948743145878476136714773918305353143222
66954538329922394045613360601782141188650929220792949664091216003590511538805649216270546419123 6518
88400989901480267801994826858553514806570539140057633449534716732308757724513069759606056795390037 2658
4611384511323064583372505805316793447259943055217500853177863339819472177438498394166462144855051 88
77066168902788747419777507278594616784819648879238392429701230219526438487691711692941913676453 9875
302213189442746898644511952336113580869952565738499513227234458932311386797831195178438771350648230
787048299803447155070141882053101412668229481600816095024682359788933239467695015594757502235926 0204
247226384941003113670440974536586103080120593089272572354285762892217986501367475075006383234798 83964
480066569636727516169718199072260193757116892594794744761248762888217986501367475075006383230836656381
97740048841235756668657161421583110847360913934500273200513079812815702225616906552683303083 66536
41347007081942216648482210415934349190820405640859522403880037807349261650300237179903148250 2911800
37747446595015659399813862386906265238206123362367459640720983507701108299079028069034109175 0963
5731456182319044477049548661871606922803035013735952241233169641834879908074808040886899822172 7551316
19587809677521653998309620348904093638565394211961230810211034710517424164346551719207792771 3852950
602675186424396926553672334478410068145595114903678238881757053538003894600276907056312702323 0141413
06631680174679733509725414626059957881594107278065966542285301608309809482798087779954151330 6341851
97787230301266392253995594139496211004195460782566744250803288180503393892187756524451695 5413764
78416716373075584797233865939263522401822608031692767084682607128840619179491176562869969 084970730
82337564779768748466753052691929850792803668182143767960730508738080830144642975982541700 78643973049
61083418619696619593220184035916356341184358185980205141363191530912517440662404939092451 385850762
70688936627099055946468937668006920468283630462501640210274379178544802485128618216125121 14570003573
467406925367903689095025923989154817162254182852000600406001540589290773206870958006174 99719
20364712097884192446667092044497498748240886590526693588948775257516401354367423792014530 72202353576
8345446812068659513932726355927699659957377744110379091071568358658465622085862107139549 3545391 22567
32806295275190075494090489639438806425457072622115936312439491645972564910184257540822004 72288846
6345128030448319017840074011676477395615436471395235581999769359010841775219733620308326 2576165968 41
19366145853311420753211952927669717042067051859842499762834604123163908122790890056026039 1472762547230
44656137389193482911984549306094462945090615911536755252862610591271212204144684177497818636 0501112974
003941119350818908357333290551140743044446758533039081986774588764665315075587303204448664 3819547370
84809840190145711080151111444662950746065233051734594525772575893078637007195767928495 4220239 1372656

825993183849635737174554050503873578054083223542866825098340742461917124106592840528111662009232822960
301721363849285104773585298392087098926316984358857422063744579956105414437052488223358026756746092
544022776840093593181777507857673345320731185308379736957382024604745009604524055600041568355046864
181064155915986925744890303471460863686842071415295195399688863994416298501262198265478995063129214
960564718499931339244495297288337833355225306650811391115575997907138289241837357409051932411808327
532105758344340786287664294881133595300781151425957827964092837812763167468852532329802856792473204
320938542101580714740180947946116048622767867343775751143759233304925499457206276842336439469327017
361084440187565356931607880312701567743292110954603746698646305896432996195799083916388510735865539
735868580394756294040228635209634721703945047038525710853133624475454200105259671217835787463359416
592332562570393312801881979934769808508853873790156788859249599338041045070956681978068909791304753
127014469119908171380579382353672715797874399564789154906407693819236783662323218190582136399034793
398119674217404866069196506586883151348340187681346790426439557385900654837580715281812895107414406
045017043965485359053827804348135830772445160037860973737431472179402649530772942955247321642858586
193133904622555731428767902253344787868856379770932207047543824471370721081728072626169220345167663
569854214600293710661317844674349494603427459097077940257119887375313991238132600095632063682362858
078987415322744621597193546312149215856310068890958077806059372828374066045173381475694066896879077
446437289032404571746893162279915260767008749579463655298108006056320535932334603406514913241150818691711
006556665547745497855290604227261924901312867775801342108401628830872921226157765095221015080446637
744032978226504795848394929080901378434911052701891586665978153952243363020381094307779722100749295291
014175775251699297714793500131896448891152064736296676121821839308489926024600418991546699738521968
756729309643421698980634192953311660150268906755263925108101725929474115970572467020836237914457638
573076310550469479661341506294987476166418457078238555743704747405720187095339272312325000336576544
821501626602351718472672153312107509640158510189813774914265452998666920870889036949102304934630341
750898351467990256972867657511504451026758358408942727333454143953879619686112286271837770262649993543
89756618638452242447394924921505485101221670555240510210372300284593613184584442786733823142617697
563642708421202831884367381928347131950877131221910120316721141109395899922884674801741656760993787
196877076344759701878701115363507042680620343222196248189679109056279926872063157344359507897852923006
967195110433305566783849538409612527795838910548796848486208679717493098452144359425346200112410842
656675868978082776276840013469829419295802033057400474913978971051226422104073255789131404774670952
633731095467107147882434746973225362089718434140168951520932937289579617990097453132280763181828994
318969568953047237039953890583965748355015081947010033649460754156809390948275449981181003115143112
371620602850821167716052901503038399817787498619635004890805220896906827949150381572239746651144204
071213280506053624460196857385732581380945079434740660360554359116810385474554139010052105682696417
436459269757301112314276156916406438293041441912580970015014762604508430299473977704434060255884831555
183709862104371824449093244990041239696802723537366980742543929025579847970934821903280505910233185
056595885693410369752178796616771042304942352351086300728128712141279327804020664614262300785614084035
259834892557120851189853822385136209728791951877465064186101050110001523921401988115010333190671539
149661273638135349062018988011860626488814169435292751301207444850693949715656963700528104443645796
540085580441624842571854483720866433386657525228581094828921725783915819147691364603268447620225583389
788430706626820136562567060429166096739937396333725598175402369018835353007990159396724928774572310075
178133888506294267768432561061426208547207080605367376268476687684621043656625525457715582096848955127
560427094838699004537060236388671367910424914919630147564672600279406939362920852680415939165569457
831013701721500246134125553880321201748024661994057160259811420538497330990958586477131121900577851608
821354656769254369586839553959226979119815105067862427873863559696351596525780100887775161394859476537
028933659176240229706578369853607110049534756757228407933874696396982052754885413863809128046556759575
867380247796245580749375238874918172010300809198899323795339275624929514306391754175652362056553753337
478403547514349918016962421277305751753172714089928417997105437997664693048399857656970388916180268897
948286649839647382240305236838578791765498736168471601522751105553564227093034129063412140503747065325
387611044057631277677678955828369536068797492992473055757014507128648776037216713666399647951681218197
508936359322145080853486264524413804231937636533552748353332158834128886647780319622096247946024355845325
305917574155275447784716651515806015968314346993860224116703396103314344414215237812127052900415296833
328358142745720548076341739976854032112427870270994658214566961420493586005178320307495998499945367755
963901544332983729598770215879840453042417236885395654311324912800166886143213359018145988153451156445

969308722687998154401637903625847449402767622314058383024632327835558970491228763755160993522863875948264709234548966040439552829693496327329619453926341254044358306491272796994144257715378660212159628384800807764860068442119512842811118606563381625787966850467909393030243819414713450444610996238141
7080458893859796343824476120094314750139145110290353458464233986653377503403288751178445621701907008
26875371234894254845267952905967289911416216871720725278954130366253131216168718400290849140108824741929033100395853328090305689816195958414640350088183835447766161764083433565762829160365278550533429201734442399991298215606563923309683123260611349847459047534817572479352289989350094495075396373482
891154711017298440790711638448822988417921854283174985756016443562226461225964028308647766387359459882424504709908678771675009139300382117519811184256499449961925019393804725337399459333773125252463064
0434299251006362772644042522129333363983888712558650282148393519537829121923251329550479427077479817
57306909813981758364267491567563803402416353009897458846645595105263773038875334873440217725854816
57032600335620490417735579097347598439475995845429765436741210753515070138512126171017094388163868180032534456078011389315457232587631687591414183933656822296246600914620551459783379115646479266635432
6278233025485821978207099743109160351106470097487400073152228766473962912778621844683555002020437071
914200728462790183186397870257027722687823910369724454864110588891669111059220294449329436270354133098805268800879346170956304846028827660119070087300282006643598669430991288386695237929866628178898
42726970488860447337676004202615377117917909677512787197471044080915599067919077234172080808599042860
0454545675142277138473782341100531182443063238871528440862687566050697234784773621962023765844110337
21590438118946982913069286115985645313989313994899983340400924228379125715552174621891287605139468
9884726771874665047852706643625748319169084915371258805414540363267478695396491037240057461302202931
99503101987750602880237975002552157499644642453349885915909369543958084528040500493663983056378225410562624168321730113237466365081832151390498019391996251482048523360352029798922437731111509166858570103260035364444751774264698915893573470565871549762563268039695816969490397599461063976343205422721311914855309460807796290809358385465468349840630635552168457034303500634102350285348776635304712506884408723226675905655793348745911332126078901928698099359636775783128957269704288379935513039269512405891
9984490604631927762990564603947687565277618898780750820211548536425379197075472907112634428136059928
876234056085997602542338053560291986990956076137368277070442866704641224740569967492098598383612829365067744980245221670895700993792811010739323086789354647775565148775074794665008785762695049132556126428060059883949951516257640827729816057275574439612018147497800451321782129786374375109974767337631314344066932216989790648141522459605796036374929385390580458098260355681952893952216695741564220430366
4372299146967604386441942130136759001693242269349130492467027078248184552311346110345348927331506001
23028583334230363420824871550255513687456321669366560444146424555698182319274119450827968870469417670296601905074502498516529806186805463813475251438433278991996092710858122008935766222569799359869922149
92546471768210127651995597824700501422174754861942960392394500928824818146877484419136141898738281264835534324101696493465526295345563461708335109501680694022867505677634457147413217775167306207782187706492244447520828000983425570457784919191688417677178688630332126895491976457394075370098837140048
70254292603296837928754377069560437339990100294852638150032628997285511301036985819326744885052228419180545540822774752760746538989050643747981698347177490761037600628289061945763967812992880200275
9672944986154770652047321954189029670858378060569502688599152022861683177931813811133700846648282316
4294019403501667489299392408705756479425136906521833097352653843359384167524753096842469778191862
434377111559925282827313295136978782140742545163126864523275708217439154214688943590977318156580220
579640441942122537014799189278853790337774328734095511741378352019197915279650013938688485569374482
16129271672957278563847386324693858405292467490402241343895188323801079314601834291621633196573700157979427390045000638416513145176559085970026470032213022198522497413915779529879290963497289851176018113744692209425310313138344961355993181788354416471450387855471665976982467247974403116060606198912250
415690904476462457128363820616674275647027759689746278418101547670135804259038575306203657293784016

4916694827135928569273543067691788670004922027323162640407025502795620934962162273386194868110608449
358956017870858831338441728876389093153740724000728025325627642840264865650196869797443042592258495
047417922792534005525247449502340839265617239093094230060093660303234802020108678668089659181684792736
33014327146956844570493654212738523641976274894176037602975201615359389448762202357391354683427259
28295090576514319420959551607261274135359833191841235784196421342887256687389708438311410465856003
88220324630865651541079929646904657770652379505345960246149402061560544843064378729945258222626360919
700634234569581210810188042948513682867398521325345198519868065279201617538965618411825224252968934
34983238786265738322488214671822123921614521633252756700170428939905246548258778512412518561257886895
45553166549754643047535031903559023214381285817927533984012460823890717054605835960586771990218346525
83057186827751076250665370945298883021196273029318588927084757014856899852966505733847710386205996389
432094133595779644769922141537865511246485379439254073621927524684823824997312571864576555150869582
41513497981570174378273366379934306509060649809298386303335394250218225663812732097435466224458876433
11812559791473648440129146862219867375818977195188232785200933278280528503438813801952845546505139323
46902691156026768435854439350762856672612665083945536858209120601078932436582291550801322381298871
46480913564402924712524410024523546050822324615782206358687167110556932854380162468346176479423755227
751310510012677605764387991571994859706517602138914640631735022338464345839483543502798190272879730
2028508468840198759498703781461796686462875466703998963042483342254909424467013293924724525832623153113
99712398944621765884271933825466621610382140069902302774264438571417574587943978979594815804972905973
777262187482791915421390156710404249879660342838987308071550425303932011381726233669143418847662675503
258634492672941635544061641605812600689785048902465073844850044199481231522643777835892801070
5287235798131917642254407902629775223299431624405682282404897958586204209590301675307700984125504143
95173705772057555508175512601790181200735133172372746220812000860440795123951426598964340376424506
08295996661560889038571068464029141271737657151348879446426891076941089531011909992999563093090503352
2277723326291470140178644514635311873837849554388250085692730878394745287492017688644731178310411019
91600631498818929990601527781687084216213818395570791840511980675977695998758137757726887891088645
165446898313347423547929805191092151446830716323855351038727187544676708295297490534595376525931925
59451473936850638167973478636883270895439596677298327096680062790539599829457773168238326073880
0654102514617216286678835870661909367729796642225593369084586671032121453105145645484863204620465
1157310033106271776366327253551051140113729479742423417996595373489421421400236584408113388319761712
5505890092454503137756058842247628652387606272469903021267047070809451241471629495570270401899866663205
179842300550700844075332796256999171787654426257033125495139708649447194527294488305094601841529556
251474049525798009901463383797769021293940853102488561567350606338634923684489507528233401007520258
2830620711371909942481755214109218607031132937255536822579473558746256777651645331092982285
7602837922593025131851658133770605209210865756174301233428992049923497335151163142174525426713978924
8050025172232090821245741107761163535916860465237641185208310455600513909589498730970708723111425470
23121673320381081548092017873914879344980317268545689214877834098901654774176228126058072835541533136905
0793139630006376970200762535050726123344151011142807093681940223698999130284742465401270194011322299
993204833287467113553834945796358368992886232904397225844938171077259058039497162595066369160424288125
825483869715966530554742543545597264320165017471694261408641380380466569532238806099596893049398139
14417781080440177680412631187307038032840781365152378659505510087403583849737817232100166230527219945
7879907436057423140992833458661530302659108802848943882627192860592688546252611811506554314391860473
868320149520141992401651017397674092260432548429456597589717717165202674986419890749336425886243
0300822991408842303703349200032109476423574937082515388359612855402857151199968421209513297601006625
2384467853304306052832345947715175211091321846929689013599203990675174666377175408931626352691592235
16675852838151330957335182942344019485759992885715896113732500733529944686451772778107293555066205
1116627864068458347421220153546184274562778135631003503800090185222039972627590546827269914375360065
8655126345316534223994033256987619903270018292329045380201646980531553098829533761896730953445713037
128599254581802272613746565690582295786920989804611674009391732335754451424181559427904146484050125
27511162224841473648793952894876891106208346787576232688199506508172349368185004920319539691510450849
06318331697956500115163300883782711074977286046415193311497771862005817211835717165889164635557018848877
330656741216711045991858250612219680110732254829518774076669997960230384720072533276005946786952679055
143195257354774111157306283794871723879901011073719703379511138790244228576611951347093824055168672

986987094588552809896555090500583947977681636213599589645466936774116795236559330196254317145982816
3
763773483041585352887106282009286734513178670579055862422877697703803358671896644007604521050778010 9
02637401436327800462862893243121698489569692681269965570096116297810488083332264011584449865788691 98
91551164987759500820116547107949547616272535974431406989501434791552148701805244068880531824450548 61
51055750824583348306015305152714103401346158717620493273682281179363822772636769508996060057645760
7434908380867249533034011936473642216403187735017426283830918160337130530819470054814566633422929439
4379129613611797429979579898222018382043393751513900818795675780849881967116995779814800468611110202
9985597696284193886876123274515246277330804465733695463654938404081977760970663913237654253918686 82
0356685427661932684390288591996788147248350231950588774756415910641899124069125309416312561093109543
5308814642343408331609704950449309811673539831293735539341187320088670867106762928026623131366609838
36430756156824337100324761286608742139189356752130595062633620498264650080206650187746333184041095
372693993549925084609322236389181879005872492386107832157797902600355622266439172544446062893294594 5
4295831001567300550754372474262118465163712077024599682774758902127077460823281087774656437622050892
2117628625949233373223067991761502464359913563816206074085843974251331593898633831027241143850753208
0538973380115912508795623407291394530386270706817801468194772402893961722164417584863020451648879583
761092985067605379167764010412278781795500182331972605746176118837796845473203989183811701978662208
08018101648347143140329254503142495220082111433074466401362422531925987509157512173913243296534940 12
0953928653470846315882150495516801442870649488431563843722630481694795792035566844577863829722889 53
534411852061006954504177044547449259708669886360993447006199388647273449927912722231658528362329253 6
482593421073552499528548442312732204674710780642436699584238528637432273244201823843973400032241859 0
192380305090058722292896105514993883061413500649369104739021291543977494536051080648720801311904902 23
110707230770624283391952893722091148778390879044959631522298968270822053048965601639594558607553522 2
221595738349596092864920413661120498768165209416326912589404845282299360702772755091042347607151026
0847037204995330735661652016080315883563879622431208900709419217345047787877409407146870679225942590
522751818094928229533182148904042084393337728589025366508426327725814394860195937648754924471152085 9
66166588308595533616071705852042479775057905952120494899134627373339351797353749095540180502086242 2
947155610087991541470696535457299922407093258038425538974677635148089518769883646358942549428422073 1
203645100502716078039833613170002276335732205805047209990128777689353375985741664585200763921687804 8
57367539294950533840982293979706583142555393829960861877722796639729390777908217851732476108 73
55641896708494182323029269132494491340375697788000085212068994885195011870424308197047767765647705 1
66600738206498848571710483227234571192591806527116704896922909858075153627517095505284290392243 65004
82488074481318574364656698445218053666467548379873567916422013296791664305100388091714937337157754106151 6174
24044495799580315561115791030873134720990353018949994658219299219647710568822829861410142019463905 42
3285844381712408363322656243251238385947763567301206764410850141753981446342859310494968693638264 0046
41625964695149031961110485447759191706584392706760240031145212717647081333200941756887387594777063 240
9980920684630534774332419452200217630004662282023808277804197794933893918898522440855068666909872 615
0999343275594219513614894603275485400282474638185574303867224884185712041289402200141526884770962 641
2221149992288764391914730542027643814557450040765400426736303359544464270818978527417707745133958409 57604462 11432
765598925712124640749760606889475217788884675365773130888483170413084708302811779259466701208771841
28659419901877875096320028110237551436356123048655461532988282999060417451774858147760123133434173 138
771090557709367065736550203175790043062292930314450241991977428096762212425199286327402583704007529 72
817435480410639345063732675068434688183887483323541121663418804241233034034909675779416537798412586
87982886103235278857332155338151983305885045853135490964305130898096369406641813704154985931496667115 9813
0899440825457153552300652508228496172872396746508251900453245583257486877200667479497121636028208523
54302783865361171124532148648794241321331700852315433727746076806637669619861895128840910891117659 551
57364984738860695247166847523751446415213365489246727612258533936148416514385818691738416754348278131
7663142911693785564618171660966340291272053650325444763830653355045116641522470865121213129009990019
68151695921524303910229496964390635521990651394323162630365345397471515735014459156097003147953738252 07
22864326791180222854544505100668683826497290748132584808710208874950514269642937392581367718416906 54
521561087615737802053527958004468491361369174682537172803536784350361890124577775833866770048718755
15418115037141294549114272696877208861952903110006521480604793894304261211250474636225675393476818 2
2221920063516876682582150682798801607357061108055578616867049474864042027000614400977944187614785497
645823956249805444955125709106420708323908144600925117778765206380393713571176447632922162141265648 3

73947107451322905073720554233262086352301211192300993282164753643692379063250335682531355433034789 62
915330449231153809199575553294498705280190335116740752763665598147220612180438573003072978792173568 5
0001256233180674259887240109968969813852397306191955928336948611903239492535944158365958161383912185
41195151992655070437222451106336712668962567275866773882387907913364509386511722013962859447865442 96
2183266784517002027818841924009364903662723574437287448563310872878958458482352508201562742207923922
0339204508281946226152918446070619758228521338779696323678640133130410995564537477406457775768490335
37916732532182625004401500646241636404631178277966053575667603716137442026691617218914291639230484 97
3825428942219985454786894825670545771208306069664015747011398438289931933635228885729811548286913 5
9603242851163621922207667981900638840484271526086590715537049718593352260571181041504579347353963276
39988723256063686084010315441364214878261119285418384995743044276868385145914918918742400661902831447
9859224459633479953102863027801831150089300076576280808707768885656711361061861689962499638799370291
13619500621550959100266394298058423177896496506627565297834415149733882552363364465203622013216280 32
13500493619597507026917278323701383715764323028881013296332873938245738746245096895082238330844176 19
2408476051027246860191447430391510803777481923871052911279590551749882293907551274408036416932829212
5537800884919287028546754254666973573970536362544007222398956201306760481133915634972776056714496 40
9064045114809482486851179621640442806897191576297535624518688500272856943365288001318441212141123 8
98385195278511948146790166528406883821869586830661295903977459905614870361228908984113820061585914 24
712862298600417189064530100820327940885580385760890512269876008642460648269485048618629651722184875 18
355652881466312752376870674667526917244167297354569673316677184928439431385995773850485061097313805
782920944994446323943060687595811900386026190492109839871969936474633114066294511147152055694804027
98735824309185973826397713404114160166237752693577223614775634790555275216648241460998148628132876 63
118752410714743298743627538504577774254205756627393156984043856772914383783590182351736877088048034
3742863236659074522885395228357786813472195376605008486196896263300503309360409978222914815147525771
83795281384891568049192181238360412271358296116497247108026125459205952486511422783391156497546662 74
4867185218756181629471196471380668777153608541786941838614660748536339552589010696623778000065583681
8847197698244458732298445308869902993378877952657197280899159793941934367522718663437829079368244403
20624636018669904972319031502436250408130553533830653161092378952733391436979259369072869740426234 0
424758703379170022149258343524101571864539834784545175892241236136735291316260171215544108493033216 44
2300759697110588546958695711726320379851371929401448711950153715879163321253830793896944127468922739
8610118372083164339171502864690987328481720738738152340116479451230101019666620364454129562919035
54810519125334387131600152412478550452245480417085800974416436084037596380188386074895352662695035328
164809681679448801761593992935630643145711148351667475654627759421672537822961337952004829042288165 9
99567050760734870429085908490684909529491046326851763652246317013487998937688779842092948512962785 23
0159883015336126834299177661463925494770519302050031055605493677633916328395389557886997769714313544
6101324961890191705701201820660702116577665414605136853434511733028437410975267518355759247185151889 0
9168989486576041645332124170280811484675907730132725479400415509728678967187961801220443350298796 2
196440483278863799604098388253629382303958396983739496101712358218421774397427036469191125810751594526
6355464675143678868979822902295599747715764306712652975718556151103574766104209641784124768301584033 97
93601121187811200823174037140757004092710837435401089199345494983756076127409717699213954110952101 25
0839813654956204515024366843113398973858736113065124745242321542571509170983114140086026489053937071
2774412440669076834511485070300361178690524323205442686532685500327030306501507748047387002402672 92
41448002051530677327019111454898739634929248920628971290478304492685380028314875358105997056148380 69
273730096861098886378902955732473773218392968237265964099999476795576053071821618693945992778164 45
2536969658122450039365458994432191272916867639855789329396609882487227058690249200178490271214863 53
95532805946828846993357856689685299143803410528336938389808106541631250749494609447408886469083652 16
57106852902937901140010171041875820439261982433726156111356858417307628652063100312749714469978191 03
8819926090166461797540691029725658474690450719252449465833526417463995176818616905673265931633412458 5
45178258082449007188501785142887176672083115995559292672621178251190445950693489617572283250711147 52
044127676575050686936709803366686786798668098545163564504567000982813046712442375802553358494967 8
34550276310756161856761102420062290862217868011243459156477561363731017932523468140907001105009 79
458634572441902662005773671790461232744208537976097262686877009464227258685007116364595723606381633 84
7494349975206543190488268758253505156910742713458684179456955827177098027528168620203131954435974916
46865492287630804662143131853417398650352264519025805451724237931971335479894431301843024011898085 6

828428763561151135925450517590066090293175341972370373166267631045526770572841082661953953967680235004676393232238198895390409926916785915668921976462053713869576979999088841317689151137434627023755761356023595321292951339330688041066225809559753015227590711430726120998040611204959544702529022458298103260460360661079078461456247718072122094537822437960892084783698815339985784383584762331111455294499316358956545170743494100864563756823652452255289027971719998672996381621378347132538829807661287146255347835297146301393797884322985295845439519676803864770811999621581708094434989580382507071135827079833478098566103008092483633401066437851717050086728560572566582490060308166592502760359401420724323535032907153406437209910495544172192961731918787667540913587710997530683942283296170848065834349771181400922782530261593481394755173603558408942664447493309584619986206812924842999006904953095601991673592700342277058077725794298919248350750002625357538268748323634227672480871144139303257644516263630141577372991358552647618310615475055054350039788791534532770215960445663570306770661001920179321401714796971673846973333970705605859892283092531292526494279536183670792879940177086017608475303934791124788612396945329823362750327417646243217820505863121003280810253530905228112133576906734827893771929083668640305282799470624862476886704440208595385347241370469282593722752895964155974991677578726870096143793390912193869911363019731897109456037301611109766624424400181780650555724623398592568655386116826127043340700980518008868971398949213194807654561609544651226431496693496697436169839616874112409269250879164951012251867524836356160571234863846829796456667649848464767165046561266990865485403705281050282325415823196482458286149790041803457596958657165789359912026924047544686256265670714416277114320705726230204570586424586485386428718325923588272100501781925910321862102525429061064196493219734842292946721450880276773631002514561458975281845672592050077906099263317935029302633914975478998055915983874072282012111603184744607313092642236057201406831874074143684735669302813859684496367816355446904575253185482659911617918864760699221326838807888021781507987526270959165282807676731436874072662055402771583328466679056225244150316045686489418112599997950353070728399541880040621837690405286046378220683553744365548569477836150626359899363478702790903099746277218424110017648215901270567118208766882275764676994285411305424284697967119281896536371908112434126531705290231148917416269759846890675493815136439867883056142206994865643705634568169510209714342381265370529023114891741626975984689067549381513625447682319382165559881689756544989847223360804023621521263789857002732047970983507338475588088526560046011136366410695357973449086862143285777692977138838668754917368355359148530578245719109930298313951375705252562099580954017897695413946151720261764966079521063305486458118903323277535560804209288079548131044308314254117564693796449370088051788439064650598699529934562288497813679005846246906668982348037672283914141463938344197050525552745661524307031168939958640952184680068090113619130090893426821782883757056395195330125180182350049293106972725805703196643197756434141864919570951944115202257960157942174332998712495398481643215884016823801567682205889033444305769062383726206054194708302698868088319400151779250677517517145937223847177220508209307041593117362202000388130607888400911177739664188836733320446529646445934419768596428126244512562577886323153831909565428679230834512402761688835965251028829254701745885087785467432335413143581398900523422703880060771431783425268029966525159580526739682566297578541127324599996348271937105702177276790883005843940769066347716939000891364094376917089427005185877101084611490569126445840049207083085260645566055541887941017689160277283372571529263391248090560002823379201775264068935118044977919973237802038054845153464214411215740926211753177525342523126563278956547994952499966134186685611371726575357616124675639363634585290219883593583139219249139341864245413593442816603840579430540585830595161258412086641797040450057090151031429790145799585671974561364535372447573259717624162216656098154776510792433328455973035022141801910437784872406174681283719961462834030396624072402411784837905118698833839179026793076495649132796657819771695645745933313313426260774897136719700587905164110576086803903926856263487063454055157631398616763810774144512259412855075449424152959294857398986305684715535148771193322793103860622876069707269223883921101042205418231418783870028474888838905667506331222092051480787061361084283744060089044614667973715826272029111684229324748247891789685877780597760941818644316340028850264536445513550671213401188690785557499410205012020584369359438338431421187984969657966712318296941917117815804943525759524060183758509979343171308802640021542881644344306720286303061244985371567180909067836741275202011134541349983917111725353851702142430673210003144137288710554407895824702376490475232030970596062076120274233173017656903200367749269422733032275762751700794150691302352338229522938042374299195531100137570087357404896049301491001310514828563869984229241736475578552941533379493920244023171942

716029023127159436936461304778015704697510260615433560235322727132552378164940552536518894649839034517885744396354358013434976027147384385511984781089286682294577257535978429545434990952690771686980501260973242575667396516641860943350384149618387370350938070380101695366304616092407294362211337355372256317992452086818271670641961004506900017835172681539217865847481240698829943944692847539212047696704008917516980044735013401137800552105630498825434932067996417341838113208226066190871983360217148256562393710727708004426030257864375469141406467379988133306274980494438444339372858515420929071406931327851505346968127345232046363566630091702633597632388614244380188240491100851015825229325656727939009661715167571103702279090057724322645193485395815336762018043079066346938595228276348528073739266954153406812865943469956911867276606938315621924386564103131340589115080721932867391688323814944776071992081035475388423453896732654849684408063510671575237120307503288768536916612388601773535404009108803958927512100254971669370797871866429220140145588245623424144670313230305012810324420163216305765377138709905272596849408782981186152588849252721860328952218240262829832327008176345561299714577465838754724296746182466438490297865300776315937919644254783948828268065501763316234010146327094725972018823353388133530254653590463483105173008497042615373676406203301379064787387371214645255533658355828253129869085390026366890725505682714100041735228984821717599940268074649141888703063814711532964678955931808651223026636997204711317478634905277669414273422723279580870002598280525801378538783200311808721509846232270743162776152994081204027317878669976955481915380842773570993281373470561766937072888062119550680890081599398264876451200332178564869843209063760256299992888973076124774122838326961611075164891522825064454626830641722718033834317376581971246395144787832000935133331865522233895566025164708101900224467477939875010877614131867655092406892223028644317156237562055325335947588834818441103183260566160857078032412132972749166000048992747414021583437012481574191298877094711169331110411304646457319878710695701135487660168472395555808872908971746912203692511897246805917112574573943911456518061060805637865308957847735391188780952243969780018543644395448244065652502922023722984608325547054940682218788034731789918341605402685996748721800813207788553382430527889525289709770802608507881752684419747475030089442427002772330423929381969579912766676269025359769229524623326878831389484003936817044687218510139571324767540822323043782287150498121221849238110607200441017372402920022571947628031499451547833447030334645316373131527491626927728719658075797656249029624121342739474949499605804582887192221824373601628086164468294217844866433081941654909805035061939353734418444988481855950496502632226450862252087098394179516137292615631666411637908625996615984832695382064061026825170404948988385791924216865098291704758395293260119443105729997009488200064628250641429808857846119843850931743499331587540568461844308724816903828496944549149121128389917442698035544256672059237594015085287584120562381860705431189016681809717891902212293518874279092195515518088868903134477084577771454928537845296172487465733320431666064130222407809409924086148619661646179157317812411352085815169918245541054412567621643344107017351203236836922470237452231984680649794676700844147762711278182037067394773027252712992031143845135201994634708981058146681732870787756102442048074753084204104430163487263326788345044502484231091775516736707605284105175214522934924802842643883484869002099445286264641842034722165709568643477981110366220426052840320341215341862403961367963057239176780803491697960775267805585158444737573204098686615089568326175957193479108242562218826890023341053971891760363056471342188593599166100404311958899686854709529630760786863475188876682139900467692737094847486632625783263229965264902904755726556426281710673612277932681885116213824241463851419800618482560202162079647746012249996696722868524506028513800617648263418596708376361601771787875847872117643257127483452764923176264625743670922662113077335552056833860543070541587402449290035756111655557169985793131000681933285380018287774723250914115819850576954509512870720057652266342717714995753519942375852010832755725591013419830661178092084032701596302419735914966068109019295387886496291209154791384092762344623131144410252780158536443513026311959243846145507833436871321090511487271230958757797122207183607136023614162793363027620066151301431575584243644725271284482860833476749112067999001847319319049613014178604322552671008309502966151612339140072324812740694337484813194585701844948519546090713962540695926556353258428949857222812264509463919495411072692219061756817721293282395098163294236973124724084346206764151658372429522369301743268413874102094132231590431123090085591780898098639811472423431259772730725874965450798846085036494035

63606421360246367250297582588214239709069638947515852196010056708757617434222006881840186782896407797
13421198798942420054262322639109160808328221720623832181566009563766131150703525394313704384764067122
570736598637047057472995577056332924928706676715784263974841648189874644026273262918091863456951402
11126211183077842318805504839023018238559619868972586637538368548518808900275672391487967574758714477
704496390596664763884055513983208511051808694673344214696438893651987429294500796357933677630658354477
9134409437498487491781105259529349460886960396179237563635270568286632335693875478284960491495594355
81123376296279491108962728456306695903129237389498739064645482345265301124597093691663681984290197555
703961105018093963707676759574613416736594918684071597994097749212951447536435570510404907182268047755
346892192192239665598892640138338542564942877508240456558180339798752807500932132516595562648968432677
4720950845726942676196203246524253611380812855410809861388993231752761000593682748189193057268792705
270668165094738441124135225224621640589669773776626694722306047925937658904054510890872719669296026177
564605692234708307765972494222517344900495881032587117349851293340748288968228684514065870595822088
54635662223796926845772708006242862470839100012713227466932910775957841160555232253942755096006076083
70533448080696135747986610034569429050487428865415852057182474934302926645020129015228508576137455221
097273911088254049019546790225373255943701093302223435333679249715548650809056910531509201053297700331
1490993319144158254039376710385615125897084131515152832179811625094407838446359926982485147799826238377
671542281850696667162662017616097056709486112509359420925417076181444365168205906937026972848430354262
00039700202864233353238003083672757694716706320727156328814513540266545280537056163375657565561560477
56839735821127233493026657137376157808881484169546069518924508977614582773056447115603672340137375122
391133422213509952010356177643077090804473048268999177976468760034480364444186413463436899950784555107
203029886333848328181072699200088982857193684138915398111167635237013599604776794327352194624941833907
8303417727528719732163352397478366615988300018701354800725499677948156412225073198207773749493693405
159261512147252409133124328535226609509917886242062141707618144365168205906619070269728484430354060252
1914049775512614504446572569712315957976429582179683132072049766762865704713117146597168994141419415577
855279213291655341080358645394236043976346169533528984633901443709771031718852633529786009896930866977
8435263934184369703188574390628654710685190800238247592570669579691292282779776008111989619170187077
65770471548040762342014690147470123007236720063747914465248848800218697254454637056860047236274220199
69808221422778472139305509963930666588351205734209536232706833519053743235096424169280243496246375291
3196824007038389209946820797082050855330265006841729064678943292489042266623714939148718520453330509
2819272200267835450900672192934483571874004864525843619500945520255338530150193608275455081483677622
94317346668701839896632736870667173889783817058514355616626575305893492837995688371566190511746934017
985538752506727461577171542792976654112430661688579214663048265376236291786363447873206048116856445133
319633775974521089142064262107733716871629057910904737837639993403176101329586566017868610584132025977
94391846880266009194120701567637052752104460254246476622255379685622112998222213661949692780512346133
589388009378700644588895530093096780802205671222917324496219466633608501509594916091194431883188775
57272880726049204842257269502347772736256681742643079240723243055492388314054863945929521673461205977
3473310812267074435854434036905135486125846923522772955595950816539124034488084615734816511028056596
9126047382200962429878618959522873019383292859627993498447285095834161821543866108691365440642590891
158969812989744271470860637344088950811207925632434391216048670980372692982232485582499570894311083
76548403737521383697754037253509308684024083186592189475434246566189331920286973641687638070559427
2649094054318608688358706239930893248937760639588088169278821723284010080077874468552849300953516
813369878218813198796091570780060658751204136810551501389914072160545407320984244571124078690272955
6371764996922732097345339032749384224691777564029121963794004392913939936963834312381057271231601654
623818608034169377923236080648416685167008845800707900539901913898692886788685012546791685295729795
09077814007758372919595259237989529009537496931446488560420183072483668164534889890061641848721314
17619792067384253885282960239842855401308933923814117570120808301554188871910111712203543960412588136
888048929394096629476679405622656481939225363716863010600798228698905371847071955248636144242598398
054972266738139927123232697913526781947508754528595707182151747793330838053854225259358687900111
20176971506948468723239775690621905302772913441798953328457172256959551393984500808185505028461731209
3463323897674396686784116298253664412222305420632638997861301160460642764252472119953879776525470977
8991387573337095875444606881810333079159233516940026805099690092016813695028758939377149493311215724
15902123622515497269878580423624412748789865593145938269757693143055498179506531441866832732881951307
27519956721753823705715225229178899738983906301739917299478596473288245148057317576623972746812720692

835468599720491582006549321426373785365377765776310542564915637728001898109944412679472769211509560728023628284880601982030177055736035503433134769074127602842407027856245018676049368092179666630506285791925690321609215503829815738436504956833388199142037563207628943768460247661039435202294481010140411684096352222244652669840429640253001700640703723317193522076178313500357425684523102015448618474112946240496328898364055154538023000657624314766841533398052677981987237595528463459530388125240783115285645913826698540619890759859202685372792300841475303461621845748488155548751802807500870114860205663005110795262621816509997046246093820030562461035531104034228743366878696989658957146052636013394740295502774288600482358997616032607498570471221845586671322714100697884695817126271499385892858592146253686919607865356133568450075766467433310863247115800710250863570422004275510279692222982886795051603903278422738873811830329773296824160426832369253343394804561513027747565564224353863128341393926559729662026384969453375872939469025962873887401488070946338060479996983165111192908802511925602858173049910841216417996843252364020412026660339061356441401313893221202873062944532613198313335654125205821195292932149405346882300337137813069933752626752573427205471825985193439712284754974232282540407662617308559179774871262988700226674104747061146869802870273514819887933069078040518521692872231913175583775538329306419343327670587118572766871456496475004672937066771990706913870008711630394429748229895181509410741915838284983000890515633749970923454812684118905572030913809374492626325991473345522166651405838474159317937747013883109292072972574807268927486362545501022618036545799496941836305185203348583061388608915374757681450689048305047432310417567254445678677540565353324624426384501125401588113376287922791445769993245490207100571938902433231581806774444726672317664560768234838627921390923872430415110888337404826020756447675776852313457857843464982933624074648014159794645214350928554441456623672600705710744294714808993054625246150539546672945745477522309885938349227321181401218199372897274783639130242229542283226581272993978869758174293364400546233397984792366191852240226254562010126296227425623817530051776328637553954276860419576758487868293563801648475164681085067253073869350186544318606782117480706710238606132897325712447823979443239268514685587071759326786933587685471603448961677611716341299667336456508979211075460183012363336069114496418838793412184219493537849966523797515391763059159646610347152121465642745239718539650213163259112308165283502948681956571358874767723962901916931991677686458730587552887475970901591888584134023149445351637158204611939295386500630901968781148725200246325005859509732656697594080902488527662670444267263984945494090633358854494438550708277860432637669558890094233921334532437458692369553584084415785410486788230887659149925858776134937893933817239561267326458605886099833130238817706692933483404804894481441182843463752808166560294729887698489519112897279950097630327730517769376782997599259573441523518462839444189362992090902413580600242332685520325343646493589775270690343598809038877648352811417289852206157665771897459969312690499427459585774890071501771028005010125400749028182672386637611059367221251178910066446038280924572444314173858083144173658768637249757501724180505564815645888269442413625697379292255945902050613210392317227946862868956199674192340073901316879381804182128317285099359198047059989085311525506061112902960745352765416055543234626101670881285152031851635233054684501630978768688625142095539218201938430432208578808474074825651151685245792053763628583004724889499062322027715010389354247920672721803903231808454935230021570322404830357516834614190541583770461312945022060404923254337208233095668957898208001861335005068753785517389822960684579261346012293364012293472374372417841737611056719947498813255663024174855112544613530731382580017759590020575525688293535410332940582476334595489119148002630594112285629020991982970892267520245118222001125715792407336505746127588382381239480241104044907188979390179213491779857502875171121244970710366288905267450506250939260196539954670766856145659300467217310496756990455624866338937724735088968724553086783819397974149839080870071248444061484882489613102697330743859124987903741870626001051421583904088761237036537135202092948787650548885182853428284100383985536623133764576709237044770544344802824908863222096586644985919436311920583162054531014457848223373995095568172734145085035650280593064829271852974721285323881753240077625654889901114846834623218046070717582378163621157379393336563531436126557882430018387059786827453486232134036924869442254632051824297733063193782880524736277279010235751583877044573638023243101059133025223974603254422842530476392499368741225941252534427457191435790426198559803235541075551717960222573100794037929632241778157107508804949631587383212867557526881635115628508491818467151197608709723569786036462041091972064930041231501888377704583113976291870092380526818209787516650895211993782704945513992044479894337884842312224876272598929978954825746643485755792646037586224230854016062202312395837939679061860820207384862333850063860371679539464128279817191131119614782167000830808950785445812695

481816638625695092523164821917822134624141759133543797902834142377835388892059566846197981052319282576652938328809119668761048185889645442440220542748969112069333534552351730720139527954521612369050645572702560858266064853839180881791607076066138641953390753463104231631434614551495468392152545717652301762790713467029896411934810468624873109129058828693056154855010986571699431794832040020461502892242832539872040350919383645420568065798193492377797399236164339232844122291241613102368123841230940005955594392123890261014800241184367325721856692868521171174389577355671369440365534188266303219673712204778461396724242393438206557571014652108390411058025499057430927093659718121325372945327612925571603381159932729512007950061113205331463611272220893115014719251795352428207046383536281323733050393799214257143099602033906880057835187398188931411813030607371891553698731169760498551140763775109513049911398393241826775067861293093343419446004710929787141898483801103110274188437832920482839062201145924541976522126891453178721214513312677987845560445615959524975452985753522215468171148984145511385408141055013085489623442362849791041944000953541984782940988043650418337571950530816933625465486685067659486021084403208883697917857456235888145994833927128325817566478148930763018318845031751877025433815121721747380864741836917805118542713928305925515203696154426557927433409571740879029284689484767512818289936936963800520966258096309522447841141623347063886751542015041317535385000813287716280810470018834689655554432756580375673502026726529968266884675106657495635099985961788943244307515831054977176345323806390058204773181162089074001240149200801923869544920213679302425802464651769831056714850197856958739975556787502966173286707949082756613216383025793652727117235764937197729881838765016004696019807626071729733871958649870909996872471749476111410147825901880118243536102152204220498197618163530033988599978726056037579758680307331475224756400757115187922016448589055409022006921571006776359249857239955276320144421268491543577800552659548499985124797809019990908814255453142011174116939154833148205738280414543627180464686626291373582349646828752208699583629003637969213406854850887900797060971536150493030709714479356640149031120553080824095024438419875824164708605864899073940687544131101744503519464872331767462602490066115788861097162688503239042564216657930197142330777144535538786284325143360212501899094526144540052685213771876767494322391929359590651440122799532365738785963366187982100511224848175402924750136695005047596708441801263734588010662253040018506699181097190490881302291872876366011259816340730258132566262078742943937308834058169822940480343932468067552000853043214053837036811967449736424643299033781535525502252468254277644827260490649082697721388698375592442072377285780670194840754825024590313667177787679458280971200747091410334539792304070212470478959724163063167939042624636533158246802784629999842955829706777036736138091274486510961319478375542147342690785221422069774417333930237588915451744621841465888174063321925106639986263693688001561290539774891786922752271866000000468688606943843341695689761910479315504073790699480541423931403427721205641761019184651983162275524847093788750088883031786357307282918767303563271153307492718627418674783688027045784857087073472818436550765235117062567559404561475108174628886513889489019616918735866440178911396503312482826282127732341495662570381436670394969496422017426965254125031386671549257792425824718393040622119315529395247295659610883491077311846506610853782237320730728643368167640663482712659661133551940546215024279598826033934377785136896658920777477038091230619879767544853888494348861517902298951335515131781694479389844805510160447538729346020810564239999563313294381811817949168206643422986141748886246889109104015783835789046470338506242254509152617858172096015495904190190817249863332365323996203529983358128818444003123166844128209215710365500317793276716554859267887642911944388521823389650944951082992926077565435181399883335003165944187490157382515153491173025292109119075949224254748307159707103394639477290177111179548387532454308219571097116313656585111336167799267475002254567461727026192658228096930167652635367978562181504406855242520263277431207892459280730831609544932281990578760422714125469899220372255860610883491077311846506610853782237320730728643368167640663482712659661133551940546215024279598826033934377785136896658920777477038091230619879767544853888494348861517902298951335515131781694479389844805510160447538729346020810564239999563313294381811817949168206643422986141748886246889109104015783835789046470338506242254509152617858172096015495904190190817249863332365323996203529983358128818444003123166844128209215710365500317793276716554859267887642911944388521823389650944951082992926077565435181399883335003165944187490157382515153491173025292109119075949224254748307159707103394639477290177111179548387532454308219571097116313656585111336167799267475002254567461727026192658228096930167652635367978562181504406855242520263277431207892459280730831609544932281990578760422711412546989922037225586061088349107731184650661085378223732073072864336816764066348271265966113355194054621502427959882603393437778513689665892077747703809123061987976754485388849434886151790229895133551513178169447938984480551016044753872934602081056423999956331329438181181794916820664342298614174888624688910910401578383578904647033850624225450915261785817209601549590419019081724986333236532399620352998335812881844400312316684412820921571036550031779327671655485926788764291194438852182338965094495108299292607756543518139988333500316594418749015738251515349117302529210911907594922425474830715970710339463947729017711117954838753245430821957109711631365685111336167799267475002254567461727026192658228096930167652635367978562181504406855242520263277431207892459280730831609544932281990578760422

406391113657529384738732739647586951190390961292144657800965627823085485257649831923721384198581880491491408122029836664762563895359153095424421023811400942201054358733602176560751537841989836183058408095095792328872313144825688115900293087041948841952225155771142536019696000711634744743935005684563443769003357417602825549108885215323925060874448597881672789669026317200106720647207404594248900508076282339720351383460410116855895150189454592283258699909935831848388833349590571254321971399928556698368262705426927598788149893889555327752195716198846758409266923712272429452524398252383641763866889502393648767029318151507672585976784849233745844943089231935310651523540784459458922923323904494552069195194753050359746278833569846132022212993897490693428402335529626356027699332292725089430202069727851055809584430078972050488004412923489474672898738915026122885291287429527504355650209629422473679346403525512695447933149314320123748438652466578663381046029077025167406752254703544717818686148529224587986189617232446589528197151406141012192732756948471047837285769460924518169168799535580404984124110491975743929251100593314280976592191588701463865514029775960129535847007686116829226447478883090358961861126606849828180729560094573218650735063439570123354825915035368957714669165095764454033626512511991682040885348821839921490374688320638921958396771807212607327885619513135576575243413453665708486593415984435464537694535909031407965508106049468224721453183834665013178906741103928046183900104441926004517726990294651912262530275450380603248851883604793463224990548258049577601974085448238309028870415973120470316393483654892654723025152908040772707730361851825196120166242493551339087586691042329502521962227426242566715929828930404314190696579041201008409478265978836503276031030218439483431325642110534772075835031460386175329382737617103295213812756699397374441853317163250635583905004766440504318465550507304089903892699362886204694026958443150631088978643382598791700659552819878337554177791762887619777502524906729638062184794501401437110271575943564088338258179644194308333085998477598891768522529027457842168969003684371389240369032563217458039652881882758186777509035040778567728215280374417828553924630303017935575994696195281418099816035856227295745369637536195570980535234513266574644682623554839759342756845834543058091598382508947389287692102976887429987189541517332043580651915856132713516811149098354192211828347036572886744881559980262163347287343430001461760387865900136880255812776602634600144920817095907337266610360763624230858843852246934735966099379697569178349256036933896156460246531209080086702727847741561036000878975336766394492456229509234066383358361325146393239122021141047583736381538827391201590659474884818049489836365636223485421468944007752508002307778637642480862819028182034711831743730349334426177975436957359771948128215642985465056109643559380311477664303553133888829485169813786387513074910740463176728759532365533291263947640487356284411750797139047400551332393896195551285697859802047769760052520606805483520547158124294736280121908831151208169368170922796178050408943578359913093476322800801514996988913569688136111051402341987841705950776564064824631188859014803377611552376819383063571431940964257422671435498184739540779481350877889548854413001331746125100043073325401075314242536427798010605783400730462256735948868856191266923560123298435577265000064743197514702226705282803889500570021520539080570268516881790966535977581263688859147694197974829700113258576587857266967996800320896771899522050292034907180241069102000562865477296008692638027291114694014711195506772843968772487327158127480442859788159810302450513289974417057541535707520610369102909431370692300027584895321251197846884642301068044903732891922605837738861309121071177058775287031874653045525558154038878565241732495736977488760767011950492894183245918218070770062249502878998405996609728435314033200749819370748032083424556361863165398752492534372358414756843571193494334897563546491917525117456083081918043863357646830719241410750049886850106059961850142045861653689695511475873960660607619571162625250236781435059299927432365421787159533803175923627989878294913269018960179615887369958353635150305030483258616920935222145942082074130397798738379096406212477970343951141630488008217868194136392839249637892044245671039999528975670486179028806273423845473978447281935512811607338126327917421992832097218939696111281497697228426437027540750347640790134533313313245649630288115713535811281549952388263090680098257040325224821418954573119471050493392745593551320420656215368929493664686409505631095853659861276550296528504908502547745982156692951273711332174559075315718029099432588340538699481640099672325312246083210899317524899234438119482066431676943205728014253352052621878287214890835065610874192721644374574088327016759943094565014599538832564824040661128333329349447382298147582021317975355055545776242963255122256361396446854997894034439749368161016591941235650173277012122574966526646311730711573483666975131418259115846143281085787581342173761329838207675195559755419571172396642221267645474652002119862782287820430645259902635461770596464704118323000965795490248713711791563189956210816754283526880911798918327001510089430933502265509880417268148908812291287O

829521097664154447618330981136912789774771229802021332127465898526346719266245553883192771449991408341010595248725690306547673218555781410028384420588904146821096690543375552065329134232041111494441007724598556415289012162258426352551147078595553213609880932895069644837618776325259427108701314679076750276325987325766791190434282627051475245536494231169042157035922697585321023117194458974786817232156114440928449874481225591723161679444150344586844820617812142689172348256874778766278361683677432494087732449498986910631126003804788658181342462107208455447364215184728768745937138298428035092082752117689938459491096246243754536774918274654184673475927049744842547339625754198994080811398880961788701145201432449909586952926214021933973003160651735918939332358754137954654082138809127473065435451735664490932758490061726388976458421845913459045197700840345225599009618879193398402889543235627906082493689415463236116775294038268346235519428633099565123696980011564888048929462661936465231843409883458602730585413338744627054535383502039757154252212852521282183013029933266179041222692387061468284264574806844585557582486688721350254173093712604179691242764811997461140265341380050202471813955693977026651364840094792620145897591381816690481772316593805505270690768397746184553099126023122300839997984107604782938514159887872479842239091437645430731876545189095326502310947742636863639679565102799543304550196299510289841429099115580018885576177703347447446784876147552096699444506590337393015003131608383205997298445703580164953993575687340656022199212181567045592086580305446081036536888051972130577603369779128213097605368580630079515176056476299641062473983496958999491685187097498944409345983593523263446812880259885425319651316359589443410520606815724356053980185968094371494286556694414900349115569797346153042368660689450470004538919216435945629122204204541554833978183611551289833126210250769674369298551683203785418677422376337169514555308472609142939192564006809487739878753639043132845210957787024572168036765425369755669337691689372159682642775374494974180054169944270856774686279433457412781934597508929237014499253445514353396620616846162869463188341163816873263566096684426618168508783583022017268272795159491962988946547141503006994178790856944502732408037784271503413853735560507743296153837658200559340272571019331880464424558912948593645383784405015959209169180992093220851226729247492051440151722161097921221026915723023656934985512430923066030593657798502968717352913344997241626947348256737640787833029427750736521959651753951173284253848099934666697011854131164051892495265549500814393148264881444646402094078323539342361234953898624815016462628725140684415403232410303812921202843427583407715871458381042016886254561952959225028937503794382926040209740029765913066860659527553275106987944541433996555161949869209046559152382011727678666162834438312481666252038836576254126699845465663056295630052414535081644907959147990964778200244813028767072494468710070397155991137593470149710862440766515214718993306301860213050480860703451838006279564187122853624232806654123242592970910770029297212199474442622854330868412733791162698475208548230041569057210122026554698035671548157732519046914043093788933407978357663892407945831053809339861714613321315002679579725818962290576615119916240999193263215059982468159612035086196162185140955383507737097381865512965135202631895293924356239514539031754422363313065739019184123145894299056068429253759897791824573968694733042393692337352111541335847620381814269245862260816051705773722333299611929888870521300396966800044633671809692370885305036087584320402755767296700500709344562922107881691647199891507108449626352546703160579177421169382077220932352600571823423623934394111679245847481169471199891507108449626352546703160579177421169382077220932352600571823423623934394111679245840915260616864130025591430568111117790828777361518323078893856675090928097083693313274270041040190294614437259108134835491707410473230297562216932801692770491933289319141114908869724389042049221875825979707675643926556940213811426712898167295742804076469588336987035423399064120619089258734901166405703082400764912261285010187293510024604618222568315323826198069982118384010719482289989607372010907570340147326678500305625838998049807198497734231901917472465983722720742108557695432505903407033405804822422614308762808883420167299418449561628869589213472341857679721081980546181583973331783147978139288506482582508818965761697179187141501706091545003996639897044841309212546912320047584424535710041849477104934778444112579849619665499434168902643712951268056230282082708780439999881177877526438418231405761652466193118860690405132423239531733139514493709568694609307469314164276161660931859705889566352596023775307483043131652545728120657912367284982374554553875474062826027002266247481546710531782391142737172074582716509627640099158475043363431826546354402640405973284371165072254032651366553949818950099347937981985806978218655021813270393153600498153697632986505536587743634152179587756124518580348309948337088602548909152582823779836683487395570987066276325192187914536028557177376711378299615779605186852177462472797744589456125497437428319048153508095360334517620186222460126046248802226814489030208699540534068141543851849396590038459117458690409573581466833603032345644628063430524441122

20994971113470241456023969311815207755289287487893639618928372948700793311171332579697602283983177455775458612993110518107239694876579428203755884446207770717323413389041519555404647557632702765338010792659045019913999111486319557132221245993272985911849735013224173775276724983321615373711800569092997008018124471868966578772680054555115368197971988268480789452275424334870367231884210352035048530872411544593847549658620092222075697993203380369371970118620686785419495255563324660911586550421706176765066487720230673352128712295728646285932926867706130769255514137249340019279652248539432700596196410907039162084403283279898661939875924646768939114950318280834793998604665837250695021325692175649023887656152401852170122116212746858948178984266477447787288052235235682285201359372244127993663796742824483964378068084949230172094688777742518660795923453717290802613994109697676546340943127143436880659820699238896633992377814747893502809437366027590036462416130091873498595963428047847369567350496512427843588542229450338001132620039715846114551960415420912023544705338201478038741000722524862318388819378019216582103744167097585753374113408393955228578747068218429437450448918247514388781234555863450776262702177165763086242091747050040170866401275592307039259902570961549547425743327576518018968833828549557132288485916402098721897474398984326459126549923360313093772591162447164874142268157619944324616057295606007775572712147755502821957088598761030816526997434567736849343533075310785509520368342153820268676367990408755129976825548332690005219503130028482569368881011500900708334010090541605439705010448801241231425172792669780942466246160895311361541680530976880069062255312749586495569879991888536255300611324583354828575063694882255737304037192527978009324019657545394260194974997779469639167367418736987987825059437117506136845125588074655979918322876715428353719341954906222489359562248723500159655195582036027828741174520515745480974327538257555219568073477789127242925258547537741631054371227392203054066316539460792954280119497228598688626209597057713657674576946422985856740408549931614689233548567863314418192212213686299706540317111527979213893436282996378841327782939509892054955684786747301183738455056131478157054163011814060518125831815256609932668565644749486427236111006390020153193412882598321073694949529160862702977939042236362515914082470343247036819246398271518952691022795031493373089983779927924543947041591197331541232099771817833270880660153630328005323212834939228219427985955455416679258510853206403209848850627815661621488128466767270416201689843688069485991007452575301982376453842056047226797984563260150554934161892553326356677175294303411190187705682235547120159829502893609563368361185608376927023150693344026594230419562045967244077011364731694969106130497283217640846953335392064815005835018255851030854903480838337481831944218813095847742203769522647144417245993959341190705441636307972147768559382864112094716562019410130765693954697559964008297600535418866178823111020172715573022556955133501374025869285427797466984463603029814118345183219329346621191049720611973683629567033419427791676238341546392166558972067100211039085968668390528173719652781392134501547568786291318332843345475807220124754674876828918923468803249712157862218584652381791160338876220544502610535277301691773664780882417390884452589138524041891705393013605620502532289090913146599752230465464962648727050504279856355249995689435484656290533528389052407476828234694425124220524321460496911515995027451192115682913619877757442997508199457141978774047543664667121836718639985938647588048049737957431031355610692264889332379418621034303009967129575625060330359032217667494155356337239116889032366927392379605559478222318350257576956904849957835534199270800453710290967308048719319218734905095783279092334097901123050953551778787974257032612661514733097195021283664494359457963959427314484791826536859625326195043897241592694683925242421860380728712661664237383521768393446568942250005575813151840308505972561946962356965072584850934813782928372217068228979426416542578445420092847486454285138283727783313351358194082795332984873914137735931487315643413640855773893828279744037399403858089054758537646920265681816610003762245923078253690283197605974261345582628952000747175198661393484522858606709892291139860073767216427450218418990687512821051640714206606385958780283243197758108844372613835546296124606128103381311012814748306492500156551239064926376056315572415059444644757897193493099102668095708148423810488418579255005136768429135114112517579390222625874008845364163387127757530258196237227195074255547573401193163083529254662769457148113386654846390212303691690580495406784173105594659185983035030658313990683169768438761011995193662536138889762261650846673375947846630420393844658903846488347282284523795317069941161364108194469699672444074692839236413056793392301529191083607535780895440722657842135084053883419547823568079973662618142186804968786430753139516367036626375443913518542939738639303504732241669528654384395492271047001209217380108357609531156944974648759568577232939594612549508301794218645124976069095855328481413039828332195744156994123083932663795175897709125491779731184593992615345221399614109834910618405773674148244441335724652915031005080446257502537452754591

85515392948604233021458280713117375319139408935171215518043017046220554479737669667478931730962714817804320524153739850169540905116883068125314868090439523232437602316225250438836589868570661653066181904568527474665587161772658516507341550659451756136678425907159518526492326611989419927697833028832577941111750144741042290808902366083391940652111393237267613597885389234282889008977305473363126803325171091298392614849700054268616029942960963848864440600491029455039665291702342363466065042222960210987858800694421569857705369844841086867310506963029609061934231606638748990018882894951256155911024404462820531593770968426564803460337817314814053845044990759203550906445909909091445259725671495302827264521527396612218260413461475010528649224457265765745290324054714370601234431217752065777645527190011539533425054180753912191505983798526157337051622954411970307888743173998914579482534998715896943787734373767516156639546476243520818653159751036668970777583283865057523580501714023236204182085387774679830295084423383311037458065364190794474970628477065476794829621886692590682176906731391606658167858284251386246833062603366795467912492230323889295687728947612151536396030940627653119766901091138048740549377875158608275732363545668580674906271659721599029921867086138968951373706831731568009195548277920465055777750668885211866929765211039077863184433695906723520283999964323963876867320361379164668679503112179126791204942268631969522649415260318384601653704509259257880042440594434946762285550854525556020694226877321894633390276715358202225470331223479856682702825245988012346839537763551768912253678812381891995463784412703324187373763326513043223327118444507497424223782661970461040111226532758895572384985057636480494092480401111627708715552798840790361257278835912414460810333685676978349582569733383995616923989000805029335342990956574069637795021408519030064495477473548411684055102877611086419856726622678722878593933583970287883595451263588076265414634051480829780026991158114186054762951288199430838665320484267405555510253322746134221993103619477720495243388211785032250462292214245695555003313775842857102650520168844871279318563741260727741141786989397154226727632286584861696775818739529467092943452537335065005690136073180257767909915513450348415839846752650573742397474714748125400935781953545973458470765007447097325493931035952440870242766698463084264446386736380162865119392598513341956303926559571167112304497992321173606848975917558263166223902262248984862286579357279624901376403823104080481974942859270246321273900690507498877317321885468936342438942096492856644246175264232707272793308155715588664841631953161411736909081320190273895284251549967224303489023763280543729161398227252990836908892497388522995923775450126780887326119546163620900490010456572723812186073402918544779953528995921194551880521382952621774080412733037942090367807531125676008042604950740619056487808050234373558916155389927465633076200800996912907610063205373231269372796200739054334374622102589957213978891909703175501973559378692404636111602720633383409032239894189167224940658460386888359912725214200892792743098661635079572837905143855138729527112983667896161918139486407368579422616671178599051403823472914009097922491302499787040038054145464002605039225230449217160958905202817816159901170435828913409135889439304950965996051133242944825098365107498514890151282305847215258025185011999672909230195341591046634975148097657585419544419386603010239328940819094792784807830452404529657217508603472258594417577247107124427077986907209955806822251978998647932984732830779343237408852481802412599360762807499552231046321991068299808118204157702024037867235236304158690964281330364662533949494026426868625765918335222413541020048878505699747168524067286575194342570625734736675473650312557208374721707773173242611282037590654692395022906758282061276734984710635197644664442716720938461560725568203791375209490395388168627203084219274933011968906502185542429917425792226204508851936642892877228775513009477364630006424109929900045932466773531374440548668487587771578864754288253125710249958355230509904068164634131457314484939386098331347774273521501166697679503234920870266706333426771215264880879536592993735189924206722511948073132659233435772079645297777548674384247689261606817339169281246359135494609645011156588572925382446911719755728510786552984542787165830234121111553300074663529554050482452345410485731877910034762330588407257833498442549880584587627134845326920402256838316423222370776457714050912364134406361959220500003871490019578335980781008598886055972848479116532070874593435630802187162111489589626808097330821420606076376919309905812237161050106338135993918359318751387130030449261801897108928202426100316149965985859667959009443747212333443895178821392131689613734560088472118340553704173908865502215538977424805978954172343536586149015842836947490060430889525606111387554385849627269150910862397554805901362380403582626005037357451248710774142634775125097739773093907252823363212002648028822972706410952010528838078676946203715639576047661145005808047717647287800131327741324430272340885129896856038315246574743679823704601279075780608924227128322264403079900615323141020797123800533820530512191173870930198385367908173470015597347513283321602546122502724867125834953157390056206935329540

```
35971715394692464427438802759614864941292475277337413495071487803353778663631185623061534867577670 63
25490186829580038258788485176455630870877721920831838898622536779155140403649128193327234646410410 42
08811093126098396184018558463702354359467248574915309080485941831260967299406674844194400842598321 75
48033459861740020043137850828618329975681776571031196315818659950490672379791787367219075156406534 37
76447630621705766985670874922707502808767246204225353894474142413436311016643663969988087857335045 34
67245406692181042688115669543845133919605697691307452133941219295087691112172525555025022707891471 738
00685423650303237374867787025583890589936305021008716646978571180516617315670294795858844262719841 03
55094621510005467265322666861744293511621047567390241977309661472422087045241117074333885178666262 283
89759505203128872018952354482173947821942443997589998350060780412387921930656793257248728701976858 92
46523250669611473973008800477104433865437226884395682500885716481309468446127095654313955754758050 79
35142675631381936217024229731878481227768318957407048462333383621659586843308579628537616013136764 747
84725658660926676629257608745273124622235504415865970636998927681704349941168129792821296922692726 65
07086672448008802330978282035593083888597853534119785021182140046842505467402549771141733336689523 04
91683144437629773482720505996184279601382438264090054050050543899562203257694031141072969385541238 6
18453416513571633768620415488226311960317539535157965115617591768572008167453622412708438608227154 900
44416068767350715278673616499068407910205043844601203826219496186924400308313429304478409256311863 207
77512643059419959491771706413089068677746385439177938197576164468111066389323583618409159532116359 78
80906332951732610281710148666948643961131017175536540332278904447010019763121234107646363624583097 73
26893259327741919571499244985964697617304672915269924740555769815934966307737547040710403479749755 1
77108493121353924293297367572202934597575880257640391775690512155280231452102931304996533393730232 0
09205054096856615040586803416379342521973861293658971856933125374145784827028060598929162654039560 817
31874297587191910693275712179952087320826981863598460067276433876482068156113679122840973794403556 13
87140597390368679930474226932350310653421466419163614019702393423257526942990274693396440815930157 91
87951544290533462614049155315688450125585550009133220611127010184435006952174617306595986416881093 8
16304249760877061545783954493791871917977821405022579049672212207781602380149238129400854632371404 05
29727769335214212268461859782117443585242059109702894198462468999523242239816555904667713228763619 60
18384874319465372981229633527406286436910417591624718206354197435685713007167543681062853835873611 
41488327923104060613445141308647000247195647216798552823676900196362140435649853331976888064888620 0
20461779025580227094968342346108349257247535445213392687369393128985048595882232814751038118880946 0
45143978036923529233651090034139345846785041229174135049437061905866619482665643182166566296456063 7
33475093289553432517497273791362242708239356455824432730660077315547130178084016931960752234183992 2
70011897688984446350552660726916423787900804794161768433438322845853768413811825116775638388925839 
40201206135674469196707575278040693445818276915779566802324958841160289598378405115255979783881841 0
29014784762345089670289667288896476577197784178918709883513493304458981708646817971112838182002 0
44197499117215410342051654270270319998466349146952076159865068488522031632873225903815064835501000 8
45177843754507436145059221942402418726282490356341265264311926550304063202575366053150918626942304 00
93331024828960919247004411577693505738700428590484669736423070558612735640205404826776438809032580 278
85087690994809257329017331171653054678392387839859920844949442835455111107556714179415670619711986 4
19956408981974696503835985600015377631978659426847065320729130183465878244925078198783011235042612 99
13080412937786717714417436494170413095989427749615238523119521635415716204185542617257732897542789 6
42524558369008052069719897834053279524962832641569341561399857286730987422621402885181286001908460 32
31750133307807483123616417826138051177933714470414812118995919098022374678777281289045303579959340 05
67204305644044935478026963464639719156600092636864217347411761939064145402317939063173828518903652 39
86290070998309862522489234371226060985810304095763885109243895097633649477485841674766441141210442 699
67768519849816506753795395453287080549905298268091627341027240261084195694627375916305703069360493 04
95517602916771452567323503367130957078176961089025569013133837451158173426087208091976896237615824 72
32799696762654917696890350001810871147385743498364099092793369454718156290345287008933177864626001 8
48458350275093946565743519796264693962789671332758001033621670580288600203271745531773549327571281 289
72230169431756595295821313773640765032582412095839422766174898757281848476321808715860746198195930 145
17767537614228781538324691582039685564713187068248641469435543418642408986534022652979350425103307 4
55647445859877631882708589209421996518670330606554024323585305741249889718839646582694981140327296 8
04009856740944722209444144102681929134110942187415132828733107982329251456061078279210121801031119 70
42776907694303544975283057095089336533333185109692038906980629357163430388150748514989827901710659 3
```

6441270747611298518681777313704723445403745600286219167913221365782707024079509967914180334821962558
6266951974361124884659465827287656449897715944442118234927551749896827448384642155445225433577710157
1115312646908838680103921379271888579610984245440176035892736009257986197434102460936289896955972601
9190217380700975580237538083550621119923314740908846814351990335032977622554102622476650244698571713
5357265288218568611239302212235302190516277356974128634983540839032773548105080711638965572944766845
7921741120944840199655614355306893086120849486986451710667370769181572544604466828202526485235358699
2499908059551927828814586941368229897692946966657523038260343104885807605511708300919710119849253648
0728361569767818774221507110373995369276615245244965350910034528895977702974277885415339785881066071
5689492621308277949265494362107345696787855068738555613335206511000951672238444435335158667179835395
3576451196680957452740004243401544004906266613796295747065274075783714609400233265567720783707945010
2526980479349759953631214724888120659995044732454052956949161135437651275923571260142803889143997132
0628419546819688588417529292266886442621434063384323544903587183699050534709475844958153940545298839
1465186318150939736233214054509690894535311623502316264542428788032188619172536257980921796925106 97
6631553486672128902553534276632117995418894408023111712410789106391179138011169084156147104326412032
4352019466878197773271630704885510951900436354084713867267715378928165562125093199330846173552270 0918
5624196261048152846404552681817698323603329448717990508744856244052584667057078202967384386993693546
5436456444723004453622737526653145299219622309881176456806767139785437916584732095046944304384883707
7186501843256927011517134949617369807824269408040645009225068239337956837176121539057082385854829100
4571562854052309447709233679921158488718865186152442294590478039610523547752945184329182177647040126 7419
0255625784771039687550336529962214488179352640586781333262159209134904418941231255215399597795265706
8257462407649548661139248146700657227719602495734508054087022130787994543507255603602188951099 24837
0331300766185308525391274076920389698099676917560701484757577914252713802209415938774884940496468312
0257215335558064888819994902449124874438702696481904561785671518215321421458688965726342555779576265
2700629297845964506174896425300021092247981808889462516757327350029677147050327998757989962912792874
6607578196479484752392035222645594063126283248417539754817457010657092602259317232087040932749140097 9
7818697026014374214786317938531308155711617801618419885007706082881144649675351462171955213852392511
0419841081641404822044249417827953801735743096369443649506351874820396050910708920984431416784064461
9109367713615057721430047516429278462760612587932649555864471865902984818220160210049977198361803778
2118959565168922533934182635010371363211577812146023709369106321911883916952705363199942535844908602
3980831429872436326258410691536617915011710034582739884502300190025577477431214334215609102142563004
7292952168051282215647987913148361230334611027752033829796984764875107530862096284647678635312023068
6050987180875128226550528198916999445054582528742547597064034022960335685388694979938282801471099860 87
0595684841215196473343476633762743513560052413653371901147912029922253153318866025153770683310154889
0999536990171736942984470666770482793224482341820676742761811151897477494072558489155464700432
9222039895064945277075532932299426267433298828142825568965733266786575619337195939908066429191475031 1
1328446751487396357728793429897564133604265814305092134813679852888292572515952463686670526471898396
1963598492113415695319863804558148479870878695292727918404522943010294283671108395063484250630
8555865392327654528622726648834185074874139385493074173027955503102908136706943028970120253177018 4
9418209149792382963165602229836198034820405795231938132665460972135883930448521662872143525751137236
8450592929824956114742268292183057942653346922593252263720317019199984476757154104896916082005560929
3880438561539393886108000388931415514251988072813467570939331543005256903783714451599730234429145056
9986414675950303181271527278942502383466726331733374558809943544395003677567031091909407161141154055
7340884834283353399319214437106006859366026469934657200936031623444263821514244283917159614658131742
4731705156325348237931474944920069795480164682189884332759152807095398821268398936423181065489044804
8927684892185471067697886130333352449811386163739573102680251642313394115498106267049975020136128537
1679450641310008468510052112163995839031402261910807289909816307020635413111542266607869548717738 20
2141054740140240326319645639300962185517129143265244274594414418275602144551534404540028743249577705
8449408098403302725818096319765249173570710494846684172536591289949476236370059804890396763138240276
9423666262269015147894711690498025494902525878222068537260457330359730907758353931703088870854234953
1223031802292879599064501773733442130627431431286152209051808992309090028791205122311868460241332 78
6097500246750791712605577404575292564364823259159993166440590914614802860025482343871969033 7398705
3600376311002385030885515627394262315739071459324738698502972134778226071168099600604407499093418534
0057217199340913082088234123050761235269621245455084222401498616621573160299790401496699491727462378

267301789409424959606884377389474315405894732714701696198582509128215714053085784661918613228061036503420039013081403792861284302253788112844894688305371033031235700434623165907687630898083954159168520545880094762371192788990746785753669928142352470441171665252891003637559735588443545254237465176941851447073145254768769289970059546134968039738232936238998019862056505624606681310225203680583248433496956896531003852375466758785679940094642537266615054675709129876236192670083598762678555030286306687993922650993773988460839066820820662301246129973037801157482964498260524705902013885624414634105814533631328256053205200592926480830313212035078776629584645441891695373424281390351629291422504839036261444255735810748683489251853965653149819630551724199950351323821951186967238515145833934908285975599196697217045343043273161935718967626863978978868443886119041809441858258695796026345630375524775831115431147620359386985073776707427651082481928976810133856997861434244300067684394207595549574076093250167823992960045740357758333106586041469085084839959816766251290531530322439401172444887403452240718053522187971473853132168896260285312534875858556725137078186863521169633110572031722370604678119508611426666560265806290848488172315565972646242491374667624461437195284569352215618440020898130159548600299025188585073537307145392372312318131849179367341453991341102203556541051010772435986055631748277847240137983476330972453826070078510252638166027436710934976867796097426453923241963379897682484422098019131330548596452868614246290873537053411643476969647296675258300786930930798498427116879044083381863267759733803325268242243329808073950641806716560271426821123031234294855165365321526945761064682470070767932911011196016546877992087415062186924352558605660814350070546365728251289908986529882762656629530743770361058062226645479965700376271442571577258126868254662964687239956406509728324077614389187704249767623422145130917936877406629168518390246803136439897416961358829185973655389064577513081193455198291814912845765152299640726298338964187393802463224707363927685951291328962522675690312898982088924713495701429446868226555957527745822828429972809777383212732530514169924652225066614655913575985554748942601571238087794284727911437223671115303827477533414242407179488602028343057759385860474399509433265113648775432422147266313648590577719435372785810437183379009515798179631135089599605570568421753214407989555923184808524878083675500683459493934247832856563524764189493364922492977088581624231502049139804359004981004314932330105944902474718476398931631684443474047505681172374643921613431838126854709402109279571879936931386000038468514368458211668002182820129991748092639890989566539405829588860028274480269991698010642404924265810759212724690630995245607421661169418389720402845533170733152315114205675094965756120390076831567476272107045735661718784029624872115917692632706339138767846553619998598318230451455400969742468503775012008321332069588601973051218708541843880610883814334713432566689792627702443419921807686310016485733840408240758539913655790811783908542062933068122947293335597163936794620335391527090063720694673423381513834269471410243982243883086676898962441571506447479056692331808524642956292310160051118640956880068052259396549137978135636662786584825404606873151675168565871305305211587055927383753478689080818447224684035527700001191254671233038886156296696015872181330247282478791621978491164332860672224463528831053208583604805147849852094422354984713141131881735244076916938817007803170683363615105313355153149059846623655392600802125833632096679155544462962234692244319084746391935174960159592862044537991279244395476191279923041584432891273052169124081319234088661680918543382424029245446101708827516088104274721924920945394697026860285263079403107069099130922043115964504298191085799442609001524514242833415815488227675820036415251050279035800048053428092336998208941459463316935938990700155598350644690822444986197170443076286932014459515648487770125948501151405397979633930100438405319067907436471299584718332138871109099823566722126380548683874189401010754001378810183931591488087478658787950015155396440761625775081863582179310117726144046668884818211945604405137924883197154571243841934389672475514442366589364718518226071883638750745929451981123641178097230743409552194140524462200902221666865483560810185144317155508745203070668135820388435215498794496039494729606547080639824562739470639740640513789295335728586689990085990869067039434348708008176071405043256903200881536947167076099967680837875260810732992479498173332139531838778328235916737012583124289064735474215108782551498414142350130167267508217911100255792251742760073937313921421039260329708107698190589880389161813825799144443221052915690578874758211103826605498946693764305728200945740752925336116365750936006367349850284388726581710725994107849303081301175864861158329066057327427060306999694824005868640332505107457077136203520686864946234411905631680969892406954212565563981942471888922035224322609723175488853172826677391037572371010673933766380898115584267604685203620367240789149018792735561807661755878181538307145722038817729718759389566296149750567853450813459765928283718236719243140097410543729784451016257543958762457903820890293493150148104836451353888928030438603690164265

89231622072189666519355916274979550480092555159519046186067918808187226970297962202088800248778711779
1258102117894250727613197954310962466401977209723226100268637455300186190885982565656265279366085407
8228115251551182617048217569561541627851396377824799523943593124697524369059218677942638333073373920
9231890347396557573966490982052618777270846169558736145592952198126893316443382397314238472615945308
8525408871886577640223851051087973101722522845299769476173758249952632471411817301602018331119181122
2813508856982100481816114616760485187369995611471710489695145284969675812583453612978034773231329653
5965223906335742142020499047977949714036872153280706435798771212837235867186123284810142893866885703
7620723484479675929144219538642201210820779887655110250311529371693291996515388780425162562057124240
8504228844997762389469269631528657293741343016936007603532936970641782778773779306501056973945611621
9240336676403090783541455138316434889449979237557818750144552064869347130864983300738980880568943846
0652073487595248873814528654331188696637177260145990129962451213621618212497585510924608606703292781
9860265252326322826483572273239183834925820212759389405751530411273628188361027150253648154993700962
4982612448826291798672824976494874375237721882327023286078786741446467607560018209489067346574075767
6455255636242430212091679637261321178552432544516726617845496108737903746066329417692646345850239275
5446300306386677951156431097661700950080536746629214903192289315461344795220517567305078224066541216
4309774725130449522884992542381654379640699109303726768644369877159674558387881782369993907291362951
1356585921262470605123799207098445306806751021923597625803820917959940362293224160002144948306949888
4798402571157588664260148826172477609227552782493180061245468044327036949428051996932474056765621398
6896794541045778067938189867874389603154930808754485087084139693776973877043177428723052213171260
5071600536828028137772342999917730050094327364174874089319317228940480334599855226410892982215449118
1471364885180832952437115609605107576510978703876312814395958783094617349039834758336946605791545
2058959144056918656906427650852458957058774828812041326123928584873523560338104314000357289123787862
7432502504636500743890095572740636949446188365286660543726152454172205829830850965936482841958696169
1157259324198273583245853831440167068161487347508612648751989780776187917496827001171248189184661389
8797807611495726137823984153462245734739133766906177880605463474924802150100010241121217139454356212
0000272927378809068638911983351608116287091044323730701597837680528428190952153900899546814560480934
7727238999210771270883119994083118190220414788829239645428650979881116428598216955662881290310706500
0093202373656256754844374179931102181762310080201404616684899134498942412244970191755576492585423992
3972482065041525370392288105886341191972034439414870165759984553384786876567773477091766714680382367
3134016398284078802053495468329590014417804615539401292235930446998544589414479097524012233423549
6542515475897214184893791435027789414016885281649814395529829554011283072905955477612250008820025122
7256136937374376926594900873746044344756874791717024790073562716781833766636490104058900490891903746
4562017867738315369910228400082287598197485352472052014372421047928489767599705577863311114044769807
8827521309539908080311709187392213784736322633200766495163505108258978872205344514715051636392345586
2358747744662242736212811092579582876119166254153684657318540677262662818074645612365270570261399929
1616530519804652650305545484991794037554110839891390539496061499562413086576574402038396516635578773
0617890614207972461756714788535806936170019844579843509116551020022736519845348479824964461874462583
0476746482617361233386321767888312504912779200926688484238181854808972739446180692274654904673210947
4947459117309631426559931051181446498491646573488929909562360569180745378234301662743162359500345076
3718679986930718629506983310875567687675209574031434859002288112378236242261019467416930842492955818
8229097115975301249041505839344052802235065103086218445972740352939276989646953854696237284285171887
0062273437169764549988286823553023276435037635204423303121918594735894560426683544961268990092401308
9983378701351245239487953845969039342470609246933983359445142833816687683906014541970792247310862821
6859276547233642121807873842997293257587252274196921100888994834956422769402313427533912447443241958
9510468682513841571172266308178034083469444691605476793293894204339662086472722713405850346947669837
8481116592561577153522840427648249729162557314786408820040141917345096547770301284755681410163286181
3732440985906856266371605889070719967656969369557771869700083478267984657239831280624856641195835594

66864633682410147066349690545555892651629622100768416618973454423248558753157654115752350824419774405089409900945155868296661067096709024218856726351302207694616474196101248350048832439909968585541868766773096136362855749905316728342790555208749417006678209583055699624947773343363507138316650091609944586030701991206362094726162816281543050557973001978755133788352283359301917015024377492382939563793589629549538010617350700506211537941161369245401779331106731368079062915857711534352906341740273374652512188369391593552057152548039141014116076074932551286532242006682870834932947027112382304481910806367017055551681660516170852577538979942707296686930460524896574146620726150470861373723724645325437825575900925675384226193084201930141705544579737519753004704168611858973324224553893793995889725739655857866212042113770585010041536324733806348493087817055660615334112886073767434320714040687517684530102491894478822892467092476454535781147272683411981251579374793859013653631922285377250974594891650089063531037329346048208127547665544803346897509941733772273008800883513861216396087695080332774463275202112283366389075701736586883407143249573763761318902317957201193318346840757130035560035891019144267179484864097507034891571557880121672194813174243874698484019368355064257538885953635406457845952580263489405368485245661207135202821836091210728244860357098485986214991107205603824591948598298235436647477244318744362593852443193498859769508476998303011442779166402353624989613466561951206330022700964314289141781565157624136606794086442773931508764532782000051431144169384502136024538595273582412808228405148007249950509504215819802203211911703806022762285315127782250798281601848480446388454238187782171045347616593762607297088842176998044060306095071630593792353372760325679084296637149031844698075924229306159288757880937545566441609888103227332432611609552152055865572880841228200008752923083828398024930370185292062238770987704278101256226828645825820748937450657346805895735712702446993304075235454468384826117949774906522988063969154916194565753417171446656667761558913022855172493749891423154404203600837602319326891099755462658514424130691699988162808907485969048306877241500910924822607221195366278356268880666917145514720406546986724461341839275466663949242227578726053861942662037939092642695130808143167659282504876472104863583938050930913730692084838813775627306411491551091108441252688428224477997658280451670888789610015596232545784169460313922987805452872155967498040850462270826259860992156066567760591964786619653301861033873012068186269889831755538563493334962036254588905835397049964181414344002208060299861413468584239139340153564232472142530142859120565763856935627762470877172637820785953181433146684288614405778794843941854414533901189633756637537059024278787829869814914979048998218102052906879424909942849374392239729358949375634644471174134618118763775567553171353328694467886633685252997791338884062520535576727452268759094250269880262302920259198007187271534390507570920489474278808112161753705708177786065723273944368967597196132409349773380717748506768066110002497094831775850489613168171047157601328115997979669711796489374936580182829314688532667068635647384591938163006141247385388455900458292731183045319492375073949535971393532734157355851621085639294721898095939341482083584596933469183187941980144332181463352839107766288844960868696378265107375503636164492166918087904910333994848244220774508737919354745696156441020723233200904265039497045947946721723223141046197781792322796891151603338708596999475128428050100075079366668644883362672503775902521102625881334848928731366640405934702802062967005570773406220297557733498918427780848503294924150006563621022687299216940843023786907112148745940890504760915559737701406543115136419939101532216931977506053626463245629378601040851025972253251730926509659234796949922483317681165948060880008261892370405949807704001739674589507160597168771331001039785153353453147106859570944506057335919536586171963149700601821601511678164458777993723274548437021671969832403521589516853551392478205591282796153491663015145805174884114809718670698427226473253962785688562186013575514802584119125786169815402760912705074195114039656379289828828522740860145039583521126538953883867793995172958629270947841273911524331129259464000726134649249972681810945151440336063005287128811763455747440049685242334339885902991306423665017309231172549322356936462708264736539614701936146564880823663856871274311814628024296708416102187589860964969235031298244682022862410158114833808695873215776018223789618015767799892633527484089573104084610685377186963984413185846115041482873050231180669598634179338953174698177066866352511778744575853111495225289209150408014527455591384711330427158913553411142349566732404385307357246228684692228408379373448470602916936013993040905065459619003081838020572282563082289372675275298952076633889581100393467050195502721525407078422203158690015995130826384841974877768499169239221965372316654475744076410100996506171262795901780988192588848779270810165459373325352841280200757510482591165251375783401008951847685249101862211735553486363969841574019955602865124223618740583409084470687897154212595303559132707207606150897381979948118563835674506257294695134369167891826306650795243169686

15582636437199760377444620187910161354244770062736476565491962676164446807721909229290808662281129 59
884924781148671534257155367603119524547561565236967703853704768866046980213852480305611763820687 5750
67581364128628780137765140619072670023309784587766918228333916512907870760866937517711689174162 769080
9551757279698052016821027162621360697060416091561788705305407209797684930266376389686627360501 139889
883673359135139603154742889747138541696939293183086800212328963295993812716537221185295145016710 8432
074694204902097667007422827466722215683386341302328609110460240156123049845591409523921046097565 4364
834446150855037499705955336660221346260483996012273273641910694943992246524570005239136227333383 2849
7275573933564959431552309012332305634341523510569628228677882109308461215686611444387076095828 406715
18753498089714331027439670232072868301053055854740611207109035935222453189737748928721223938874 43165
48020903900810810885266922772461931438253290972586311246157795502759822070616687080487695005424 198755
620466992746639886393484868401422335287282144403954784228917454765827194543973721716746000608536 8620
66597343051524283327289391251077992281717140591722360207385278926420803737929278643557802115835 75735
8359060207378135471081449504904591832117553015450198368358598699232679936322436043562460202847 050065
834538669263343147898299146785381491643329373377064728561197775471780210572065624167806204924 9605068
1014505544400083691585386399193466330728633212102715925478049198906708906687151219434944169901 5818
121629691645296566728882261734646998743806066636501254919352995156342670614836862978791948285 8285181
2784137702694022167558068283567621527125100272253217550117539167618570824410005697863552015457 177509
609839035001282512873029138851759746097154637485158162912190028421493728409410165155706168620 646222
073668955439393145435182266250177317479799689206120653252754184139144761139221521346140692864 19355821
654479670253694488544417123943334998602980270985027141815954167802543245054531516922268832227 680570584
165710882489400810598243638806930494014003605405882514727455882582328529765554366006097614188 1745
43950683518943064407677097341582749754334607411032688197363251029598689127139183273239406546776 520
5701268560215837385465875625128673719730520083545205711511563796715242543756123193671768567009 0063
735425030945510191851196229951403259743792093543742091184315511684985132549158654381928399557 383125
945367248250671147389629672990878885848882599701074227252504035481804962852257052082363485122 6938315
6280672836714743505598392797326066842950726351382715619746889767672591306512365731719532432 3101
2869561072580528336315890095826698417353963848720165200490649064035498544923722942741837841 103308077
7386940393450466333220017422240536339407527505982935749108698818541052273129910222333976612 5332246907
3008027004902589119786609546141221334121389196263882666178655614634421081326330771760786638 06363424 3
19091832541943227473241152377048775613809279604835764996953687485113088070624761273257503766 59107831
658785768550661988397945830269086027019641122022573660586407392914387722589970473598764473 315527068
4661916708199094210860851403328060178785143014996799473634326222918206903244530152501557857 6015228
847367151823494492654828857179769926316352753845576557644067754489066824899661345595218009 7814220867
268863056429154013829653468573489967197242017176041456299913025141444752488487885952192226 8437236453
29643388078851106993397696127782691625463097323535451462264137652139297506809924776597347 26126909397
8722139280745585647099981589429668515562704033660587623053403109834593229001269832269205 02220158143 43
0280800022365801546328825529047068250736300630781189568719809407671585126103344019382962 71808600861
169707239670394318126130974247111717901985743012399075403547165062880512649200956055134 533362029263 0
6649809154087581763402129960016194911103493751576558875248272142376797264256426977368198 89080882172269
9265553049872479047981000650468434265519305954864531596583419325601995394272542463731715 15176352439
51403807424423857724765659615245674669931628602826485504068591224634076966397694270136 69388291856076
758200866943833756314593294171455199666977282218147062090620349763171208646603060567561 4933109167359
686926299803365159852130088802404618633186087792966500277922134760116174813520006025989 51502189632
6415385131177521071464107072577934971822572679005993370186596157932780066176911181676147 1579646487
410695224537611456887923115088630362384071769195998009003565807547270107199166447759227 7306588961119 39
4275860246657291646477829595177925217284078607443295113156581166361862904473631555156558 23373410205 4
10970553981055118587774847255317701694561977327400076274662916015578149893849915480070015 88028727 13
9647788555334050152578884671964802052467242449395518643725584409996006306300289818578112 7353194003 44
4827534004296065595089225535186015542228917642946353097283905715525192200671822520448345 69572830242858
158777835437195126859654267081696079823466161792592607356189398955902895422982006901135 01199927833705
76175407125834805827806812389428561931111141325049331642605280233039814169006798898037 37603209951643
848905869936906083071012317961526692453046492374632284627990824233479833037403445017864 48039881275 59
2259020465366084691021379484236888145262969686621356647670201426725320564435031311744 41377135146736

71697265452265465106144038202743516102254601111621957755376909012981481368832862481603668992773908334
94051471433848747692011642619481923649494632004463847468847911151704484391938676530311008242387049812
52452040553796205818316248694503295350919858056458989087442568312292543450747373415195274236476418024
81995487317377212901252421446754042945855402423906471170225700446424875936387663670677479620244276809
43778548257665128437769009314209866607101870293389533104377328530343716889518480254779912733313963367
67250697624004443998428715482735400213622800363878357808619303281309994905965891889375075383426925783
32089037583246284679499477201616184924480670720606423943232192320717600372523136002600793811788852452
0504577314925350249739942544051399099724277891851892389555672490883322532107988627081564000322527803
15109766866228334677738349417461225894542098000029093296907437726013869091027914062598515250597766818
44010781670669340007514934828540555614304755391105331437574642294862097446790844758764683631787927730
85598851113550957947449542186298667749669615947444409273113100649178611380859054534176264994590167819
69714986139949226972233845270514380938951350464437551225486111708913968808936677797309943848808619331
90910435786072084967307382593792396359932847510040727279386267167044754075719202503934191972545189483
117279024614049668955179079963610253764890964845981670901713935120678283957996122631197331491377818343
84194318523675386629019004633433285328343418249210719160927673080132353694726484892764429797086112565
46280611726046073321917630474121506403020178932057956876051027752505304361222709604675845312738661652
14241940868340837589140095114133292545770094731748116885364091483528610997103416723555817860282349820
20314998679401740417191402839362105194806048190182451800817013710421974592127980404242689630053335368
54941640086392177456769534458630885282629821658460164507800003598985212901239590070420266074498878506
27176077622255060639074531947718892509905810093672989195409957663353865837809804479335379918771214297
74619764094727212140235326551777236163419660743763915386601945017769140062406841273162958608063506762
93775125293436759607342719967513520158008737395483897343990123825656829078591144258825821366169201402
99004131462136806323526508654112180847998758441118362909790891983411875376649604809362648932910439059
77256329558138877674469546181579787041915975583350003772867246073518535088549546420293097158563694978
576404837410555809918840209343335128195514555185733348881649167694427784205554336977311920143722541519
27450597601379841443735994202870652555718937804114892612579830361838010771100500423923944156922779005
70279474013649422918479369253543248404878023177744518417795835582575476184254957539587854945834308480
44173087984926741952893230314895899160068542287957892948159000687353733333338638139084209948725511726
11710872944088868447850811634558526921415402996720988693715886232220314855817426826359506893623448134
22473784039460092325660075472125674141034218490131189639676547329167424718993424330280290926950752294
90779044300923867287743936255111313628761591973832413491672226122826312779381175080740895894397190025
42641266494798017644201645415881317601897265962461441391319206785035246383066573204708920114691713372
28786303703845313419426414249669498474404861773851352937310809985947281511173650719139812693074826906
14039286422766278494329558938372148475070773855119896224911955882370450645820056101702053248444902150
22945617655618592145743990095548229413751544136132915043046914179941224609338165136262790282788421381
22077589420438840462294331727989259418366829367925960422418459417706530195486439481703355241028740744
73117150703477508343237333133258767276368346858444865625391872394608273247135004489570868031503250386
76401735315079400345572855037001845602769961506156829141617610046161740824846267035518915255188248217
26007259576567889125678702014978967908471066310748764673098909147197987925985906257336497349830132609
83098734868162739358057907354908246830497284097732381167082491516373468087051052191917620541698826254
76054458177111799377677869654216992578557720426344244304207458970339043450720611900776931540912532813
92832583120528240037464854590928048140870719534254761388363353511911214414314850552012358138062649231
35383387675809188923797585515732036568831762424116916752381458859207516403523668437267917590637053921
97981126597708113947351671962999970520901709075898569163890664270423570074450277512010409039481482945
74438097462603510578981954076398706078758177407249118450697884299413834206081242839014881487259985418
14802929247922787832434356105549136410091714887706706782011985910958090983868839551718380134914824992
74914269855259517765361262421572662448896091482979703084202105160279671478561594064613638247758902011
02519923421532100601752302574212237542107495918728675189552155329945322689425188409422826757744222275
52820761560727710394718256680247106606773831206303146628474433862044743526565056892871626532908328372
87843996507162442206138294532916060436087256305952415328812093700992298960069813962796862795676351987
41824018562931984922623336431899309017048819952588273388079532658851449393812411543273205891645786422
92452684117518808184050956904691381344307784890211487971211628397734430548461498079356243549671412947
3332666422

4830500436445454702749172320783995650868018761732703319065657954753952069292520153070643504713068812
8903205385455639879921010595667298287304794661005431635846234487441555407133814323441311788424196019 0
3702868696242276246502400420813713504640159934720532755679503533906712721617790603021623997804185763
3481005448388703173071625525647995999986935531968130101562970978711673069649402749166572674943320813
2314613886925533494929411831638778899032594010934111572742980133181457828168791351424395578734205915
6145877368988080791707114551154762661682081777574724842787917129815482385204369591652405437305246672 8
7313030192582952498763932098573120916105613572201763927169959868610886760613384364961695186548486046
1684848240743838118074173422244839478939982870147342230578654102177292648209140949850350132241515599
9901630714781485270516144322582547801434401095336602393626424931385229405475364168364670415743005583
7298470401814557100296962346985059977885573699802182230354923831879762989415697721236384408891789275
27752847144776218920574254531510374799777068623624218990630309628221177321970823032887258358929
1140389896916662207787683260123961741999762365506363266016990661168908868774133378091951614977054921
41711819110149543478693820620980827878957313278990600208633911278471536843043814981509703047708868 81
6359741925341341221913962874578109934248327141091787631765420963125071366926365882135751199875401 15
79028028507806698333841833826673945536831965657170319462933641092726672742449927473229138009835744 50
9134079930856836048467786653542105866800521154264867497239537936421566535872081855412598194547427516
1184169366255632419395158546750889053531412183366239987802211504245660108159974192025443787
361022671294122829888805336794395032940480496167711072878810316146773037150218352890241464817543095 4
9059448875410425016804069999819671937982082773346485581585910776178828497505468567913990401138456243
60403574402461912376944022007604214335733872603925372466408966491471465473072195292737417679613565 2
330678042682000242307412186674866946280587887873829952327805010706610138540288011913855730911380572 7
558936819403093807188637937582154443516265599120308784882053239749518955971581551828048148078313105 3
50968235671612355096316152866585785901952871775464250641298494510573035332433489097997209457291700 95
878408322140440501474114381709565771252671577981709767638478642064876136793945784423858603564966446 3
14013104866705613462769017492833229904327112081922956415894815282655842219126189025305138159118034231 0
8851491226653917944817433904322700506874288819967061960688500673328353614680498596081402879119631127
482546640437038478288400798157913445232101668673741121294496813451012237984294021947946916648150312 4
31873969119602156712114227943068201040876756809373898205464812477688201880717174141274937 46
78175132277109046730015309467773069223290953495694491601169046818182723477816197512572785155835698616
292202918847453026983015663874917702194760509417486684119727660444085483297504989078258061 5390200301
035218027179209950817839121233970070027832906193713368218336961540826520392716151802279152864076 8315
03299740609280483105062551572262871313276818247801169154937862483224017238370727160886419294211817 18
564674811856940087052996113141059458293870330528620799182649324742017955124423294762903488349 5272922
83080401222637227611027520380012093752088562406453112503726401539964371233707907434951203202853502 54
74827798093030202196632509329094488190574510298521728306996224219547101651939329769370654576333432 43
325643313639122108352913555770081265053979316468149451341207512188305473965395358780919473745666 27
9031868161713641615215540774535438041479845458406344747450856802808623241269693993112296405603273109
2793849169187933359614853517001814969826164800344027822813985879014862852128674617538485034806838652
01680747674402114766550964387396753796644295809577591541926150719665373410817064974822204191 35035
2239320369277923907369558805779975526019030814355084924759818577979802981264125192698083107065746 94
6935112856279759057800341932347600136961144729013113729326518877314672141074127522101051515586571349
391276885573006456566639352535450945737896835800270721080754685197902156776635596085899524492722
0497752998154586253592955688586552190862034885742944354883401766168305526602458359945433095297362 0
5642829474539664101759387807875790401431956726445025653896405200714870685370566271174667651719181064
228046438268670641781701710145579987859492981028674877417901743550399316968352852911648148786128 6
28911637599276847108874366161776239309829498234064137828071121741237834222414406998486328823924065 63
7743898084631704339814190170417704585646785993494117367101851048288345847591129859911326180631166520
9771093260254577766404781139798120273962790521780796565934383019638637263605481726175440262684357 26
45147411143619747488532813525432313210406553747535609140513357644845126503159949939847591233867625 89
98862826742421410656317028409052554785278498062554777967309562135010751350710814376712097 36106
56933898112877059684395086822635241957989987524453151326843577125269551165684260624979530056413423679 4
0326869432022127715528452294559060131663002332978184272936970542564089572247314796084227990643121 1
5572289888413406901570607916985539706270694986922613155059545812406654276265309898824692284240221 7

051092088452264193800405064723514106544629397362950062687069185743503468907870252668899031 0590307332
020997707019556267080078421417500402487764369382704966754069144895634000929852354991296996 0465402832
129951288171829317338025893433608886395618814530549363321022467023712349079740681256872488 3352801820
879686837674893286595156350015843734542714861534249712522615178086471835031991909993217819 9155353873785
623455087143139050384328696513876903290534015238428243963913256867292722487084779696809273 6560421516
904023007501736619499358544714404152214106474972421615230372466125530179653345435408898008 6901630718
915880845552219843157422312945196703758620587762633366848507527528479034425016537649163317 92832073
572043631496293326217198412041496939002157898758405506883631318978280327068182062011470970 7824647760
276296351324513522519224278604476601784554635868956418344082595532543340887101615107036746 8245292204
678084853185295603026033369987912529602307327858469371627037491448389163375005262169733233 21524007
087580984856462897791809848253427212024812725655515050305700418119895186092167870927093724 148047112
032092752163821969300536677508468358938654528152235347557559076380237852151946018272456823 639549056
812593274728954569893447114180981315364959748515410685460978633032486429877701863553136745 0811076763
860473346402957688211528898135664493094910435890364514131755241478099235305914813095314712 262341354
064609059803930072954096921369456449857878131484983471061858858025833538008663482207458754 7577692067
410067517787021480249068099354030409780869747523182136878867192069420344186215299004686685 357497787
101475284582152153509827098635277035756293849313931604238716899073982395618988722188420611 902770942
990322757366871860797707077215592326418136016971727774070728841345717732062530298816172640 64980178019
269641550387293286248745115669440090146910371466589528076907506309916580190402824912396345 6798392695
655715947613669438739333006837383081812309420441592959517187192492679792971295029491581185 2694465448
881606448909108574738497225266256135983379178319495991506543998379398350900517433878817182 955161844
732570694229373536525515341711785761849777477973316771607149942145263817047522727295570500 1733744013560
297812168718861307295720086827047281509944936557613530068382030072487238918604555823405676 3187626236
258904340796023955291440818688735039163729934716281576586283465444767189779664009434202490 164341084
863053640849409128877119965069804578807137977951591204253802570919860455967388601901634414 8290009729
713922827786387126308017669213704934976904380819565188012979091465075440684531805169404962 484549525
186336324222226145869950092979045626886597039309839695602356438968632992332376590811424860 358035451
697820676948892486638330687917659426382009263778903165827050291137857297509740049386753580 5163232169
847363385841973520487633945535946427392885249745154594738849977874365758947077177943562070 743
257228553610149292967565026585100600321814615361292859106934321487868637434297702059825451 0958054174
455625081164881956344323355550915407022873450617663489800449281114848603366735880522498027 6245065806
416616460933096583227741201687004618704873546075946002355373436366858829099159321746543217 351
175531575551018630965885602744732881743617645365592624617643877740379976886080184145764476 4303571298
399555691565594231174530659483625020603643822173544096553539509045743693865510431293427166 58105214647
736285604104271523729397812487106059859320178075773843670653411353760558860836202347780649 8797585
317840167290597876570281387580260323333798832030389685399914520291291252264234896619763397 1826810602
370959756662980918534819012504031589962520471707676639986835315319660868752912542311537503 4755005153
674068684699567677330577510792350490345778778263396473876429056410443331682245650903957006 685608
251827330581904243226112894851491882214253968613100337137224301868395596441734670841783204 0075571512
416808375541314824792400454254308125367729689704440769431797426815935920313582151463967892 38533146916
848019237856136657063609422132962567182221408537658070684361685834963756556481521327880020 8251696180
798484047486754286481247189232457274880995178492002381321495055976860978043927271365667914 325583872
939628855575319898513390471237808455940718448328361153361578025705836438593371292347805105 80916017579
237173588327126164560656794503480990799720508422730547419077811064940216016666295094593246 968082788
587382689608487744561183798448657304543205368926840349142345088153510167857570666874956376 0666307293
636497984091052840084546233261781814979761412181872861268653460711660595135477832306274118 8124807717
164665265427099148478807776892823457954861106715049473647399736480665299537367799934023532 7496480103
446000856397693373322947146973027764094077539628839012582466293969389877015862089379918551 6933084
638159965395913392629245005644067609742316093027155970074441832783863920336711950419266647 28752278979
549277034150826391547896926883308166047956420911237931251245238262891323911425925924887235 5135341403
716733799306171199648037407664440361082778150937198926678955883750525547609294200184307444 3896493496
814253542442287548263467259939978795979725472244117287600771260884093950481192159674887395 5586267106
549030213240224770792652497769449552051080564104822068021729276356152237738510636020972031 3789483890

87608773044443391070377138713316374909282154921296609435599576542505065970239239800133971925439883533
58193260423553346407090728580486521415339035222970604712201304941234553577895471248630004705562889455
50446086222664852384731073317223038564186531180217935120788367893163243192580644850051145282276974 70
28812975567230419389696691151684034252191266686056130038452053893369112100812050252410879522034623 66
01390629409586806381048615670349189528899594434218075363312957022412607656628378577631058413324970 08
02109639513491608642051940551379803738265147259899902124751359728276487048030240957692462436925511 13
92235317869482928421855572437473618265779560035189015876875981442667626582583777353798801168581116 1
18854587199032823772576597479255550151703522520432959456689504597798593133546030580462871210509920 02
63591999022032802985513499055895296203508118519116239331768415451061061516247882198069414695314021 63
85429484650993057590184254772587685764147491708005076586416337695814660781872244505837942437576287
08545403350110537092671021601688974251901726647710911877340196587613192382440167786694473160364332 35
91053046392692548647340576221166484269983303050481743973119350053676164505084476304732563764216661 3
98181721671531112891024558516207319700700311254050520823829089312875756421875416786156917970090093 8
05014298466685212001591511327205923213373624510092718572734941352894812323115992346158969097896744 54
13062762687256330468233531371596513000406153320056264213843223148270354735863509198844653383568400 48
41684653835147529872668521047590091051673159246264853570708194303523598755049297733307531722035573 47
02365170793159327874445498164459324739383779436527800310927753543734929701120308885081138511036298 56
20594883469615636232082831084250287376136412864750349795749821634469594920423203866360959579652731 2
50489120123599962848087517002713559088212824459212365972636236219317174993062078006727038520443869 4
42023962659430711733875448439649357394016544694254389183159395967500993298475531832019709586811147403
41146430774907873234732383184316977580069510475962534780392903312278972028317103070441520409620275 3
60616877509538749135648654899352851081375466230230104416440607874337993886594343197397276280988725 5
35964218368862525286842920940766935310174676061436150402241061006575183857818459430973816665274228 22
35952326563156004363460606617992954342681278246980231107831922175488282484434894215590509624968095 51
68161417840810580681730090026315272179607472345797447534390082662384080704887676620125669754226324 76
84144360299612706320364247672556623958413633466963424018339637852422544743063280535146449208028401 169
35071719251343449175916344648154738160058964567608405798016590255748976711877225395270001783834618 07
68335742506196377939729071866416577565549223989511605044361810205941685054187658471590296259403076 77
18693860546243793895291773829741484639269221118535479059909597299760747387296264727605210459511287
43563127096059502375731806195452598765319888191583707806094607301046153140796722525577786092730919 0
24691929884517398716315076857663518649379762973806603165031986181720651187867335919765239567443188 87
95336522090886520080140499811495657664501357148699739221946031656890119202263137063954921500008096808 0
49365156179044413993654732119422707627168066232679931722181257457609152455167050976549279875928496 6
16737619368534354686042600792782105050293510602699110772925833223313894239808716540504631465861787 62
19949974969729059791579184648037035187388643820270609804934004556786093076538261926526318112021863 85
92516569258659483274483164346145940369617235805971795888290583815236978853316823013314522929991540 59
23022740585503740628535904597657859645738065216026315917742888124801830766646377077199854694714324 6
04154308710677882187690707969413136784605770183828392830947916091577622835129330815994526976068402 65
94233928858711054408683063663858102047363209449653498877627578626990535391553279118811926335106919 31
27339766674091486139745614525452756883805182825266087134962482966457024358189231686226499499218634983
60741028893034546596253808696255263711592359482638454560872583686037619817067263374481287637268299
88777229445404077067899401643812500275542016983023069545116299831343107188990986254101600293552441 54
86259137615974787091851534025688811301989218844566111649106326888324234483096288049657974271034082 2
37366738282577250767715305180616483063355980193407109897163098840874672591044598887595960530329248
89936385211939490968283710864571352345601580438642971035538466830596898796808358734978576267855658 83
83649516272354520708779533467658672049191672087981314914169891375664645456974216663388028870931159655
29089096858324662094236872595520958765867639645258106276095395529149176518173767291898766779842511 92
16807552581600469267289283769650286492087987822244210714046467307485519743713976050374322345155253 212
12358708516162852145380083686124629515480394533296837736424862964570243581892316862264994992186349 83
24681214946737511994819504507449659704327176040133835522360815340004795981936852230925881457545979 5
66969476238949572473252484866070654288256287673592883976141961905913290839946916384100532188918694 3
25657206677652332824731643807976960444572969446502834306166522169870802617575705124164111092073577 26
20397041168885940376105843313029314482906595568429648729637716132195073219971641441702565468946101 2

703082459723841828360403116603489153716076232863966616561856000946715549151974673084232558676741097334029846169675642107632231536789799885806831989228308597633750929464903879107437077400846349362362400830084950073558164966916759777865808111823047707424696436047327935182038471988961170359008300455685125280509551066769960237387823537663727822677405205106153201899838061149198595287500266294554227353260581489488099794079523817886224433013323645543257410388370423239809214161496327553956636176867524381765566251013374304648082075528515649116718283454130472387959848889991562759190821521959453589439873919878854478785588393019531703471240250072072868196663188915409237562982477373635896293037302748649251369197889067635824615369075723811889000386834089303979375199936538172287366775385114171618214640630053993407936721095094123328350571426528319496746948502122505862740548110939587405372888099652279949131350411485737581899491522734135204022017717426975610324053900259385589578491549406875910851090004456698779029235899957858043995947222843355440391384551575459050101223677094900429072516327250643891141494031439293381604921411482986961512607568119300488516042892565334373068623996218120083759114317308941419899802933772305065225027098497205959635360606930115540542358093562221999017615513369475682897390100935189886891023056203345606748159556373372431505131046388436846166022210580766055016338443951392753905047198111267789378425274293071714287376055437904153578146194855984706040401016336531189306836965939759008802914458513327284730880044871403931718394647185632552751559863968310050334405605278988766114724308920739067717133448999590499556286687043138974556386419506525626338300722560532502544979125893372866069366527475695458554937936549466905956264822166011097222716643362714785994645999926749049549615126423085297640404369503078387567066522228753365772170054417615227578900688449222464725810068854831676168705370471402932937942957597129456134524552754565682449374143313858049509737206075394125741562910110232605185216108762961134662379674361123425177993400596014944329479641646003108837766887059200488620180196337697614505829217689137668547554193811607470643646183550950427836763884472620716138197119177044448797513778892289958847382008296507046223127280621051269910394458936078904429066128912164886732932772124505955764999531645688888539374049657657168809130392423485693764450199912028784002735463631080280448393986446286631590478040102917796877741989286222270285329191455035549524356674461195336689703928030763361342278481300805902452322986453653475937924700977123725148559789622335055037140823738855989996357257681528249857323029102096633655298097518691642928076192749832296521944816960437154884759085527235864404802017581425754005297343879461964319673492934702962918697328305692012266285835941429117643270438839971403798486296507440722652641634634289829191540694582238400532287197813012034284651150002141596938756059895846743378832151044581733491029314084926196254430106947558632673613396013154936002702882265375501691319481721677272774887979071308645912517689622702343482671737444762540360443612383426368529097934730262217743440936100473438325947055617962424701479577599383961282821867040465947367134674003803628906734350560654196291543982323319166341577267125366761272875677647403587825136632401429351230992652418610536526239068980576795900937826721251220551131583478239251417189326668486967410493386288058933141258364873085596852364870957352273551564173160005704356833751285864867987776268789119701741989262975037390169668114553105639638913349194784914126002887272022436295541133021032888312521790386091534116785776233880138563643471230253133125783492149626816843461805655064376864844321601600993297935886697132634594804765801673028762362637540464177371171233375552907616845760984120314814906712654478813087496692652495072837633958248312795542416998444914156090821234201446635611514398786928367644038199996073613063565876406833411029087823685230451771621814804943262467842034037569121810020485713346838360419049331870082565113342259651395183628517923265340031625303117768583043905855303143470009499540428993106200690938429859894462576424364247550200920959821997053713856754024239582249361468818357839052925627662578147628254905215311884516726792585106299641121890474290326898290300195391955649074348181246684338893876522912442311981457466200937808079651108208092025003722176564662265462851678069756165122984690402876381965360231356361364976638902219792361723385491819809573522970142320454913947600203098311826513470831957908274217797322911001498110420929199727079391310541688430564766806828814326393122924847528035155632827957533408431088161697961015384887094900429407026823091533999011018517840853479664576963995643862325907256630756039470551358975851270259376353252345454607115787656777517113231407198493087072741725990383479908377522266238501618900013696653310225295738651936909936258963337910411775860518781920708777656099941098051551760524338544158024148300373807294222057611422187341912017380974191603908969457081286919618687718234413339778059759399170408525740245952795358955168367104612241414848823507042098949464053346102981481814983849628748546950040743248430100423702377026935949778090939009555164281254168379512226340810340240060173185622836711787912863505765122105

2346487547089228654757979919866349134336745731780800010159228032458041560620405881497336905468838305
6831118856426927446668256282605693182711497110790802221674273725205608523980970519287617281829491707
9681084956342092215680052226576590502708101059646896004191594139713443207974875200619703621436531076
8194923146657468333019425617725802965628710574635667936196429335090417591547605643722848231520544883
2642676308711147460636034732834323009419042086748123219633555980892981390137432172319029440338021386
0378466582478492823716028457090185093998630876897463612139222874464372822230689814427108087673984745
1986859733656832475850237902290438743386565016354309204949751394651712042399638048515240438271400499
8736609221595167943665520971624671902878497224369847942146220930654035515146353353438176212140974995 29
0813343668874219608546300562941871851989796549038922609940064581345769898902937525260315947935205 17
2873837166700721547961956375740701285592607563335804361178569570321751086097332660021100882712319916
3759340997123032026138109889964222097317552741425536341306159729484372048567305363345356618321960256
9723677437664041985539084586654773133442430617212600862976052967207859322562292355846262846333189292
8319836100025921871137039971505577493204875978319767136286320184428233653934716363794571277158840984
1768777705511446946524042208853820282498929639508467046570193475974601082109002237981389393889173661
9958702353222196294149913960484299279341165867048778571113372653492856653689288750896260840586049730
8118077965006010068109796230455954801189768561967624238931275652416559831945661471675822057205150 73
3187438649959207729217154115270702515737429327406030388997018176392492844496144989211996008741455920
1197110619178135600831179394370021856788080126118115799464147173035479067670391626448036915230016380
9750954986478577153019332075870767280738241627299798565721981668444519315121472153946260624154 74788
4492531474522234289814443935660965208172835201849104468501236353466594439101771286059078923203698177
8144140657657953141693427566274427564792495722561771122441508205905726084814714045923953079597060427
1724592752165506005137153967772426318254339431649599940488134896294439801928050446990307433295869 6420
2347799440608555298026016876274594816524700132047509569931582510565956554511246812167410441892014971
2917059113204648692995218936078725719895433268702764211207971112580634371790702853656868188130649315
2903586491231366257190281160804499253327418869618739913721183963488792038092148177397515121555 0607
1421237847824951464949328585000899517323133417944919571954937381022516204335585096404429995065494821
8174918298209802850161357646979930679132224784985129587419897623402048276020205113048454320764
3630546829858997903588405564610067458997230028648958952414533037922186076757219603000199840534610423
6967394846398198903322644713641778288583186156410899810603325669588014908754891185529444754877770703
3254281654048701145290259162535198012387414010436036309202528139960818578217507591992044407167990 15338
4458640154205422766030949590985253171464765665886681458842246276815471774897706020161939694778352 9779
5179206393425938196355230388424839581039275670564005177458415453647792017073426820789537842419318 71
8112513400191354907522301885366424561518864030265041520622587526974644830596060008664808467397587 469
4031004430507110080328575618526003370683321783410069730737036403212982503954591566451625180163585488
9250188649041767130862347416451200097805267165914094586669529719432662613627554662763974798831324 20
7660024708256555168463454515732273037984633497435417605339118055495848517100306504714657091740008 2202
2033342084271621025506998295166962322792052067901249991059790172386421758536906421361877237109133881
4995600185227362965360920637316839784737149841162314047954571631028508309649294656142890777360309568
9141375737242200843195367673720751672340211743272646707775707205664465381708610430491301993304747985
9635164887964619449295289058257309790409506277206668335694427988054961983751273274650760182121520 5334
9220808519176266170836135723334251006968310417820719671806050297732577611571650724027365119900861376703
8988665032755774389422758156617307697218364368019421958476811417575765780225941229369376285483761501
1371209444777267883907263765061893274497410926405869319895905855996960009720463567109558642594222507
1342281083161078785456208386552686224968775589567428400965170789743998364521940287896992967688552250
9924548113167981196095844999451283017439565864554253492467331927699229221149864442884274282189623437 7
0714914385776080580848564273037171353764245306793786945790575233507643881521810656607920750515725 2883
9918515710526005616833910853622127568842615368679981807360170875673642170133324263530619346354543601
7260388552556774580621314638205488997794379948659252807324771270170153356672430512874551113052270769 22
6215106526147981396130120548259553984982903822165854000278718281794527593457598160626825669353591091
9702531758789585078624512381690228648440885550462338617680937438712550030626767911446042717502369 0482
6053496822003460862797939752823723519731070856623247325548156691201846615653934486047540357405 73532
6497356594491899073616081176313352491857472402949142219105534669423028663739063083795795265229165823
0260997650740918733400133052343101030377785078752063516583462774483536836655902705583494796 12983505

61239433902475739693050522951308110467091324988204831537894226132235566343098114183098668476810282 57
155028682748383397571925873474011536646716826293701552682244212771105239091796931923912773977176880
15422468834962014009973227575932939177240864634892836464443538761811383414398882353435306237178360 37
72709222067901491315749461235248649803680609448488571198938411919684715248368996081472224187543173 353
33742369490081762643689316367885345454285713963603275099228957125803241463056342872931869971175945 65
83631036168351573519241152917458568654837479206998158899997133603670512517832216587046941565206271
29418830367286169732775954748983270093376170910380551593860779518316479113151091832801275251593651 5
04886079398845860993081193493801277896709108478408470943154894557705504550361821811546026502333598 6
62621190754572843334065465592362626779679009724023458937667270157178745301067836845974624241194146 0
92702551816535744964080370763592182048070317080049503051527919804684148029262375914410938821758454
22300258710509897434053801960809670600172944613159435393190215524986391770306032510939596603820623 56
44413794178842765240643899787215289854873908254194811268497400341380587702931600234975233143783250 6
83468323902948629612891771070449877062308584555830072538581936999562749473165520895617944627647768 82
77511157614704990194822843238680273470746738628147526753337201210107164668619616006337713521378211 38
85721215547739721693304401947166343082628509805531003676041798533591083991213560096404211825428838
26080594511633337459993309854009095675531282250406764600734126635440624826166056357691252799813016 15
92594268469950047247753327706499938464441139295589691465220429908122439062860004650357381269521702
25639216821773367320209158675045004788981687036460961518460048574441398417260973844721274644195450 19
83355932317688295578364580208982952763154907355005465411113719476304362617329785902846513191614666
55167663506411933457586671378181098415646592962365769083627160722338147885085506388631774908723491 9
198767989674636254547126044095416819779922071042764025154843319647739055810496100016470598951578529 4
39919179851978060893956786471973638900985242234000544223763116403151864279380217952656842997814288 38
321389598243939588770155799078349492067091706552034967699860541559021057651477157873786751371254730
44660334994239637877962711768362368962445967764401996978693027895704351674431509802289865012877545 03
93407397242671605002558785498888940993842091001681738748983584562893527517079117054900054362192170 6
4294043278062798737866767302228520183704159213524224275962837503729973495914312511292396853190129753
5611629672308486174323176389298817540247039152280625468479103097962829343042133940228841293104759404
34608356683432324956136475425798625445544988963516713644459397935735007027731767348845174887099668250
810438991556770998488740178829907494844233167883018060597371518899642950324199139680647782140200104 1
1177747896885824191590883768045362608975083421104441911728556321464612372029116740614660357396431901332
3015925808524191590883768045362608975083421104441911728556321464612372029116740614660357396431901332
67351138318086854427530611568220064205077627624316309076339574680273152917599788010125354943612591471 9
887523866639374852539797888793077692813208745702729612199390812546255462541090083430520403871083838
5722642176756302904930171692951152885790854789541342729679084551526624977107434221537659954921693119
0035675588797196959360894440560279853228297459230057292902250191869049960515196346658738457662861655
8370926229549052587150988780191535985441716721357307006359746236465069994385958653045819455 76 68 6141
51886375167295828551947025681486752161213458539326225714469525507007692700883053550536951638370248
262056345211908643609455508786374555688416514747774466068680273209052745681753951571320375462170698 26
26148475907106925055688718403630005305138681095753887480633463299948769553743481082454131098108373 69
0707510452638146157137362030563741206115436983956431136731850971749634340583229072358340161829060 87
165871016013042159191713902364589705902734815681072289867195302579683764563989718769225175933155867 9
7334077373304321675565339107575423891616070212271737592070056307136843747821213528898979866821776283
25725333451583230392747618654930192087965652589286081930704035731099693210894322008519944167611717
5273539788820266187445622820220531528314278483033868602799651751797558198833310256254382304160189423
391703863783579210320377866821806180986326486741966200834537726721380300048353496696010662241386427 2
5292530005372917737061162348654387360335735131522713930047525770309217990269105683369868147007004680
7131900900638769068942035418676510749733938247585963862891901184985160576720635681836412579224498
2802899653361861777668677814410290398847347627179023468418153776428435790683635879056700585
8121342784860173429927311498528294263813065849793217196027188839888813672103025293737306872842 84 252
676818495032793295747070453752303208640889188204692025781143747067742104393146452786589638070408 5307
4482407805394314796823453620540986524544381900969689729925320004693722765192645252347104386 80616 284
4150263272906125719272577676314022420185623514890590306131224408194389634712598106722230807362487403
88234464280483917599711890693049486869311881573358946333731378306463075822060360727221651203940 83173

6077127229108945530997277130413106141567730248755030303311974593678235696920692900868406551784355069
9097961301208591342082523653831972164002578438653155268518045595065276218297807292700162623807539845
3178749321457176744284224049806304873363455955714192165559906931601168545680475785694458258146525410
5690942130415186078742436405050757001169784515992851436363314191985622039283636376980736964248023175
0529550137335897960359874442590363690717246275208403121238030892613188023207103014046119559039673478
3221472762103942851915451503469923928686060878948272040131551785488924211858750326012076622758794866
1078061994316694602312466700366406945698803378914916927909305638007712556961113855111068230718626350
8781301659159665719956166392695240132819371226867383997102313023718061442402816750002785237013297 42
8005731412337710767305263002918854321849520377262517251966548985863027520198580555528655701741242478
1398441147945614036133723592910024028800169542825276370017260558174084453430805114569010709200685375
1074980056229956793937036094216558524012682608620139896639979818266838146426887629454986869869560183
5872133246870515719561716040850360702192659349072702510994757252219108424455952078308514342748979583
1409061138136873218656247805199733098940110047570718189852293784438141434541275082709849791964579292
0802355036436353509601251717716805655499827883668720305795332335489227358143909560512229294254511059
6156659899801588064005422943187694927076222111028476180826159644660270430972905492918095775775902696
2478243427196842521066737089539128792695710391317015524198956659379628884094286905195234919075493968
3374338510867886831129748407742561428880242025456470750857403395398767464470647241224405084157459877
3169274280657938451008293347136974573171707801215046556077308798757870502442018251306625132845793 79
6693426746591767544732128729799255322939565415828638562563929622701616958143610479646337016816903 83
7002557364944013958190229025904302917933014131985196006539395930151180348550630214863817390059278 65
9377962839746005016600245619242505593516523938998578583292259171164760183315867395892269197986771992
6426770722804451657455450128217130748500736779344547024871148837188476688243781859855683230039182 9250
7222124723395438145081249592057273858216386717412145544407200077462566779999303388581438395224684186
0609946505511749493126247475454114900899298880244751171439414717156204843166161483490190460011909296
2556851628776043685120922176537252030663261027926045712112346384308976910057607603820506269489451831
3368129757008494650362783044503424298558033635619845445054185239483989082514808667971595308723718 73
2952461752619649890592054696908404522519746547763406553670508396129526943779829719822550740087246 75
7960737559298851192270040140899223099769250729082437252930253655458496293343701951694483160099816953
8217539750893931830881834902561945268977264011060413490845313145010713937105463476654629387332788
4965077889115704433899875968620677669411702353257219697006335598012767963217354050170173873456002788
4615557553918888819015793780554417315212471104852527959766608726189792991456157552097970404808675605
4469428312274540256323519115710445542963152252304360836308442210335683371003474828646197343120322 02
4724439322933802978839273166096657341966481397117329057631860775941914189828747832911366843302535292
5249647921101964649065217824280216584448011941483756087062646822170527885558666093973084921172488988
0705529500413886690763683994300818877934807755411804546951933743693074015004386915629027469361458 81
4570456637297627949440616093119931117418045204492835456146997128768235017514453738928383768007204168
0696395364925057869308093258643237958870379349339609878348151314266450625451524927558
7799065325616332081271863536440504910181314864879728064836084966650248920735697536217190475720554079
2696638443542621309943524325188309713530823407873141549480966574879196207042981321762437810817966000
0362780596168645858331102483433125102935266385776539713455100522118161656726351335384136775589213883
3561375461669015061175377649637276412757297961854600653058815433746646375024676618736828601359389 97
9661048870373212998988077189303455828424798632586508329716478813687836934043158699441584056073105170
0807409062423229834214836459375406668988799024529677053884086424623540006792050169402576684226733
7762347007368506104194711069435804534455704891685726842546490009371256247610593669638889731278456 86
2443799144139951569466810839263252231819117286400745587634104562885751255056808152522939279259781486
1747545269477638044769959550506070304057230748023470574234467241031229653495050651701165431324285219
7592232503963491802465431634661291802225697714089021233932878860417394101352631152036911145392003875
4109001217400800370764068065824627805087515704573154793095847220829190461794537539554705
5895534904623810081663731945502358481019922271929120161775202444668605900966401426557247684365331867
0352206558014589136521421488427956558662705933887491878639391231094985612126299412921957055098215904
1459138611252665621879456917858641408418466291812312771701856076429841403475924859795364139295700295
3996000447652417411806360908910703299357601235674502894896672436831132348273536381370078481809220454
4878697136394407140683810853750237906792549199074354539111875441697879774458807061279467910261059726

785068754968691910266428987266041554000355937062096146877748021195590129744347561903690150322987650074
421165940749484642116353836077421426796439831027568155546121057167915310722177793025457322874537292988
919495464444302147439443399152619976466175605004468761414451971378481762117897724143554673546702420250
734783642218978843024064895284631438816343507529653333478207439874478444243948343786215800052959412012
695844997595662195314594638517065744940644541377088385322747539546226742717816410785971506263948229112
743733169123087794758840162452434673160765272923038336108403393227859230437069614233618515120201630032
710647339236015940934876141393157801437552099930571439690936141780868692008127299950126894063381545452
942618042524870705529369512012093385006182670089611255740262928428939779819953658533467689013525133133
654478486211389757939535376384770437896000543746315267922612655403500132944795933203899504403691234100
089565126418333271896351165130651176712022579372906101484235246743785474046961258221106399832604515800
125475467792932216010039689183516686733683039313629853299887287733133651400992007718115949585299647810
866067889568730412979681933386864036542998250248976083898727879968322479948612906215381195278117506503
350076812703463806998532134523960704085038220311383731246735440854003549805598897628482182502298417500
192421381995295183553440312578431076669819823628904985956939761472019504074153418288497163679555729000
981157692899029453651754415265328609725311247060977431006428102157231991439287247182840939967413832600
975913701920603934524244628182094162053668835891745861519946108292975861163326253936348579165089549700
475561088704308265783353395487206043619313914287218417528918004502462401321388921541354063163302382000
161565464533991219188665880282830367309947128978081654878897206258768190147486576432647454358647966000
050544193140078330581576657539339176601496901725110135193032764125437908172466689998107870743298828600
063210542872923064618965421625785755601612552340300580197329431255534421488652069980970043014298345000
079053809234458243679438749162814834015294287829919084621132239087623502463789189770122675476308791940
532414911410912534877347045290222856769737570216702723515203683228144986530301933624735824640022653000
178178623061318267347574556271918313859370386942322406078341585617375014891459225910591001913262226859160
207583999928414258360755023703675143747250061084565212317820690269323271707170180717500766710702438000
300882134956211420966662792576294735365612231196303487297992396129561001441176843099144359806556684000
733183771111489825166860528824315829668157736742930175053096196635158767553864128836864401680329010000
209836414539013618201231060000007757966076771445323744903666453819033035659640537748486377056089195214000
360679643287026511419820587140996218849541275905528966271743716378701546311823493958080515162638789869000
011697057433252523306688114102309063716096539338291754247405722813182485286562201408429483589030530433000
176869938862725299013743012988285332041492596472487084288708681215621556590555205118766990299353492885000
869512934925620588529975945981473505525503315145734981302496356600057156212941288201622999848145291500
802727452943367134613398954992519726142954671341106550874868735678990505261841864946700362615456651700
859007474346830220335306212633198282054810388329437727848015435298542340690991512207114081916532842100
628682826366356108343235566621845589654824351140825271341844670438854605640891533111831123919180535398000
781167627676137036584284818073139200562490471623282329910806060606447062640354285919007969734730188700
585654299212914106508310502492118115256100637422924329390473242858590436910997808736870161271565760800
625272563041379461686532715908948468280276438781968830346901398630993534890439939614169462770034461465388000
544854517861469894836359302502333886136605788253493820429754205348196605394372634779791355298709044097516714256549173418657402833105367868993037656181566235182047554994194349715215489121452204670790290437621284540363565004264448587991544652650458844150220834818998082523762637834939192990877418633119502333535509750955019850442946460037920770326472316538371677976322551046150834470572786498431040691955211902908981909550664375400366598915712960903738260856886220943158567624141694627700344614655883709614541838367816532646717988704751624200276984105810499347974174742755566238344821872394925598724808179028869791780821130782667388227175699536815810068349441770141103531411070982796796045742155734677372931318357420990988172700102702705816105909256197738725629549670643826979478631527109963323153897802141958462837023585522802977582140619125039390009594720689692293515487099579621681865172941430023651007141027558943126234999452621742518340731495182266541367071120543595047705298405516414408231604009414859322767183356104103192534646106998696317165885028193933170090277543809750296912928290287182718686765605650103883625974768993319185150868463510202044431914159077236468302553916808091377426723323139947128849775977953977347922967893621930992511200663011566173906575736837676369371892114223684991202147557197223055741005752542572542538555605056864605371122826798012939458441745580652122388934637377811748771108535856365104804007023513841408394344914958733631703374524744207690310465894028070652040490102610180630714409508901183652829100104263112612172301607303914278299826054825930

748719707894559086334217056537691155991337169641525291258655071927483459371130752682233144193075059
9400673536365037293550076798042151214351972532560562264394541411316747298778368714099675896563070199
6956082462801450491199124071218574805370693964038921234646978712255066960696516150681300606294107400
8047075701309161735077334755648772547369122521509813503319299338334382107885543833236187356907160805
45580984367005743508507531994097659653368721043493322280618834992004827873811252750304920888817484 64
2831901653900630464650291562089053229473199967891042999177453412876891030909976718814803093271626 98
1236572060433159649649342053559309749955465353415998438238620494250693932014266637836894812021997614 17
9860830589838293900673951455175735499971707538924803152941090118514926140875944738521211593932892 63050
6958329918100862084040102864899234632603456423704248242570639091293808017046596536307241449281837 55533
047948623350633663637588656105890660335316291117659790023653012172857536620189010164981597772472 9054
02267078626833877730111700943185140353776219454816752649657531968784148514825893441405295839963849702
5395976212635978594569611063706013322795334191053037954042597830413766751674976878746529964917834233
64270007454754819494713598660806915666645853291490308232059890281205878126815767430930721017613582 263
18999288797323093201014923212637326715179649554869924776611843154567161757493939840161652290784 40894
231508766870544652757932388055491266616937775989539255610804069381182248000513508093720468602003905
5009141165394481995941732187190340971882559072629584283719770085817941850701023924759988289613573778
7564358854963425611467807833405187109551579999318051909219022665616595039282619479016017023122433
05754431260654464625008686622748043866744194420153942603811557827540000404020712181901315746401093 42
73357483361014694036854512564432034707528448389317545617646315722092878102799120290218922382824 70350
5463383094194477974691308829245720502922062498555235510565416430457629886417680620174113480892342272
88347454320119807266068956925882293270245447753472176552839080617430025428281598287674713598313924 86
77545318468446083804815081566354194625658132963595505948290763301068156449665178803283777234264957 3
4762093975759557930638671004774393400649834055072362181198948448212717277851389956849044762700861 3
12697815775722954647603359273556343018515256595829201382091502262008800317497285389982150390653559 33
12828283530229104248499010140960808012922737134515814582959625143817175466341676306580012467619302 73
93974968282716096494372311355756290781019612424029811176798261867140483954668523855807275940560932 97
312405555373044799293256493279811483183320213731115623240409254441936217129357321551955582693070 3620
86939103095626257928849446571817529163361050844013835636074650721157188857988513496298042756138550082 81
66953039086989620802903035538006256321403668177908180211648438779482785129378171151953129206025349 9
3978761754640800188549112650947737940338081385165821773654742122319328208269808040149053483955039 64
389278202924723734827169643746781354156230792934930724033675169925218892240566961469968147387885186 0
3142897633797624324269819101596656181862939220488095891535233677225687364384668169767417735744924 02
7732442718521724582490379444028830673844564001853650746541756053710825468179256936469644180207220 76
79501529746750884313523113561625499529176351416883845818879384276920770036330816674413405717150430 76
445505570862019621875090213884932155884931463611851668192193331068710415721922258451362322199002670
838022234873215757971119113968268038488404028145926959323895964188662679614930515982037118628310945 0
197691335579388158951141532725042249535888645052952518856976647677540641989646125632782259701702333 7
5585208862221826002855392814463086109070674171612612300225306229432694397803268357300881624868449638
0618338129063172066698253927339740049478569587351393454769511348187322641752631190577688592199823 73
2093522988172495982321802051414645423317460267710479573856951746726028070968152504373358982055047400
8030133017558152271695309675197201612009205662308775428710696458634713742806675167831937351325652214
83833173672308119816534223980262474376558947675692163436877669156494799894490895333582608637982325 3
9491672689941674998347704690921464089968375824950582908145331422006226370265889085675892630506217 72
50459027499099932791976237866665291918639558768793566387776474276695160789608393162352529278325030 641
5584678024618159880514292691440698652471948193396323136854634186509092841382717252169538362006323 00
921099620624942506081118148675129816086548637849168389142024407461253734991180744446800456578072347 6
2101130684460779794221320441751848161601019084311857783769230285339399275611116062550093838015935 11
1113590785216204804853869143238122459042997729469643222737151895258029733660453560070753438041266 9670
58679143693280921830411392517937860772590433010536938605645312282573594317372233585221168154304436 35
849742772083634227879617830153625028018585784844259719867132834248512769481082148289899874541722097
792403608326198732536158959601419384136517625882323166497136618714988209913042815510162243911604524 96
33842565407862054039684759841372950931581487773712401871797783804789899494365429776732570157053812 63
785222746972467874241300793642328497818478384187095000920272327654649817697185631159468301209971547 2

73531735570252640974294862538148140078594209375626383946867371632446500669475670531599472687811256170460060640744455807429019997012621053694428071409221661518213079348698972837001195293081073666728548798578714822353168796747873575266193854236240007139113056755503388529014237718541648922941556716384588614111063333831204110852764582840260102555584547225937596187234636430993963801234448925565529898272029900367843915104186874958244295462621261555251986745094446529022196496329554000070385210563219676582482265073312536205462602686845226609688043838047437262323316166015912793688199594050257199993291767339310271001259536094694379663585968083266431931649658496339772919441451837316677385536453182023023814826377065301139112891369134624913279126032535345919916323452775814247666027954795407043305095730585710512198291209443165335943240846813807488753872970853754416828066098849350666996435066972687636706605922664955035210430834357140335183877561742096308666225048152511378485882570025040515838618361645076359197566456471215061989852062746107207788534090089741333788845290560643246783723784423811624539607897311433337060960525973019965093743954562466866281243275277881785764509786865491389233961453938720160995472773168875997332167711844199589484861426103891518757536335313900844721671593136105659055230688227016390060464654237193404309850825077350154851993174718357373044915069497572480079084269359829143830931389855487549423227449491627921911744168176225065513529230906270328623727172713736747285363862123221521565981401434617442086412238248701842176211379848001289184611502913407235104009968213635515588496259347022842452965644631458122087796414813974052131898287854284269657822434892186212453402142291873824879831283655208338110223504587132896511975297239395600011425296402196067696795758147935979993141849812041111542822165132392550709874816323371019161139791981311334362875608519136823562637758111195201592830109802456170110222573672111807537839397056197834785899801039586427329468431331010946879646342978777226821944480569992455385460167533539940861811762259294277647153412400002769010274117619239922487172129321988282132145815583308103276766568215762232861023578888664955757065519767142309504205762010616009270364200234045097208356485771816682415119269944850681518629351905611712803653926782136985075087749822073193348619790072589775826641420631975955986314537097065477664768007258785006309940108789455470972063803894836930369700269745829254188935682769767592328677671016980350973276936027288312103987040472950733573175722327139128868230896977588125791454937934298535944730061655931505590245069290180294939653193727641307597943503018619518284940068279456592773381062149016449883428475015304083572143224589265200027521484688613520268320201235650451901911872323559167037232979034597875506946927721571141718237996680794722945122990853445312956919781673382168394091646111762121075682647338261461487802212088546109416072495014616471801755745231575725649165520169646297706183473704619619470719316611275377707519894464689611124640677993406222196506865913705310362339187592819657416107839829879513864770740645142244930292524076236866433594986611393309682004315013611711744028457058929342504889972302803590243885579715131545883284619046631488813005446165010619163392539632499566891885581207683296137502319540263309630469405124798684956587267645287697099156573617747339190531248684494625872269120955020707217071621879691877607055804810151922950743263378200279183115536826437678875799119811679688739631778145299544256657730927567143291248470230227864752245335402514079009491825462535291014393528966648242984559934255759177758758242352338997254784466334004725931357292624483994770117076154813188069760615022889309610656882030057928247768456505759164018129453869688032645811705112881260506082220110294813788676234788832482528250394668865604791472434959362408575339314941285595065802576280008903568814923209519216400318809729991980254670328632185674487704766797400805244494123019257706168607933337261667952372310225603456073494572063560313170703759627214649019425087954259668365673496784779640178627750094170839259651792113718885026960580859287154656418538911511244828095751369332743829787840259129242527015238018394282776423077865998007149900112718626725697779749355858588276207844192101150122755574177976386270585270654598893783329649518770115264414216448752086329424108541931861114096827628512901437848023439124804724666828152772743115489646256708029122941047036654412796194587245160059110000895993476482685734682846049695113680191131432130230724663416237466344128795678348610301168491367031251593211223143289163231551486390389204907304675379060333848146223790476336372022317683354114324733331141893942473793198755133365951923692060556541712549538204457474709743381119982520639905592573907077430609154483495464543945550651751304659388351630848247441099051994962367971035589271335620109785056162499523189050215581468446772463615546978316832482756963135717558314707897091728778337180485198954598438289569977420594382809953116389448419977603501130844851928516254905890310872332963148825592074274014344023881417245595907001193665470967389125872702735685273077751939688620387064305373419963678594085215789772443560955383209737122234993636234092989901253196033994721429578657

87475147685543217321671274622760803323300056382327022754529494217257519412510539167218643358594453490
26876922082387331430039261354663005729636733155649097959804899247266938645643746202224968338000050557
89826856676967114280187672621778655766491577003524611009327946825636771615396243792771294838137972340
47290968522053123318201578958447957484049656542625020395674682354733532589667464821188622077302441641
39640057439589934451240708100231076667283429860057554936374749864908555630488045734980897457849263190
77514473595857773563077254678529823332546870200957197489619002324974468772535045331273092423088905700
88306207272855498567467690484861180492665365738112130031825472998778742263224505235417218301325634179500
43964983867938294564119675227721807970790556444508380435789200441015991008710562086545753586198813300
75442552127302894241653075803033080796360397960666424281579324486052872495607487409036181364062430200
17652262395586836828920962749246091188942919022126987681974610643447889946335490493676584304851434998
0640095449673288773001522154402956812345344894693259234979857860792193479042487386442067922841925130
27300496390538838157937479116299593337347104425865737219183591342311892468121510056417553783567312752
77203394204577330932293933774147462912041433642375845322780010418199175484164690798896663803490314000
42085797512767023436973090178204120231201633237068160980961937623835316642814670856607220894937814000
58508265825612064156690803913301454037873470086191634207868965841313273363314363382372598692478567010
08198872061494310116009326430205345194368673098889365873618527462830468486508658931662844174281581300
43991205834318479331438251361257686530937757477371966080824138797890956863192427878213653414535171510
20124156321455772612571711542375752304356111855891966314442308693667051199138153403226210621594397420
71207654652317651989664244752620471519096989174455118437433125604112060004810628341774495189990610220
13412644534006534857615800636404668827392202226192144757111594145475705652435388670821749955628890890
01677082397310205194838871835498343288088860711036176903323107813796305527389811291277038682799321600
59040531468963259863919429152076441218374053356895819942652091520607223478701115078494599263794258200
73810045609373403738470052604324004765151033974426291589785942161902765461243710073141533133956067010
20992357055893555586437332466239368273381376660988561386081756085525751881822982365205930398480268400
68925648315727203819634275024490538138712272836538138174118903818629370668796552740183516011067772150
44274876318669385166526891909669324124156357125547713861679074687690502863083641493189655956235420
78624551587599337404868888059363509476404640422660902373943469381319809455398305963795278500402818801
71518968731583106825414735475048320937668798878682016249772079654622965998697092863444271187840432630
48465824167244160379004862906335227396263264369569521863745855477379277081083209980065603785849786179
28168238018443737739659675829132200601255392869706131183075749736498210147313999513780119583630465780
45862188194414978493700984027560068549802633573502769035013349159994106340615470787332906037072311610
24034187055078837900089817694702594740812636634023320523472594946284677202796252113686448377842127754700
08090675102553014546480725459627611374316793083227144449518215553202530688993058531948318061909762810
80166772303612166067813695171764875979064176720782749004456671143903026184811709882218889166496393
86851319346911259894986247734192221039293183653734628244718989578879861281511099659370100378344603
66990836605499539314209994234926019517200560034985940409633599105472773946979904135787022632069202540
54984165726774294639612974391368521794850340610077285098742812276901141368164024867528877543456347
17285451238985915716062110621038611504145324978791301213806889378088785741135694023601086117274490700
68634586796259736100199117029869639675360563303585985059024044857052314430822798558580421066706978400
66858561399205325248703555458537010107735008864959163441191453995475570655262775845784612584150
63107474953677019045088460451789610356300964498325679803052637110090348336832738009258557839639769780
87574550409776166609902795429188589308917950986812311563053892116814233795813108294191301853876649830
84326660489096880186076989969469391952355412235444495109149042793265931689165384524647612102977323800
04949102697206319731445768722301888645472310352033608803273451892514604283782723573738137990445139640
61458772415780099182945276911489256449554457820811258549746299122723264649039750796423060026194526090
81238603894172852569764474274292893781644102259782780808093811848982866564507504699035503682364040570
02878619264007846083106256689672105151078759980626053310210012235808990878645255277392745448389494240
60977309768103288181048377558490535754369854782487209796239028564633703672244472560265313928132432200
60277494460178010800643538046561772418288441537932797330316564262549644572452478522515140333292180299400
36366892004502808793014583626559274530369320858199236858240582449286797047004892934103674324968078710
67805287155715269795387317616139356132503003098518745068872904606280714530295331444248363527863091514580
11167703384349810903471380868798951289092704710360333677107781408348446246860614725498836547671435070
89716501445276993860022792288731241611888686514548757124587583194866819607135129780973428900934829900

6091613721718408056889128483203073780439902980119776744595260872450178788482758220314160796646845338
4013730036339352239132617464228397146920127293410803224014862978907413563620435519583015124060878235
7990545993705598339963964272542888444321935083783960444126803498854986780142412013984799469472672513
4575043284161152831893433625782555576248604469892081082951581213080747248488317379714406575523709296
2046872292297505739043255293584811980266329093403989497358092902735650265209686282787669265416618779
3651464999633518918591982811237177050851290889147989502296980227753697818326436580306594608092400080
1768907232584144129719282424720380048659076248329843266612530268510334990265765785799633057285781580
4481506534147291034011865107527575914621385957555798465331746524212479703562965764849716996877845984
1432813641506265880323735372672051002039187208654889404916333923844480571756388725093012313790693034
4640283694891453218747968903689080091076964875226793196883973083136328551644284854543895519235162
6705524166101161919395622321540447458844931776033264586483326999953276131156081986730317847770969885
4732069736111069687352300866613257328235545132889270735840381580895934095400477668633738822098999477
5888725251392894710257611581130581373079256950891110603633747143004818074544707123585696701876163044
0248592810294124481653304300197678125184887324291465465374653084078562990904537378476391634106971344
9483164149431772592573598987741410425852650646992257533386670229379992990248338449796316516582601773
7602404285435251468929382667737884010050407781498010651565529747650230253048184829022351663490708944
9876811196121510650807804854289402983031913436947886071972309108790654545592900442696210241265089862080
0237754863792190058221375417994513677375502934421394784384540882496830055969807227427147523461863684
4963300372011058488444262822847718356695105783641870908711811976341346792304827999192942334443433745
2657118519582758172429556975365547653558715371587886731558237317912035819434368590246116895493584548
4381021820980564566711522781111528663148797912241046715034541226359174023105072675781565169079497535
4569546610234350635178926282007387671578418448532331264748169641045009329526393910725514026380972345
0742145266346791248799219504249506398350575631670024222097888845014235493326447708542073295642006479
9904567282882730897363424010635598151271655857087602631934760357479711367322844254495461412410310441112
1365329590731844662678118662740199008057408513360217113291910191481723858110359763233621594459006860
6885817458272106636078324721105539882285311162308307722493207187603142835549725999909954386747919041
2064095816350306921536539525593982910836444057789476865590642695907855889278014811531296052829373963
7830522396368797932913415829562315501056197500574788258435068389480272013160805449241765541341553672
2967818672262975219357296762159745729982784576999581857012470410652755234709316760210887462018949
3099905626803547323919174388035285239451968559022629912340556236681684202614065946016614268981636489
6322670567145452761552084031997755218112222830694640287254583090788009525538267001174808944008582442
2344840901443930995076045874992960929194486787284244655292260355040153630485727845750678903342063775
4856518026058646545350492315463660672149559792319961283892566068624538219662137909561418094819626802
3483737048644539098576041171310701243314722770694381624261607791747358936040322161173190223629901081
2414634682390910147478453481264736939378034508669020117854092269189072113908427362640084022560952795
3662226878036310749929518963349372878042462077385456462979446062669081703284652326380242248858174
6684478301866887582598506335809530473059262795878826628135669574185663518366267796335366162569000655
8834528478932612307942533321209343130986170013940224515939863015330027588574482615552011632165583054
0110902479913174431860947888944247191765658573938336152938916463157877520692609899223778427121076226
7547136296772652833826045803154881842918345790562055852313846748688098747574182995143608952757114896
9698465913305515718249017264063469473831622484570295688213478321860050219575949327957537140194691987
1033919215034601174895493500576483118480038321070455100031972994606266908170328465232638024224858174
0354229851081369311162482037815682402670540265060927629574130039459067445279197996647299332750032387
8534455218203096952772541183313194929837454299674345083494569207945583308957219416697425544682533864
6561066514161947402732330689180554887144239701482488141304333221160282289288031591327546040639314
5650451024707847827003100830623983379035059208538782367438487469071591071661383527097558515843491321
0350122255865928574296051076914689252902235938279127110071779878737899790206010405379271265542027857
2353850665444666703866631450870991655715371206920784428728033234558655314358559330901816838600053
2754965792572606903799575988813473068888074744754711193224014850571325800645281485106494506217879717
3665367581241855617818186509265150896785883853734600693514976694020085414241195861577381990262189019
9575558106225483616404106207630901758560745738847326450713338575564604173253371130280137894079385793
6433995336278092312108397002330283646942338620299103313769745072519281930448106053614889812723727553
4536451517984386997375729734479617541226385432501523178083972557268780522568712161813534253902834781

01919154203432425934609133113642510731939943370384323886137588056827156164854936538023785090393483336
22482273182070996845329664200662680372687820264867057829851192845030221581505303564175689377542106
13879430081690861925874286987546755092349447396156707592501916836911293037748049551990970398233636
364891350584303996481957236211574077257682336990702374463935429270205346044803831023144481145913795
847492391572943127807412044060803097587296697310718198441066933408260496984151577165929551948658164
867577942339368988277590035789421276160470473421167218792048878694233196384054499475111593254796172
938498538240196477081852094864855924228773249668730030102462610446676908525456097605252767542609722
7580423153215767353290753481536411137564033449376447731570714677623582318651700354720537594043178
4599133598339181781909632989815139125754319138983372544052815081545703102289680299577227508331211659
77042219982689658324694122451128329498987297002528869278200885612715994977513562152414462120118572
1525246972290915140594046753192199028178185996457720580220115601746209906358907518617573000042491
6720092148567640159697615970405736942590356827057621726980826125845805961670618866332840263385482
426103859307490073261464768345904157879793049982050590432130023886393302978738736196275533218595801
6622163258466407754647657936338085449649570965549194681612051155694391080796813539384605975006717950
63102561224795271996046570862538431272191945200310509319123694795458561223792958467448322080038519
66315323828450706222355416355752016048458595787727964356640792897731517789254324601637419516533248
86621259981179724637628056178491675839232577223568439231390463625063640230283172300137855383978440
960356833268427992665461606721109596063883424295043002336365108733945914246608246878679142410972599
551213534391932339520885700159038044698266311069545853638464299547406804067960963340896569467612350
11815471822156520259380056259062150272901224260404147261584034278162605459873385724563714779216047
48274433441112967312376761198629786698630802417950047309905486807868392900997525783605684232527345
49388464642611743784590176544695409651594708729576071127626661175786960190328974286263834806460258
5183651959147287102541609034171076266775815046516946254495799382899617866660985727357260685506695869
037572305499557209062621713947039685035239312944870871415247642250606521669707955283041281086860497
922953659243154172599393270748607956674145938706841232750041669086025761487588798762359367347546290
459683470578182433923747399924993182137763038375299038515104921386268559486228298020909860409840116
50732228999532240562234847784269088277733300025209570716387891375728361113161508282405867748061283
099765881224500975070984791439925240141622405328590851784060267753381922826784952738543837535945
16073195686216856063512035024630992837418507807604204773843053254838763592412557531636452293163517
65222823289644831910269247416363399928053530936659261798644249549488461748132899568137313948309599
958112473554883738711252190337005154680402367783261544240656690622404363579199589191618731985304828
00619742223209883669364708401812321417476276732394733060040517262859354854118824613991225459936046
89697141341652030935025735593612614513964764464902106275445737497714471550805588286003135966157597
888082513408417512408423142188759570263961666768421788501511665029595597861841596054792017855481164
654185831131412091278284596904448081819980631438903805220749709996445926804268473445415578103344932
59501563201963076214180739980626708480643032178002476090143977156423120072262363543493273799157315
59110065247277485199710198477976655689449196716486168670790375717066483562808032596486767340458420
865018193702426952536481695581459090936590780768681243863993425655353304650567850655485557128211838
17396599836335767803088710405729720638481947048233307935220906084398628018386708952979494559033909
750568234493536783744414698853888045281008136062558329519311211934751776284520665192222527386696926
0025665605713798746774722632191103954463938461218455857756926921127446965654071571414181979492296144
64039021523936521713197016823791016253096307962867690370366999872010105519272588390490266619397164
348114974239117387042271429504980391844092035353505646482647090883162717270439141214238422921600927
29123600670102952490482688952858136084943203538041356964515930924338277301069829074003637819102142
0109190692418721602775558045932649401521505941423353135286277882691268505700877099441767082111178033
613231077884547695892356286062676906368115480861944886505985634982420785224821027355717116828604374
793001905324214732698076035933663485592468191202932977148092123328418605006258537910855539500693514
721418217342416965828916871047738311059705099813494096939955335172453464547253019452500219704236725
3394199259591390300437260389765339723331018927319980366786966546139807538001500749940589639656960444
46341048889101877254017581364829927018989318743514757195119016432624565142006231074765452195350624
409076226563967703182232859593609028162526206279685291067138807712746419025705602446123783078902648
4860064900878587772245828110243449387066147744535679373207145334080832496103686003164871160455763852
9640009288990565903880787201538168686057845360262395376158436587641315689542018451668042531125090612

805758605185126126372701960423621901669912907552891484255000203166394336803204057114116042357980
358565956312479949695698495135867617171614861342118764921998574023604525815531400876092991964744 6288
133829073216882269131573555149018421727816778521303836600376356129007685445315638810905871806940 81778
190789574514335959383903513995923221581547874786726203342273037795254810134334488701126988252706 94572
970442925473858639589403162524181768010338837389259782856379513349072256643982136040806076951229 9054
794220774558697799330344439044199973576079752080203525292863656227166375301043481541454332706533 9165
099594673587113493338216698948567055738078220021731209391530078265261286845142429072659990570958 2548
604303545232999965155578146176895285778053665489485926317291170626203829798010084121593188186962 902
828715797871793688490402576691969478911970166423079447116300479691660296336654116567141292665386 416
765584258346626506690416995107981990415638970961657836067803458385887549043983668110794625922809 484
398900347357457657221278020507663612424745302728327791975368709460269211026412850520464771511319 5909
759784752009540677372174114843995901350727293059963625355275183651258426047228081305017016364582 98879
529638747352394412898404127423076692657538919129379270227358715890678007404859216483963090839234 0398
840095128732229830618530714944412455289744543189585611874000180952752096285137117256845726219543 8787
859256273722400118928592109735917744530909913760185710513365126896097982796691301664713645669707 323
708148465349387888981009948329822204546100201720436127512031538116582991015118694304911447593744 1511
992148277698846672390919809215085158244561052008871854606987301372553634632995644554638726445232 6955
782438168955311089653134069842041218638500690537990290112965503277304964605482196018497714995876 1016
018632187928577655514826098688914664178674335486398857672164079809930784644700415892529812120087 0754
694044486902285347709395086821323407338738152117640484760483455529305299958930207165021318455760 8516
378241629071597940866179526322650908706250000978529459880341211082557760720141887727490710128389 3632
437344755327661099462982029247845727959599586690604600010182553848674565658275750715858729686771 6751
114872652122065893051470298169911459357597062405238204272016760908097193644030104714270189011229 1978
328160211355975632554869992243388504519858282629103818226122655992591597082919786233193166881062 975
254106841171086259870307144086083889081617340189452178676169102178400070572151153318183463981090 485
505984190806537984093196765261825490144962826381313686830190537276290221408228253205031575814491 10
521496934008005917184288674742328188920822109870751009609422271796061186875231086513054879407303 2475
585515857271295685166711502658557537215249702073566554287480481883816894380929471939249707834269 11981
621047130830957027914404488752944652228809164269322120891437353263434701894141865498758085443897 8375
427611160482194498573399194641634510588275964150749708686870653330876746855170863672200413355069 0898
227033638087248324773623842767612729982790004723503365691359648505894871179717735895375235172704 5329
880897687060371716893813114926117990763868175722778250303799481053099714133210679114139690824978 5759
501734734327486335668431549974255534328798138728885884082866555063031169230342624006519246182085 1294
051384109438338309943373235611840834156109525474454266787441539452743808540854781574817180554292 30172577
082542155145523116898159841576303310410248678593445139563372056885889069057119083999697995213344 8395
228759228397786569316970545027552498603759381380065895257056548715399259424999717317167976251830 7807
180065173051529563382623793212718062695948063416496910211160236643235890477243477665172504370188 262
115843764422666620475127825231887020178980272672802771182277315210303587264208245289884331683157 9166
649925682161144990342466951532463913570478076852689899848932052533721012467448559451168698251650 8631
506559023350608946133259647585740474307164510219889646665875650930833004568187257241680767059216 0435
546810092925593956489533295027971567218920273250815062770907173871031323842496009919676665726212 488
713499205697235869332407387811177055732346090215798472234521382771579580347220540258966441539053 412137
447005089033871538349390783651145807920272101479062979899235730340243996976240607889732184310882 4493
082661908026220499901128597429556277217464611468993929401059494209733817594382878025044409968753 401
903886017717101245912287676884675715236965521220057724245036539737696902851785827201084335983398 454456
817840625749043191708774382410403186714142041720368114298950367725654071050128601831884330590260 741
359144689996884717833470408291468124154743093009981538425850563276047390876890492793224924095349 9190
341621154460821389836454362734525542137157891521320603765643461287733464232652749078838192386444 75
770595009119444142257604748727334642460795924629817064415346065133084095714746871552128608657847 0735
295517681275823209533816373144290474179840268935951138362336190365221936869489539292986105606543 3057
970271551142636084308592728205732303824526633753600405372551293149122262029825256244413425 9456
551577015191896924072986258052706984831289548079625435319741712858425766114028200953496918021677 875
090640963889389104845046640865750372940522104112800095473196600465004394421684859422090144524292 7222

558490593663824402773328362645030305335399309698777034309071233257624941496016107924287316685263884527512670603005707641095261429896648818120396063873408498555009417321987796675327499620118697725690044469020624776565535065566423759146917333027248915434058300807072214809831677481930217368453700212286491519944808643918607136506738526628020901568680073865848269567625421052986411659338692482583337878752134922296988954977033420478617409807162101844151617722148191100437333711693674308773266973085990103719739779496102842916186841275756991398649211424787814353108830701287103774766452424063042283708377315256119447305575523791098461773531290644241047089451669048806950914607318268944927878884930939500099507022903534353921332558947652503280710528542412429468783066310827995140399264853819433461802956014311243285648469653171701739125387897466609714539422772741011144239387958959878910105456641042681478916268572803599678303987097866634444047449502378053749408916979173853720974707345273979460572492759082492782583350682569083780835456936366817395591500548911712294589342501939702089639872042336081310929952781885040277128738574435470581971449057130415919250755715118568712451386170846137621994813881555082938848375691452590235639700631112601221180043703847770706721578829144725047038571195085880934755063748382935317775380885589411741926329374954300085072224839222875439442262626959891911896956305252790993462946153609361635494478791750377713741590706231759672455836853259984211558810316256244427655499002532750176313629431277960119164305097169595321129920452717744965463360265153656582884193638793377535522100075233062499389170886030515134982454203328601498822518924449789713654388787013625839686941239838808204334626611722043799813307423513678499084586481666628620111016787728549322725897317962900838938093783686224309281603626918073213564653713550504886916477877918584277379081886131475435737043839641610868454477860518569883643545539414674488362256907565268051777748487429731521740717222408996252027134463802893843314525169836380731179921467836533342174788805977284261602035843082892445143907954874192059946703109376776927346001157263547865330378705088255862616916954752736042627652426280382477111290356531092061375501041535163500294773602803042651560680704450345658653273748883218588221454365191134340143429808942893214788848307387257704974253324646609483535166578176707446537836480284709692610131864502931229616599437355582029288202344507282771381373531087386815666846805841653395240631511573679215491126324775012652980708686532078718046128679242880775065292782594194300758842780120959101662877504140143144565160093920421399550727498787244571094135555045015722897094705928715478578423737549555530688277162786068182464764719997298003673328772074752265359103975496408864254765241431378861906622713920337483035553828434479766459982436340653278131631957083438994184020729563312274240406329113697622685654000106238986047816686297777840312749899917073967230675221629509652878963150378284676791610967455848873593475491086691961722460379433265367175326679642757594399604997668061204001653302850494997286070427542509372077385863700415441026059524493707585003637841381860779567606680646172342950075239776514594329489671900183945085350711525082628454534527375779691224833371923427615820308014628017675628250240714602509716056309317672459307604868682487900538471580520750744820305264966898199940232115902784104323258353371777888992395174636657436312785181100529275734664184818518169944340408818037687535197406636304783372844055818192994312492820699574561423826653050981344664300969883579889993336482496472266767866094838218200691764477995154100648314950923402313298553520760127965976545825143221427726580066657557824521143511835733723773049243714259570295855158711563888077995670992404167839450785418474597970898158041308296754681520346603969878439383081839237477162789771382844434051340951168427734092176907252476310892244466478911687943251322200736515345623118550138742152335445030893883986101461106907489109566359662948150468546756229289415108923729419319561491823182561129060258364287817219492917534538823527993628588963636795180669943553947379606788386394329921827855684125586779954979546133028786214646515713991659140284587348702705261069817062568063586601281644997246668616416196987988677368474709635196349675767724117193889891161841016757249065281230163968169315003820148974579738890015268240237779830972400463108506499741059120950157077584141489180528324673061835140951749137078851259753482476926500

2368546235843601597072961828044309168026370057859215113722012165857223365413744447659413954964395531
3251616096580180992087790835257903757726750622322518588306017569323189563374733180071536686998849863 86
9143703937913796902175096258692842571691269854684920285100541554697777269088469155248117449993657072604850439
50809189432853044749398963006031851877274582105246557741081225485874613984102455231546972860561 59900
4919730133874530431602324644992651250170425989598182556523887587795429109096721908835535777249868613
66152804958762154417326193041179249699714236480394563630399307208351692249539620200838815119183724445
22752636688920992957667709566738531889661959616053753500860232336482872126727945718259271787177324 84
30958282735739819410794665642841537352090789228890021037317202758664290118383502401038262269418554517
46236005539053368444799037802314935009104335155598361186501260495696025913457693658762226557706 32075
1802591029709744929517888542021192971691266310414209140107864252386132081347634547297753174408567322
10322546073930167741895602027299203162479537668278078459710521326857264537362422553192509518617 1876
0134511243376290535554619591752548043038904417376179131696274344878531155019525369978077 93447126384
2488993435819171956729612216520534622308553410546173551602853035317060140881213637333458637 47449007
1286834136213074664314991962649587662383247945685349526493664650173730252699119087348893838 0822049031
39990505627544398890446347223256585859879118868687408069701032026820399164596539398276313976 57558675
6306611912485289248569949554524745825958385059266980566909315279118773020597897063137870893 3883220709
5089767335300855561529457849364066391245229601266959600962649236253435000617135657509324621 100978763
8785940003781112255481862865919130265466121607498035325602344356367631241181589646973856150 829519318
0065715412630622899429863180944708829996977823600837319327593124537302930803197083916792112 382203775
93869128205704012577244486157978289703237209824314199521913569339907161737486567237940901897 3102677314
647704915431246335314923116498284611912232044329798798865455048855028172241271964129621425524 4081433
83318483554853164891565559472845494159518329931788517293334982197167452209398227947038948 309387
0068880950095001760186510756826190182122598791106735765741023718866272469935107578553148758 50606889
2653011018387189835211172154353512753982643296902223394938045976378558090731716349568750 327089865704
509163882025048097942352227535687082560260318281292577270037991853012635237990106194259570 352197978694
60379340090907681690393742403294222448564906242924060904135555779781762809939784678087508 73788927217350
541572168610625243531297175287017002171389265268626119668687123822200180819981055671128721 78866928665
140868573073142030602022994417580061475433969961445419578017347130745325184665024873918663 12265912601566
0236378626652016174160913156721283704302383185801183603595637611440219634198851777353715380 27908392
39623069592894450288127077113265979220538364752490063865053846735265471279601392705742533 57239928439
256682119007592076122827272386085621382765349933921722883132892944445541859511852332513183 0035127913
792520760388229329076831775728434767105684327997447620079687854596090230370601547443724261 1467397637
6141025089189347878230418831155679580454124311921034413514902690955017999960420723226099427 4936196
2125944584701579942673840826916807034200373342817358824220480345776191805538441867061092786 475446871
1090427656490961336789669320618650990481167965986053034049205961741584031837366244449878891 0350470616
5000925499423394562165624604863627523675719584621270971010358678373762859455190356948078418 44204611
642743268607874080844054251871227322220062614347989542952826749324038150098184804657780841 61918386349
25747769264605691767553183675482278920331802726653109312711643679338196228009595413304787 06842758615
783981596678635312212649107416283403637584898673238782661373983593013623242961536031166345 79503348069
604146809599985141673315641663661063813593337029105580951861002596596933585538133702667209 30385742574
2167864290636425462773712451614250089638653859527580132228918653790607932844115312395719443 30059711
167462664902389432661717961130794582310441191900797838817152230006700944002400638759226943 62285108398
90825228755833967504535432742412656453948010664801879337026477967864356301541959285025624 393698024 44
07004148981012850384761255766532196799894997511651665256803543642604731499806941576711495 7179087061
914479972522714795293279630735358761112075829608056459270752877176117312662945676036633 57133534836223
74923160908849973395000823908022708340184421860239715622689469759011211164176352819617880 02013550898
606998035550395507069601752355717349136765805036634809891766237472777944388920986795169389 36159532508
157610426498077187885831916660258754558558020712496020356901436014160676509441751163995525 117604635
5458354558683733327378353855866576517556532384665889863983862957934979868378391202883732 21654180551
0150451839104689209505644292618713072718897597198265421314295193673784547235248191391728 08431950208
148721448681809122621419507962485836787251200849590524925366335255753971301282522666319 26742711708
740168740198182053009569012107994936924046386341005226062833597847750993619741102160991 5334124174 5
632227291428984531546044679324631658508205990084443004452923568356130595857276984976510 0357637380 9488

9043789814971338944112155340478285383410251586273452209298137895801512526795905016883110797793735785
2314075453642383267753725911397231685879268790411956752555832558943262474631695959323793280773972152
3413165102332695551004256713478165687894432590102030761299318706117131114724468473809876166132982158
9523733673235671671545614175551623880797120001523453111994541783387246086759196531569649459754343526
2414717377027351623521869084588169433264954360963069032786367827217343378723421220127791453073615284
4291764593757545276655117361662493776258341232669470386618010995264161414469436866655412146535572510
8823132604443020594782957064696420840506639720639317365073513173003880445262259603654848636596942695
1483721919405374886648224695511380014166175344131094397669636304892594972328031753099626143939093318
1016508452080323635826442101627431013661298558705422918469185853580002596570936106888670740836708490
7612579947164130996388859772806057254383677775945949487903959265587631921103227650583873043998704354
1540772170815337122816972534708484066784815303398274253546067361331062553640748023517193731976382923
1976068793186399739397093292335846525925254878475066695873672602507496468659320534760333026702587893
9971559384670734638222108716543115987110832363590150691704625433514220561500233993163621226931614164
5163467224870089085755978428672090870503505953911025865125649913517596449387846853627188016450215023
4656608575440062129328084245635478066713704565851925933592247464344496813893964095728803007741566674
0902382614245016397148992671505880621404736388771776520916125774301134751954522037490896163122104904
3534651801900090993521519640471458864773560214651224794178059004235182076670950785800263851082616667
8848558950749158645126619309838058348262969071317067310717197280807768484805796587544413799294449107
0474718297280178725901703403330009230392538061041675484669889573135068038686174263992443690005077832
3469824069242703312234425338168695545684241317436885767212185233924551451954366292887749040043945559
4665507408427025138815422580127799362494827612553010957594107015575245701740197885097699952595617208
6894340915972589634928359294917641877073511887557335239097533861390461174419754910858237894535948817
3409788639450617009227355835690040585622723780649116217992575263586771733782545081928110810215902 38
8311849529597030289045463552392776037669718046455009366389395271293756618364987721355743227583521179
6176974028937435677117102096019708609988169714823781834064514963240191341985964566571863043821782298
0536476808596413487848827651335906070316166008073856450236740863265104795655555239056932413659061767
3191577439406138380081622869424178366058122796601378688103642239344928990155827756047672145288752696
5184984097572531141269697999683583131116362301667967637406820847352207648198751988229863052801822887
3740084633583983983511917977005782448199586712374407288542554258280673303865935123488499797026231342
8093258461206187314490573933429526185556831586472346508233563111689903929807462373018589617512746201
1024135025301650473598739129966877316578879331468114620930006882230887206158330968583647938117175872
3221188044075636045626651016083443677231341484618712669394355809447299220580954003417477116924948 57
2249761131666871509490601304242787861170021809547233191914441841677024576301402297501831938044884481
6047770140408137418939545475965527353961460351911339301223329913329256507989652308090598585740 6951
0691920327933473397266149554170561661608154295217879242085018926158908427708999081541020516351796459
8157545596880694260180013035578554702715987600608073314518161940783672212238515733177317836585699962 2
1593848758839224899291106871277741693137472569565133430063065227379245510562662188080781183258673 13
6960112090771466579443662883300144081586309863172641906333040404508058228112815488916198322932731818
7999585512737838729385096990500690095815055996862377712942317619729404081394181360272405148208 2073795
1363148433542279393342776536861035254077258399965486187460681108541599378263442183893844858 4852903
5304569715109912504423138252492529540089600277653873866357415638827431411023599533922274661428089503
0235763018473899695696878579647461338527462600017284007661035335997001107429588617909344856823861
40221202628100987841947785569657670602082972972849321728359478857147856268835413960452050327339001 15
9768997592348823789600370474350944012621810109481571513305005269673475051595793024141924677188377990
7897096741985620620603365776260662003934751458951212313889676805215858321485875621004939827433779291
0085634596084787274433460556184821630338722910556433093467219264786656314689814420038187211093438 97
3898630381721295790929911274307671977729157837674854628928102080396714378507933637369500772766 4266
7558541999588569725889417188547324570412654538727929383555423041284137013213116316861676495031 82078
1233528078670360575316926111141319274287790444655907453108303045956119947788929764950706463971 68145
1462945312429436355548868636124882656124006440932704627209919380249935389605974433512849065936304283
0104058174606123043944596786199437588182107887055290972403150885180446976720704932193264273274794 47
4657352706314008460022706069559676968769935816122080876142890824382822508271626545101507650711225487
7678312710489858125890314808011532348223548575625016361343244575548970840185256248785584116155906291

5706400186193826787191671454229784270131855635300759233917143715042252368714308286973897230659408747
2615583200288560564305436754730126975992391214093078205163473056839443576608050549539129145864564189
5198421266755703261404016526464718191682556178050004366263986812632605156770373757632282557237224033
7326563439926101072777491347627176169780024202417454219633542368342108998205098399693090096683351729
1556051800329123750279741383363465570545278348537476157707297860282642316989566162153641506119169118
6958507630088773984133144332352216108648075372562756279535732619276807505917917753822080505536085921 43
7064255882040794994568053209984521619529034874717291587880433789802719917575528412495164553897366903
6679774585106049237316474933485138913107670478168931509280056525080740033602625587974658733895886504
5549105944187115107898947642772391413200859027313482305147114710862568442443056162645731317975946690
8304134034969841895264128081303923795354233955145164368583297176727033387996607050275380884466613249
5208470595623603513288763569283104042351423073689830225761893982104514711649750864957931331023 05164
0070274041872292552531255505075196263090141904288158378608018722688533033426032418874199474309487098
0900779689622152127820403638718498691155689285267885143692081119896699199175939442227659858529607316
1102727719303922875037352767460313649872860148590878010890276964810178419250513768363944327798589783
4996707510645916080799781498674621295942992949510345561239528518996648389315134163425625387229042 07
8570031005057667238782499422930431387629331075992163341034881853078945786213043844699022597151095378
4248396404519561395264994091054193480278333814105054975038632521344166525325783462410543137242051343
6561007097502647497506018793298970049526351080384880012217081134423115293301825065523290008053296095
0529674761782776499033804649275642460784400622477510229840904464576088284008438365375725163 30908693
6595011316680590317606539941304678318604500629809546411723922760643945450974640311900040461013636375
6652514763392205640853948811565665156414499546999798961551708519238964173504861644019602822837023105
3195494455715777919567588597530366009882202496886496240237376974190896227826267117164709054545896434
2513581072155099312787001810947933780019428811697138455978130343759117449830503972720050883 851002048
6452766917597291317151609041458699094829963978857884712415821518191513119364659265502246995225206586
1237108617077529724800962855738117264350439772399159135333727943712149397486397926247924574 6309268
8060161268332810801373076076896388530352476243765297729623303633029957531311700321563157739611414172 70
9684579198896782331749802500904092769839889486946584061970883711871716267893503879928186879 0835967
6744386739411089537196986182976738456954392441121744326072814344156833377839658009332416265311975886
3863145756991628226060779033373953780686154995433439666253404137239336058376135713663148496 02284656
7534731883339597694060242585033382282696716025148826483767031393661503046559440350997958808 6530647
7527207618121415394106818553151735880056317978508106998207123263571467145311627582525008062 00983530
2998598765345615877265590577677705235068007347211604498064893912344314969945629608893810992317375961
5151983265012059878144161793289005863320956390449673274096775615657921695125695757343065011177279446
7699754846396367647409519787469857405002566630430496930378286467308890740676217208162910009 2708410 88
0219803646903200660489829526544197361650808708999139972680652562279641804946454156448767012 80058394
4430038777135570780465643299132187060632208151427506224046357396332303341712920530795518615 1942218913
4174393464269860039630400508002042870415729593735309957716677615251624524092900687046480944 5244870829
6372347946634083467678892429045159840508607169534168927276857026100718961752877525160511457 579234875
8289688272500582964348325725019488704916732932741441469386161100821207301489450942209156113531966015
0600951530042151879680050092567027746169752691736633588478953104875162792541522915736474184 880129370
7600826420065741850241704075431717287584945349675596925605810680009173535233924655397658056 11506266
8571458886231368778252027075077938929852212898376757339444061238960194625480149686695376756 92027945
0041465728456706853099431195910398723702492709010433525525980501157121226328470745074590340 5964 38552
8208729191184983955166804404872154753233912561994593028105975250334168546567651849453998885243277578
3503720834777412746381874054586392925447115055543227969184529818208681879698647010465613411 86800539651 31346
7232381249441407414596471311007233355202812338522140507238615369316608946059307567586266790 75053701
6110407034090926282674887750285975237510238545267796205530602752777436781940577953384076593 719123390
4161229938965877540504209331657806959653412589292769002105833547234056106514555429320091707 450439470
1995669836627702247022883046265069045512194383705684670838235730118163228543400184304949798 582621768
4156199179199351241530949158010079374774040392763688826938414993758057175171344755815931032745774187
5763711877153224689227224249337771139575584843469314320560936259513935228175997645633846563 236793269416
5293829299473154160262278104340459712825822426701976096925423356800253834894656311249517014 68994004
6400911880740870773299643030944043581834084141758018478995026247389569075478192665221993496740541728

81911050332527900215854594317573458873276958639774133830265422515748812728581889273284977696911085525
9385652172391840425825823144703106476046481450202221028653330865637440225958804713218694501433668175
3853073158920051906575480158234491548443177708176712520461196493945018862567637414537686456006014215
1261292424645967306428464885104976111918846720498664547599992680177566433790034005684703754475628782
49348984087140077023622965041707501045403006968311235317293403026317221317086176202960434454820954348
67636952493941538476675380957735512146374990805038295646709020523046730985564355823807120013405179535
81775926611792974601486921320263060039317995648494444731611937253694206234756374504782675939732127 0
33728057813431266211173872224928063336398345193036803892593096889952092520352785719809923805762 36122
84960652327697097076479039139423140213336869039480868768474330801226886996 20788037470247047250
28106033141647014474120542138240308128347191009308620269749091536072252751286597749519507014468 35957
034266146025302815338482727764497900776889807445830108058076258028265253984671218499044394258918 8021
8806297599563142816384851120947134995242272906625621310572002786025213027165243573020132199505 31711
204193338562432181159685353143642809886601095854368526010852934463784118853718262721650745414 1542909
2412576342816834642480358183339986607357730940949860657066158440701673506468455108344830410340714330
686135064816123133500844233624141744252203847162068581577800344074422408997737952507722722 24252 21632
5270798286483423936031993600770157814548549973279149571524204771832598625574711721760004759186861 34
6576680074791388238882526059503696145637559048364552493318493757958344708256511202977130549909 59858
8936049041984366155680280639101693150794123984426141463145272074906342167466656220820733874054 410275
5527732642572666260325434141645180646462013591074636904814091394881734915019090588786588657560805463
319115395795199226915101752611355353654804495158052252430920956839139336641860218486574056531386585
8918038882901004954410539504281908521709568970985226604914977034425634152011335435956016208043 34457
0567858284751828869432645728470941550706639004202184742841481507293211425552538487682398405964 036946
755858570650778052888118437054370766859312040436790879472757699767642717439082411422515170231 95532601
492920532485233164430336141058134808121187844540168004984186371953775079122509918883959439843 7434905
0092769074304972197321668269078681894298865543260393479960279152866631367424520499357683219759 8296140
9060960027750075411611823413659519543647682659704837250343357560760309426459371768443525658456 1894
1330436265854933396448813667814189940353781251863618098614310938046860842389417657170919837530273529
46053182177484980676410072780689353948778562080766076462886434975781384916754994801336164585 53592 3
4218744439048728220232748279000438097437222409055665798279254079201827191764073156868788990869823443
332429846171348077941579260789437869378793218192762383208853062195382562620537061570533650998660437 3352
7800430297853858257772582081434930689509974690284212321480543625214410992658451320869835695037934121
927877015483766846258848514803532821005762259531238439549958455977183661774696948771337360294902 43201
4008936954352464284670390628296129253331756090545801524153362697046334156094771470387833114080 455327
202669851691655054580885262347227009185683811529132516075575147329831048174511609119854473890 50027079
66281357373328915219473513171507142118196223950640306662507937661011410909757519875195062790 76199208
415618383126232787053677949587209348085310337003707967698144333986531947300553395503613737 16903540441
2274418482508972553434113291411409921010502794060473454574891039787039359255767719730326629 20585
002184124305611014711726658858875858110980136224271917836015056300845061260958035711462218 68970392656
006767752271278115086918240931263983238895143307309208061818367053322722219967122138243924941 33845030
85537429139510060612104640152772992047216247469060199687536161293095944353196702138702622 4274713425 2
95198346411502152585217190733435287605089694898596618737246471484457540042632111691306 61080007590199
276852772312294384537537344556192707318420770035251888198510000910588697024636141339463 7364030629167
6485024313959222610891431245843108024918533766365473795428101980063946754916567388220937 206795502245
397492793604323187604167371252941868904167875605071581918462839749951776189847012014172849 2253165997
6424913961553463045596737003669827139244746172240537763884525060573831172206305999522461 3827515438 42
6248418305454161423976795807575792892955358231583791059988100013635583580253819304960874 48480623 244
2130196731285727975696380891589114151062682936713253390443220443935122562713425835719375 807555547099
528058650927489148560615014905867221863018145395527154337744115748430146045421047237106 2371630972589455
60185074706555125049620797634952629626858574191061198684337435959902767513512428421 6278738207 04073617 9
01822642116190565452355982134650098443468474989342551941906126585394712155094383577 8399739398535979920
8044091401401301969205884816241657792074065165929884295428759267653257646127865813 6538693023064324
9514872291568341948933977325772383110186072859921382743995531670717977869463102 412096562992515636907
0709797657462230224811183699337954003728904003552813838791669645146305017446678 3026773391568 48851313

37332213455591201694281994463558691199010467025724886303914313189269027234278807655956714355081008525
332873676188833143306258402854461381790916491511389868861020424410969493039946188156434692259238227154237325561865763731911383935082984473757876309088819066974875034626160773620105614764919585889526
17573029866000284492088330986935638966693573658315432198051146302918030353282391251226514579604511291362048141607426348368904872534774146049792896866456718043110963702069366180038156492646017632282354675257725100634810833246049639745818948562507159279851400688234565876643286161694994470287617278607401627879376003134030536537200163610673119735402165742745526613772464146797016343322109483927352053992161846684431657805148578748738995611735615782759272999789378304117463423317231236870629887993963796932343814093077303166738558758336589487921922929917029253219531310631375169956489555627920773263344399560699134012345561866002723864409129577681745674556204269696639791492485463176155558347884911203132980810093820200166708214261953839391351949433245587415772583691116349214048847335467048484255276266990959144485700983160423866343355234799524193321743127082642757150153738114470464896147792343239183659760417327606918396476067866322674929193323161131877673919132713156449130569370585515339505822922629739369280099891273744011072991307584839194837251638752515268120935615506896128156527850437543856737065968657129040745040213967864098050162871632426642267337613821529562365220402118843091692449620167039837724022900775191199017338872565494516702488423144667330169795649313895385861239811166830825720223337246985737787251767301674688527011564277582005939357098122586901258690127253477512459695452503898261166802128775738056368563156444219945818740281065680175318556465295582286169552862742002819640306143910590032153379795396930218519432368807946852791407258771948469041176416915227472110682439096583493681740724391256726014139205537504438778509718690612830895421445090454534852381522612136091363279625618713641431649422139355442206005338273451530798672306688013562930131650176555377047163009150624792123739193705754138726037784409421730259112504923861549381900703973226982708445933093837163148061128341179486313084619959430789684964111687084337933440570079526478025532429988270212239589071572621544919883119891179649225568725795187376437265210419618862359708076370825191601974822344820994433236500401510330803345098734208212792141200420848018050107979856172321647350440136981148554105881197260873953942549086287378390374680988320817221479683074313026938154363684537845761557520199479117741233670859357176920959284028880006291720871627379774153512958050297099442419138076287230850635785570220029013432709277729873751576162474840473928515508631642535028352650155756311959083682307340304392715181050275265003370868949842881323568496500724939884740105694734338056373384023842360725388733091113738840764500037734478470910018648045541171100256140540878369928869527413926193790851385292297895810628398059044992413872737639857194829124848347659740140414325981885024539106705917683246226948397361157181488531198911796492255687257951873764372652104196188623597080763708251916019748223448209944333236500401510330803345098734208212792141200420848018050107979856172321647350440136981148554105881197260873953942549086287378390374680988320817221479683074313026938154363684537845761557520199479117741233670859357176920959284028880006291720871627379774153512958050297099442419138076287230850635785570220029013432709277729873751576162474840473928515508631642535028352650155756311959083682307340304392715181050275265003370868949842881323568496500724939884740105694734338056373384023842360725388733091113738840764500037734478470910018648045541171100256140540878369928869527413926193790851385292297895810628398059044992413872737639857194829124848347659740140414325981885024539106705917683246226948397361157181488531198911796492255687257951873764372652104196188623597080763708251916019748223448209944333236500401

8369926251107945876744271026054910183771414939538460173089933449360476973301928779138027031350707321
0981882099042177479582446759902008357116817844883047873914753449351938140117520881370598439846549557
0550101894746331785135447806050024532228986959561098127445372003405068433161951983630318170374789862
6632707244065379439277750617384377391003701560450854244171829462232309874159926137923103063979751 96
2906214954936751493482955342557326734057366254563878208772478017012519108061380763294191048376862061
5503990171057754937943719814986022745743276873586654721193724736483688847386369550472304586168757789
9262417991924288808684663276562154894537758464269366017054904106967913965856504723142697926688920856
9296287842331651540087970794844046026602059142071572641491184272577254451851792252299228998902552 47
9313295561629333876104160849199666317442308779405887350839478730728309916977398349794368477463448015
7906910420838749535492617591718541229359751899008922262172191445930678861976035622188127806388132245
6355560680694152552978974969052102555871166808039462533165815590264701717990039902483340191847418611
1779142508795783220037431338999438878790174932178328570913622593452398799211941358296486042387182406
7417098622812190185733336815685703235943091990841310032742416308556070401839676454219756598215258 34228
8679649061753328803833759251880213557889351191633363591256530761963446885955556790319313426753383306
6316434060769725033746120456578575427494456057731809473490446364211529985330435369838466098461376693
4308461700054908361012391654874602155201807438320463794894655179315856781079228746907367749 9657913
38535035133607550732132560029191031551310728317711625077255153125730496232722086022521464840 22028427
9859329283668879907714081923988378503562205294533615680282037531395403317643158940563253112 358682394
7289210428147770904529570491287593524120513586876038284514258273460944885456431933445606103 92771958212
92194131087466667665924574913853915794463552398869067679969535565940340083926630948915637 05155183329
39400050322641708848899460174496659076686847228668101834234733580748167860825699263614579 41434157737
96972761953625148160475046445713935285259217578395278564414202769631859151199253706473854 375073484198
04585231099524566402639452567114830535968678311138004081619234069234032609897041045803793 4836010544
9292069273132391092894894161225287725417258807157698002902327692992653967222231259542378 97818796 1729
7369231269862904132940692304070608593069659530828706334988535898063170379065510234550 3598110
4722630784324378875026808665286661970123584310754871934469961351102465382307632638598 94685743530 6560
5827530017359129830695156395419433781212111804407015336153143845798731666720361818404 4659229140 00728615
7047092318388837125901137751101546728156831261731358454449555693404016062515912214 4520263040 0731678
6242123473984156636061570022957579512596067890949180714872199201151338603342012490 3343269019 03152471
1111770537674903792507047710800780724379721999748678051270543865809773836608455714 1592231311 2507069
43850574957208447061617646428970167342485313072653113141132969224499732589869379 336205382310 43046881
167270296781793531514792526227715087317841347201175082919918140024385165187293321 9223671393 302183937
18018284319234620686122487712481614464372623846672738716927043432266762812335957 2086575925 299335704
69301671377062937231618402812814665403393299764799450835563529313404481499746082 1573143678 27706020 9
03620002365147948179616695338982033036431519760589388240313104586462453903945756 3768870592 794191446
351316565638530341380551160738301152349650160262898337157266540895142764548739 4364299003 10464975 6341
864670633839370264043271999344450176536211941839809140543543120466110710229491 4795728223 004888150989
28800520298222195603137155531918072367580873515739494158638634692426492982712 4687428737 887107211779
6093428528221850751761203163287865050956785978964939706685743639441733419056 3835044072 94386217 5261
00606739944657144525650821851108112347349770923533060731560348336668128028927 2835525949 7834686120 7930
653666229848868445897302368695057318071709271048802093698860052594234685582 414663237148 3603347 08879
0508346751416318599787318213771136093379679226713048084006367127404476190169 9021280536 2495514649375
6285725553472201552916273600484207174507880795076184591360695778824861874 79496027948214 6832553636964
80557947519580422659618175754553725757800638500788289428801540451472866450 7693636429261656146 4092376
09919442302530717760646766407489013007128320670095834441486795964288 4426565334806 26985008900 1435859
79343204954700737907197624021831864502251873557681953846695713070062882 92263756160327 87642159655 9312
5292492422161329574504065382201672623986186616448514342988831519203159 05796007336630592644780 1768242
8057937751925499116138740816555196180080619144877294851166256997724515082 137146328190365180 877334737
53221390179569455469855894609950994481976623160582738973924308531049444 078072692069064031 78359352904
6138482240608140130673921194035926589229119690168304343622797153048 4459807315939371353990 8523055554135
8503033059942630753008447497312132922813900414937292573158103903671157 3439298137236609677 0703723997 56
47113183650360610405488716098619039618499361053903230560359089527165 2168824412076000293777 205420054
93450599593284765791445138948041792035850852529938744500649077050302 0426246039291347918605 29641815140

73543083458770362016261363564753671667605974780085903142936133379293319175008485509604389718110409887
75200492122237069558581249625713314476066506933678827688873103244542439828907735107267733266784393233
22019262193656446632205485521963628598886921245905984414538983552431719342491299006181463959235638
28594061690297564763112622914667465366265823575832972216617099689212152632249540972533009276090968909
29815398547781047546039704805640970194386112437832161330172173520169538667789380518472999961133017539
09536392020079729438433442713241962980107195951189408020760785422475347076597726795708601337156
1484930323571019813287494389614948548783732970942786163362544363330826939905579688104948920375857854
537640505365757571955571940241392108268760908876052263686540301870822701883547584723999228940340795
07951367963796353072634422455419128095770590315877255589220778969111186865369280723830240389236271049
807288831287553507175113342480969287697201586675189142171434255763904895946985807500609279810043909244
9452408176625095274172741560161454315945185221718557495526847527727167389139029014720235059895496157
41731698904285539944028302783837762365010859093691920154027694868144088268392159342531859686296886770
7353568178360341970518419116791939661833729700193822980150272483083508010971773091110587089489467230
5428927151182641544091066453295323935450519305278990931728114996492725306034702159871961345650020
631071336136517861654416497035736820976258374943863039224877359667110116693132791920462088898285
753887107153631368815546679590153314546236533727109419292748404169131454197001775401916619419279523726
01888253730720317523360121961711612555388978351255985608651257176456894993560204301074085289581586506
95013583730270205832447069646396922369063859331108112201396771000674772455361451739824657873343641453
51852124685840728801878105976487177630797893325464580093215613548419484217791755933093578596719469919
50561912931679934411945972942434600010285797589940496943656661909757932956680054246701383744999049886
5697217529786249994440386325648733167600407677690717691102171332461667136933577677517766874823798
1197416413893296651098321913188912306128832130617534730645943263123690219424876504426568006403723735
65200124319482373191116415861199404162345610705558803646260316221668784713489649977257144363570310750
075329860063330828997051279145819779206078069420505494926820444046342562971575340949274355797251601
57207979726607019969187009861222418896922478331223069883519302680013335158234809746911378369774486763
2078154664881263924082841865160249149598644729864447209866450518203571764568949935602043071048929581586506
39519173056385593822175143013143807598669332650487479903805646835095656163984160231551047965985593168477
4522978453271156302256796382456507087383359848615942699923530317472056202203726152708976066846029
608874471195186752852500561782972888719675246604895413107910513002573103926269008884742371565183099167
35467374573994383500092516531918101842370641827844631996490552885699393252826566262824286907604695
9122933888826367890525889629799260049345351292859166016812711585038595099200450238288052578716079997
148577951174107145878927892859445542297574892663914906061225590184674424204898306960326099241605373197
8039930958031845874187561196551694103025884274613267712860416762599716689009197446205707886061938
2471443068200786999515821869152348094620599473367284616874380187540244684311462271997183540419902857225
2670067522055708177908020764223877008302314596324376972484330228137478014994596787965245111164756288
449488239112658022320723393967248735348379683470903721727807182199304579617087484668322607548311946463
6363162955046142891818703440255160661099604396822797600851051090393629199421193882665513643110459
3773982337547022348970989382383349606222414588181571744869858007680176809831354489908073098010459827
98840671012861381855297951311265857946279763440209325404426562321448785499837045212678648629596772324
59938670287589062826992794921488808890259752971774789067202993671236787634519865957080118747717798648
4510898251955339140526404027575286220156098390974367883923433686902937949023886092769925569202312504
2770510894350978322370260097877221028883869173065202970742687092354303768898474913311508457272408
9272768529320256830382290269854983092664297981696215482643789646128368380420732092444638106248237628640
784819581085473378891721650337093171623008278354460953550001570873258537182960697550817043583499220444
3478239737270858232693632701760976748450030410906040116877483126415724925431367350659699997435146
83113000479043763389468381150750447983622497756891870633121598257365693430609810105810751283202846
4444669225835877645410765424544616023777827884230143214148376077527228666458631768687572843620346387266
6467033781045583840869900184700132490607201629106732588602061527200357037700877236393370846191528322
0488231140350582535412861184976918987401837011971124740816615401897014577602302374503811231109971026
64661414041857261089563696058312446625103344217698695531230691835330056188518487114675637471284253768
8375360270025404781526769636800281490670826847143662704493834208884829795605591530914319592305379389
70909123501683175231569389220817236677947717971362719241455588808601903804074946015109518157321992663
3163081536727786395339826503769350931962174930710360546827463851923840108589380482153796057037554136

341945317910214402772440022859505105250788530056362548796203963051416750538902815483989382605618460596925430923050118448202444405735333994846232469861642712525520414183945458833906450700211202816269037643723678632709238480835704928556564225657004171664346794270566931694659035536902250098110204615997069222696043039313415011238520208433078349276852128023229115251973791375174328405671719578654820096834839493549870633410451091155802399789655537286716719980635621882290898597135945659965583900719908411352900703212798486817378763769197501596503476230449267928002979882816144262470454993187358737961144612207504177376803842330889951248927382171911705995177134534299457297252152140383430460734029321292971835990233671675519020483679889348578542813074091711212749135188966503880595036488660019305177479737720060385947944314665010410771923517224472581698761357315539473606361022608171954947748733465753076389829979706659381130355669692838308327669547610966886531811220411325508888982062797980648030112172279233418196079841299972072008841839387221139034724731085153277483669837824867965448539605467332451782818373712906848843226097503190635530179412648238953511473878639905548602246000335769136031838256952239304394162767786502263671590542608217941626062786534137817872842038156593007446364067398966754926876493957184903132121136562023902615229840628730956481281693035018696850371310954723293767472224785729641708198589405216965981052533788923335031987258494889372864076832966453384003413973368465664299812962074532556516610754825373816739692277169563653668258137853929592804863934624068004561289736777656499264445275561622239943174891109978681413034087861609640689091934466010685781739964966919294071519797706196735563278083037486762428189532993799350174733304126784639410742390004075198605918156465947758609699994125596896228698815789024265778977792045289457268597880151203918786449716049993622264612495868772143477181778272154033175708743833318094502935381915728145418326821124819723259721432234940262895469574762101080787425847657147801608833404962554650632414744423711596789923705898277665685897719457925992692904083389327233247930254203274120827944369363547913979579720396366395782104958416314403965424093860475283325924343500781306594917949907703881670576114485647590147081936248896043683708308730139992537252983668910410331359439434771325452511125521110927987277859513827089163665909520741709275251299026204551410308034810322700046208199313497741033969935052008134906942080378722037904643828902499024012213804633979802421821068393468508849301679182896621304254933387996548743861093208514910533275877322690267241292962567364590687559148963120501742439452638939790244232370326490359685257821378096515045449893186606855909756632394422621822925196564215725418090320998044966138139531209853479671146993426949382451496674751598529004751805272660719775722496708151505803914361824012521809293563442697690775936654520908202506027804671493363267787205816815970250481840076454284365495003371942335655317061100284423030042530520952933086763760286456546110382753741547484910093128725956442056976109993020881040353186694892623960995356572233574757431587861845904831966429563222726105271648198398546337943708365729640939488820397486960694333315145791639720748072343344270697628656850889473094981597190683110612001052867521309520111063997859704188194278438731917963048374603671903556930388399401548373818626248926185175467230946628362055651216917470327887284570315345444858522369060797986889224934532820969343456761679260108472382625897599052637923259164150327636255946077462741704332544934574448894772168762718277270729479679929407037256821061295089993702462171998989446787668627345794135226403349817753083399390702966521338034698372289053247600661939254582065928970141526129507274262922793246579170438371626932075365011196015575259469406191818184837377134268344442405293060600533445888690588803931774845117741023973587752822843895986720282398743743529592115562432438928963002795068728872681616117730356952723169774369592914284446211898945750116903129574251482845174419871337174865767463539747457615954160878152194938038219063171978546364806877248861810391894489750730538580409020796321483089352318480379090666813452717823533224661252194992676529142759089092262117510817467050005956809313519528400804390075726178665775125745288433053353174841429175337424877509489933543735835955457882706037397391292269370301243965689712377394516731859670419317393074231020534492793725569514349788055434978805136093209508340113045524208837745301823471483685405703839803008349014666157562782972454384494736538947399853428754328227478538137311632469929383670295832152967629316901577016376459707311545568276634901948250420327162355547616062896101031778920813007113313496833865486540725260994238188746548272862722781054699899243638538389171009159271708230490667276596162378168640441085875745479366675438596986709554749996591202364718630251342342866031230832887254261484650491333914084557134897421213326279563751415885939383437023288367614272109109163264381199307118058137053205218187168903404084228331495613974141000910171509937355011625049869802128077552418620045870896844438063445989495551440098765192200443482688701270198303940692242853914434437692525690603785633163635916959756362855116855631624525027745376196294289043661945896802

39185158067914386065547630666538550897916719672277397520763582919057664796718038856453881886603506454316833551245320523828327827721092596297947436508273341906948411747713586694162355151541897665027365243778292737501090510930825867898462418684947221879450928673056564729372657210556932497375301720300342984629399035760405532694801975212600306698038435039935362245864967571735364486622960867665022114761971906683670007878619525727724960774953015827040191563034896276315535912935217020929815099957118977712469085446761448508350541333397842613439531495371915024132149252570111457627010326835919788551641036147376475962509762230288111889540471348042461791153854163232840054371084642690950362186833728775644555584451321260709365088896899626104066097271490158259265168477635062363573335276719538329789200090672985653235452227481654101948007407049839188230279326391449423695735288995277041099533947552827010818943357336971843195081657518177313617892037222004623202250257101959975795240224447773214620837600834855387730623739920918868352510196397670784185032783930345616401418765456938000416667218785983018332497804306684137087899778060970875131122453733179210477653219632269209206244411860240382393958269384876943864799215827507678016075065336501924781632884895067393170475081966462719511896879259504855981422537353491882022522322554527064031100450586493401962683243996750270941977257999962112631549818066293540715583610274971906518427406565937254515257474213565274061255142087368319535891534018300564376147556000590188755943248998734235441853629889772464142912985189531060905307036852909517474704662175927821270284427202763242218865003628293273448127781908247371971787331228262453239033105661231369437672159070190562786251023149650838505178495474625792863354846764750561938956048712724763153542713060257324619707305889144995762866108051940160809851371096016289384743787627083980936528689093624135397422309740440123377345283506225830076819495350573727124729146302429342011820559428587540967298247743329952532893889102882623850029186860662230600769541453344014153478027435465277981124805950108815686653909581055179251789261685947619890185128854853300197191365805093430865137339156714425310693345585359369080573111213522090149884322616396432630776114024959572757551801179589419401319774573422892233099739196245423781531637399205324766455348061014367306832579576051667436473662023462120548832579620677794618961534666628496225125599883736635615457380994239822341397785731811852629645092193334000396560522190434390795218769528629536258345114288337411880138976683348345199235437275950997248847549985348212875416021214200071674225273228186584713024374038012472127577115517354380686932178170984693047721386934362393518517720943809190247679191235016341974983001943492515439227328399895275284543090800613957557007914170816782579339825803450530350435599716301845528168292642279637951739982625697213931034888695236503388767235534591792138831157879766244044445856686261187618660778544234578255626175139151515217506997028267121482353761675339029972479438694009843980337239260825759149712252496999091625168224188302770648315381122368712756122608584023252177282389919754616966871004680666830513940546830147066324372807917308526175004054045846357996438713060250466532406908513711350478406669674081206228084952470827367848967506868606569520461593590640327826022810236520839777490099988133930572493066866654387869383628943125351751613853047656960848342689216379531764454189162730517522167897208041102237228388620965663043269375053812605807435715564425203015360659827372446319420027263684391452911352321609780682089802503971154135638074184333838437755946588993432757328763589953934333013215225900120838600512520109318668826735672604998799535122658646076878454488418338341366254221969714632518921172825009741219838894379964774246111828756492740080109580681071631909055544066376841992483030382445386120476391893078777478409553293677312666506230463491542094550301318699283587040497769498762308681601198225060378407782933137148693195576904124809600289428590147156300359952118751134960928464643882776368616444290872354236562624184913110970711588110759956848824186276594293115532643553657810786249366608887276368164444290872354236562631229233088278228785861246220198372386815821566525309519273576065755536633813081141261504182742591791584686097566171115535926504724528901397973074836568456676376607503000803886827448092560195250182287786775116835185139009239907351059703270669619154073287289116846607505209092006071456463839356591565542668711062586079966340457758882769823503546991717127874165892379796117044330665499088194970371992812185309204245501010187280970744329043394827028863200729296823007160613009672627295626979183862541923927466039000712109973496105323355847256759415833536503896957888361222776102099081807849942356230921096520200749681909702336820464796210938532210508656767674684835432116346982157385650582333733814319900742026349784281011684895754010218975450709326652742614693339080399046753111520241502483200566561580635972361819323007628627294668878379078853974048969533931470031131324493093227705326130028550543429077890064303191002285992341995905453419824838865764191083821664299146114191054104072

718375771550615135153927714008775520012284097818725816627089312739645464772598088949046387444120338340398470547482648434216601506040978703871976045332968159456039741792770386611691754030605604275947492737317580608753112966079880717230218830918163103554699678999676814113223840584872150451109777541309273348085613994313891956459179137461296122743264902894505828696018397663266768184863672978442961082457532735323785581012799169576075361634928445715479677700222592039479012456471885952733235538013204986707161550915878288956727461343961549590248105267578991639561562922800247341472909294565424144238427975134894573060583395554662066270210014102767079458435211648908816862436979656823419770822333130158028218768411671028519113734962555044156500322013128780720836321567275832894119429300942017627734317042322163016969037110211968178145961129850803567824717557225952337646404023992449941171332270648140922089039340677416590793358224796176127195757906232160753334804425925247216637653281249173787913554545318283886538707564763937340816244449879336143123185696534013864422093057439128727633815813871255067371224288300998581863210156353494022783105631170331076712499040051320012934897027213099521549239150785904214026893130046098656152361405253039272543131409786703723671598135087041444155684740934242858068269691887058701331464632050819150562489476004435207087540878211494946211509279233564167673683350164228427865293392832792845332151528920409430120008170861858410750441576216810260608335682836973843197136510829362124680025797679911553990076484038049928171802870208363215672758320810589079553764134354905293957087659913428997297701816149476080132837284371590590628796866400470614917846595143380897979017472288822130531415452675047969517343623347261533030009304974265653945794747407885636678194708758120346048621221197326834589886758867512355607721231422483978820693357982069335970925454114267856882868576218146166824675502952377526614089492621791402342151635411775702690729443076957090896064414965881671742121668318114963709144779139340867791720363604718337590738200996909450123084402978243198983074299124747509659505402432113462983344156393868466638451130417168804680082835017990694554557428581327744030140033636837871236116275322391852009315108695478540406038514296753370514492285816823175467578599332489704331947481163136568762244209211961639847807493990632550658610472649946278570911848293076400523023957169404530229774843375344969347910427880464981102733559380940469348957848319661916191568767866474420917676961460159614300711876371159818435709487634197399138085628617818195168335660513197809453225854265525165340525641898360419180987756474700933335454643863745881837071930899277747477561594007121001602129242904288437751885569693798413746195948786640495288517970299440341709225712698364347792342000445089724019564277684357400046918068857882963825556857695524334810592353696323776654121361365941658964893601269083039121890796693463878269946256898943284269479001915449176490799259672833320150204055065395822883229865421520127390385712551158338946014788267961307059368446271407317663585074877351536087854710599450815737503768721757586689747637142045085859347552037159289414490384555188824778224886065676817948448850542487117656020412562510816987302947899169290417780732082029453912387288505780471150279434067819720679870667734689917596857017096422214988438621317233301403648409062296336613973120512678548019751401068761497867822382951553014477543880094209195811190845593172841912844754245930240434415604684906365223220703918197954739023747928898430628755879895504346332729229426581981899384964339039017485919007454985194324377468897151635061784044765817263836980899750933160686709027362906796736528277031546320116423755537799298474640332323739853551610977781075212622968949860513517165602410287103772412949042870755589199253507549714909146833653584925310865474877713862288757560810792717858979025975918863519630604566663336319217407944530334592773012404904323289169886310725490859039501306666592730117026037662981068329188801540077400682229302138595765423536841724364975303391034247595466797769708002737594358006471524868350668199646207850017810354281282583528653403952123279660353632408223180898254477105204750370425226479722869915914522430007083320007429597732272579503765299376768720265918931466788798396187665084089712207162147080503329653306838237586478099701736217752518266225944889755547910790029432807377769541203788819385753362453555755538621513721579048564519552478272383904392255558608545983783242042248996058662215842368888782818750328772057840977870899101239796223592813041542814620670946907294304427637357079519463824063853597538932514553204039865813187666506718012855292092902813884694499131489621511096573538273671105194612560704832112062881259687496905332546516609855153284705020721844897915130385996182707552533083094178815330731333348324728777479051810609940065062184695791416090258633375370269503363251595038405743720504028696419034506015434148258748213594896680516971622041292189090136519426616334910151777093541878234115944342573018458460479674977341123967367460769375849063542999397454530071743091496740145218588375807908410093952512399394188780009800085298325011797155246966298052393594260533425668348417106596468996024067593181873007 6

0771656964600274918478453928375977395610105436229722833079671242759581913381790783409621408277302609
4598302411681339242540210247908295842719227209123104987777436008228204047939823835763173244317194831
5697133001082852534017809175846522941747359197349372218733486503775663769454517348058412741929648062
3788474696003236329645607187500856199400629019636318143276961079101024847744994817747303697518992283
5576935750144684547038179643560419274820296641148426475538846160928431736673260951171414551466420877
5937211606640057131871831940378249303566026421145554655096381561717598832630660635415029101213817574
0705460347589437776573433474433136145706954995855206815968719207955264670235033228932588692655211583
4045717679269786980393658441605217539839796914164692130528871247348215268404863354416036667164545205
7287892065390396896570098833032927822831246398832259368184889730076202950191392217469639190081298244
7578103017840712413718133347426091538063731963420372270013531282315612820927307873336065731882243330
3526775361685144012848142160469279280062614904237264752955289806723868980112463526170892236094195142
9831850549387764220559839784235496068384308044463091873898281103232617494249024592968705429095989327
7188672788181490220518594249649786437220191508458725241513773308659016343738990369096183941846604804
7641285774857339602432888485615948165391309950784326152761242743730419813219131097138233235365219625
6565684132100997793465867125309809163126945456552408670990257957373786907357079576233304152045077601
5138834558474196237479266731639431708110461614928063588389189301292776504366428922292524865961964742
5015639365130455542218413698115505602264569224268844270921908249138797460468842621535222232159695297
2046003562844801805143092351464906483155814707337390990940330635162847636452430703990228906910322690
6335776036486055194090278268031593780882659283867885892833981443121074324210574440779725530487580750
4382718089738160582946051048302938363211204406323798531018120096800478401312104193172311588019894120
8999509449051823520285501747845472762059863697070096215053673678007104018661808141385962780769153033
0859735979222742977968064432368930843822161613445029092444424134286820459892391441005864948555982060
0284922716267782026995589742281427014267258362020191046924411432481136367823856331661678230591013029
7723739494218220628532292532966281056278942937466150517532071023254039560695420249982143153977132554
3297586855252724801325259204962363918642824022950565291717349820738772748645347449926663833468080472
8431021137809271950366939387088898079287353281533984742606540700480084432945020104866023527925853131
9653132245368773309854167066148363110682779419010528654438552547588213894308783875546974389267645496
6213807228842572393450520834542156644577393590327231975817176591609149922300536477728381273341666225
3384147224262994241192462240097784229782912784403926399816969246024983199881028202402018949606067126
6566100746939708920646894033570492380927010701053350938561179427302169798825354162801527272038979683
5160423690238188359887210402920190710560875100167903711110517939171375466236838325414471785938653029
7056462626094815960597311128207255718281113324607610421774775964548391117971361887347487868253984586
6897492106177035032173667065708216985559866053152727023642929621060332762951293492175214297488361749
8973053872979523177137651695656076009410257209652641367228047189019484536577305232471845768564345313
3418091260257540139039411638861092776307356147710429820371485588099882807701182076886043581805556979
2895134939208502703609985291365667242000409340681562664800047750369267010067187565498302678403949779
0284984080411286490427373187832357492335157772665464058755270197331742152553436093433358781774394766
7698653410903424182885600468244271735589195629962507979082737456067765384902422624243413944410514766
8328626098116986962995732918948031309383657877203054406356888087392173364316856301974958824077856564
6911005844850632214789358625449035009462428781142479255709287588160371333704989444793554135
7861767774259300350199487881893546570699890238842859434023529397359036387799554858018444355982316
9084248835355006567840024862288539779059190210100832664391404237858347141539211252119664412719679850
0140377378161135469455626393439672291698397034431618028268439565881785670940283583877603813757838647
0121363152065502850438209268547160480494467248770511152956399198461966070480199092254387591936048994
1776424323787824512750976245752855490091969663450563250421582478503943865657347030265069800027224965
3816227554125812383002246685055927019280952766320805532391360748859854952576989950792761407664345764
6429097181815270408434116864875195291524250698686972091219727276416639989480352938155720610365285429
9804227933990098463092628786791888447458228183849154137902575761730557372190989173358087060952118139
3922837017330476881808509917110905044401302490736272237452998124794226518811858963808812967892860274
3502479061364226666970965563306010179050522655425750442591978843096292810306578173471637056821333167
9546753852570412755704075586258683249666663990627715071424544243476380734993509255726525019287640864
9276184971730458976251487164885915989512535752282229553578092755157717734942544940646365328643887343
3421753070279721078884393578408051947756982541739321292803528190432830422660811527761503372809322222

16146272802255172759402589140495678040276685368356006483751211956556037929089017497705688289249537680
0055117044514709040276477095266126178911643527084947663336330375047626681813493831684698386036717861
8075709038409994416681888508857156733750859452381605828850594971914115773519469163826632910009363196
9376262656339971438859083062405002210685624839375451664538977952645501454384899174219683121931401372
9951184100975097941992374688409542501321236476707532895687164905935102896844431703315138814848447542
9115655149549323986047347014309945309592965732996400696179056571891557113959225222377996616334929240
6900212172351354308093758013216123137754512348729033561460914816427581049952002360871359854964801275
9693337261148782204827141654286109728738531498106973275369685591326889312465886397377860527964731457
7443870375512914383631014808801025549750185056343837423264942884038079814325557800163924992969085284
5898363913292517937062540150226681735964657598594163322441515757367266219230425695566601862213826190
8801925812037238815460077600385904490388617664212001571276624408763052928397860029377083531090043918
6588120008743195074102899806569379888701230973100697017955799260447386963611607865159837648650167944
3285933431962812865551608452919059142677920493990411896788778867430392997576167646554681342473563258
5818313412266269003748336985031872001174603213572511554875102276922014334441704147593650389209545499
7699002142804930356954892006008908812289808480549761464239812417065375453403476304063168249955435889
7677978729067940476948112679095135574378201752416467534613297580009812957790997270711903936380944971
39562962587268238358947185416563847329242758695098427673772192332175276539686606177282371596570821
3035581036381518279132569446119309158813353194273406654288407415997081949240291838849762574831533937
5254667365082843965540113896630167639046301783547536947865239365547983209434681476337898406578472456
1420759944989776142875174494581453463387899796025595862629754099042522142409274144442862946764111157
3392060689928399591420734459182101298789382713047503738331667957872390062834881721234327931679078
0179277820576279424741963951177750945739527443895863353347367966070550527011421254429947908049134644
3573810923689307860356623664611507441594192307991223658035752635962493588388549945357860888234906479
0166624557209478823103870099917210865162700174127647893143063670317268396116681796735997405243721
0120623481940467735151101115013575303935536135303983876540463630317292404004393454222886650437559521
7963855990474148107366357622269326643306422466225136196475599479394751674283039388495387860866631
5660876086740168682542438590930983900950824741461151827971514792538782362331160391015979083705326
1335653050108925597369478256693522289815420801446620485501976532321021816166242359303465718412880165
644837177530378575721075721672037351922240314148705332812556025237909652855171060447447911806731970
7100206860334909523369289943517346016999107184507049095286957774131794120530566393158259808009454159
456847401673376199344464418595248458578067918467182672145793302716286484233254970208681474069158570
2483014262513491301317931897383824525493171754034351063059441518585179933228918463985687661286780829
8214112906685003925620774769600566532434851625148542827048561423339710633961326931405248211840228020
876493282440070792951867118770746416457364563442226184718142483548826605654175590498795369362959
6649725411837193356979846993998266970823283120991093412559948081987322038686457497615007315013080359
4050406734056012325709787469629188299464967009553229932888316237622770234460846167862958418100330595
1772290600660898581303058313139558858048276225962517551839426498063120045127181001922219497057669748
84459659269299760162097226642341433969809608501454529911686784527877225801508574285976431805040716
54946512152689507976140983569243094174676545181719566747720444984286032696803718410825939382735584973
8685580513595228455287596361386027589819453100170689442733247468724295891164782188536209845829683
03040751100830546676612131694944656386623369731490536304878907883287404207267338339692583482813533
46261196639727672957698744403654713600165916747714238618199164530627228981557735662292266108971779
150083462794644093605843157320637834761951700016581060210092878404435682065201945270285682643221760
71358160175222197346722332778027439894359711559780381276278065260467038575085560255608105166677781838
263691621120275954763750927356103375651797699465779495961144911621311679160046072342568134822109174
70416102540842483992404235096219691263681209164349037926693492225463510170434101546543752831620700610
5390821226693539714144670163871339195724562013513920405091861473232182936195189123645492088239049722

4882879142572991339778222478186521013391414371601077808100071612966209368046726337703019405918478585
9655730889364507857936000087862868663380797636890280780619857010023224477251403933032211957205696718 6
4238802321863347761125994354864992447475166117836031669526375416043600635663238710592793579217568711
9998228111108914246461015465535659621270420999440319758735153439833319560989417309386547454651409993
8939769355339224586430358876231576156258615874628728187813371123513453878835580421697764398525992789
0499624290653889621582122218978878116258382965907363248496977987612007130681883372519003703774034872
2504297349609934766072848640001631992960669543707142718319214099244239469582566542054191451204553361
4236488414816026053774866149864110283759516290999622092132981921641786339239642561390727753657077363937
2511982297369678692468915137165264869759651344357671228268583753144012628041840462286935889735824637
3787848447481641210733816758775922822307915498221080479590434831933876343307339949925194243372136321
1915365810872552754939034970117953105652743688849440854744660647272707809097309405202816129582
4460656291655076968023561451397859995356144918568579496089902815409931896227352107475824485672452 61
9510048267256152701723009343943029068053019264553530429770999782891887127297756731225080121569089899
2104620610303265419535588308434633118423243266793502469849057473910493235573872862497868075144049873
8143435118385255825859650848619765348305165577465354849251847062358463611289612310634756134878291962
0808029021541889567169547057841263606512805097004486533945692126761076648902182601513600217640420759
34304251549604712165063826907266881060328142074122008886673494151799724644431487852352318295667 73009
2434525278035472371332498112111447527196051728526390932400074100850410135349534043775070868269290958
9645056777679550518169976547742490749871376899708054222310370399862341314449704808333886290361922462
0994170065952755234994560840888352771199512564117887487755426906953789016658371705059760688177908491
1618570218746070392004112101273866342584584512364593848810490498891712468573926962218064811341034779
9910285334504336352935599469991079384807309383222418565543065049769449583857095754758924181079549377
6431689626073436458913015274465652114578614155770992104933437132765485502078197577561301902631627312
4397224267414660169404353621126647620668955718147149979122203905936045317129530955540122855108112121
0265729697565469723178282753593325263383311100965765815561173594869547594241902955820124457500907537
3370460335622615497103954431129334077832390857018090461357962884855271634026996506312919009426960596
6225589497975628721287757198542419674753015874634707330507378457961453545425206442308782408414080 78
4627967368738878895852044709279081010712119877987278359242339335755619371324997959387609841447798 2
1381826503432240130638574544885493589707048625142787242965210065664978707015683883864539959036508017
1114364104266278696724141998326567334111284879330939586086789075434070202679378477434535326565612 9
2901343372346289780107820115522188723937312016093709704068632553677092619190039894188411459261537011
6668658956367315052216330410779848867721149416004662110658605248504107954794838142514049963682497073
9689414424666717487367463168517045635775424231015058098647646417391062875459904737924888107274308142
6295248909319854957628983241719089619983841001815466443780823753328402254416041489599319087300847247
2855748818376975818953479390880198458062894211236113335463254004046653571731667323865207845030801661
9577080902713241189165462019870895546091872385437261138749662997666582523147148148008032379500380 6670
11859894219599482184048922604588548768880233759401421579154295287123568227193941659500062439118 35934
2678164048354162384967924220808770996375733873397271385700510172163901028140061996402955555109205148 2
3727888939586225063582005762355860460670033354514888308803026990796123184492379733390033115058829483
8020687671024091648595888839443246563446675746335849532241416518803282548820899182410294068318220651
1712210563581095080526760824300375506483382815036583507743940163080243740097984829556648727702134170 0
5589113966218009352304353945888255902667090006571188533595233013070087619024828655767907621690733940
8560706158893619301348780259524967031591436244219977851983004379844414548655165908282041411139376731
8104087719596578936089006009298571878482454316218024538611683785856234510782849182169145864330972 47
677113149590308273461852478922159576361761941580341303131918416077594410121574394177453130001079505 6
4422525109306737523223688543242791213530557592621003112211862038297502753817842541731173375517119681
6524092843676630140284331236103292345231891837022303444549741418863973986508372896066816943625131631 6
8064294041596095582815279474489407960067040156066354233056608882241092462255818735880428279346966420
6274112459752152202761318262096736092737034186306804296209480151901121564632335461994960497408344435
4531421150067268251668773960627541821696486833308233724458549241249738099818022915171702763953 9076
5960845015945874597607378223637182049117249030372172847524176818938282858524186640140709140991778837
1395692161707697901006019230526629784219510099218401232206291400604821781949015619479026466313128750
2908324448371554662484940386152485335327366354838102400072695037536724813331564958131952983295824 8945

6994701540983863588375105274450387542374163505345484390591298010160337024065091419654406408930992874
5635036122491584860237513287337313061777648383491141679742795635358264326565093438974387697125404890
9954397610309447012230170495967791613454602409207214946497104487480382550964755792337840850157046965
4092609937529297506575632445478640178182086899760355012046052898752372284096564154010430487466276071
9959290035181390822646562780069174993818897575504124552628118769537357505250127280159222927782677371
9461371921651640018071110326066546576457256839904111265510326989847762004945257320763366118794668546
8075655577881616833650598047997528938859344621827274827697573534811670385110008812060026845121931613
4987292279735435145251620662644657550435130099588759986591144348733027605783273860610819603256783356
2521398537214632713714373366852594289784092798477878664746644957883589668082045147767271907031262880
7710771780875944659287949105828127513178451193902192043621739992710417481747355300551496807137461376
6826102458229788902684662780671691840805906684920107117975503716360971232104299798331921659946411876
7390473835823972710206691386867582223407683714025128786024336013754569561210121118668569527576098387
6242013180600973001510948770418647501460347190956001644591325816011088700342409510860065966824661861
3700340454030565015603210896111956432794011332320624414296852718973390029304387526826413252372818187
3418377266831707982236681984925517311140484292263600497298366464717470320358925111903145526378203697
4827376444746579634703436277452109556074920997068569631083188119705267540760844852099074526412068309
0143476734298550655555049583671068719038492443885017162418959241597064389551791976124801051183266210
8395180919193168647150076228549154633210029492313843480405709777013982744519348392219713060823701654
6337256801393503481012122660511497687829426858623206434741211262663352782157747324524831241354286604
4140191905637144561933616734099663782714960097805768754169534454405994793961814891494772883934449386
2378544571071502666829049614037984969799047731453498520525154680296797462648196687022861642312147
7629241242965831276366150132159956875563032138349578504195023669392820440838149111506865684280960304
4852969738253800241726769870495805882387667508804699912381136464981780323271388627539998143064746 12
4047754171778696333861396561788596483175635190236538942860987981324310596410750202204906039359870915
9547794213809864179132509391514302291823591945291150294341341236215198182158312032029486003940276
6901266322070662570651654600527475224010199238734690299750611631661928185097677922609310051354456493
6116459656602849112845526224785726874706115946260327304962063873190896327472023009227936371756711 9268583
6471857815515133520340200942557325702413567497929832606616893747723790923156448369939822190622 39598
3512927748740492188486514861800676468364747663490435468522704418150329473466880598025997045147190867
2697542090927146372479398724299080014659603061266441668226040507213318150829460988850709805662 27981
5499892792431405323990348617380802149334102633155051103722338875148162043988162961449937118414730654
3972563580373206053741937671572155162520262879124347517695662745040518338716088598437046467204976925
7186263068174611117896507127338941143633316240042190022842068463221924960269928537680961589319 4890230425
8339012860852902135855253352287293692772573112483980587096220893068466462636341647431789867 0004740615
6685763785710758427494742964857966762548769795941069449268116576569257910637391280917433293427660082
3474451226468172447913454112832897445755146507864591900529246653499093871511169679515364328 2615119
8278986688971093101695910417845024882873523192384486242263629734975768439282226311202000471321 8720170
1783710975756685537139382475853381189805680552456175892532901241044606138626252972701449551 8845924861
0761575320195237921069318224655177285864226604778693839844215913918849477120401248032226146 17113859
4797565685911745747244418275564366755031861737103512840615915488212497193717302126743920082041117549
8546836398413567746088381673386741710047033045122372699632967533503566374255932788505284843 97099534
1517659032938402850692199642609108990684290371284529345494907934984037039501594365344631399 51298245
8133353138964830395468537774238675823879595073127916663391213576229300828313749950880424143 67706341
1867945790202225251977023599544884292304632875492349672208730074226135362239441584686508261531865605
7785769972925335180744046382897304361213124874166495577320583031985264922840033821229619829 400035721
8896092227607689217373213756128171401278947680860184617347613358304779955557011084658991846526471602
9062432682309721379199995026002007677928428128010218904655036864494068516374694740097967052287174465
3346534766832852993058391752919256842294613330335029926614749035530997059244430175663343834322 304415
4343703476466404927395026581091264882415178063849558473212925984863914337804053563760280606861278168
8892152483564543619165845905112065370194484792425462055879015332534325586591018391532755634325 3047
9137407024665888585517326415578510827116214091911502018716175817025131170079441408634863143319 05529
6155411318677747603095539999860807938437497622730370371697491622920021830001353391812999182960 402359
2948162240375734996478981565262268246922466182266003346563155444069190919446133592294764761753098401

445696494985417874172123136077955700462315757470710647612827340963897976977401987616019415760152910910
09253122761838347867951852419370607979165590705751518051295428310185359173183636529202913084205780739
267574115631345641060048148590656977727755892339730476017611861094666839383170513676765980608645465325
7538441719332482103500346143870064257499982174252182121892042469834596946017145410809671735478479028
96490057093695658507360279962669616846431118237198194995401755550793724878019693376504513474123702321
7858171705272875406768086778657331919180654146507023469304107602604380764983984187762433538838135818
313963322978304519273542000124434770143914102028058351376152486834654009224148555589073720292194609
96783093804725225427153971564461393207175620510574794738255630340449984090551893122525815170644691351
99454941078976605283937995021912602612047720514588368727742039393492746617423169661272448881420286321
91420227812419353329742169450946795452067395697738280628009153419557209296208702178163359573153985801
49405991309645978436746168803327624713133218716363946718093667473440335755526219772548442024999633931
174861668116858002416101935718587639139371591531076342504338406774911006239888597954611355539755839
6530539242511338515197295715072567194931591445438867480941270092974257722117019767807775541153146444
1588881354640461003436754633954381365737941755202298370463378124045276311595878742915414204162066248
9126162385007770392863484726237351441746226554888964328960848071692123310844633305533714717333033011
90172115307748181597531874032065206546663038340247240044364192958515662077201957319351948859162981533
00551052799525400109233465678597064540514291065710504302628479375935938805412008468120716572595896815
27794362993111922904287498933815161612418075418386765972717000319388966533563565704729837055561
5100268027923851679503365206653091784354680053222118117845683243680044326659025462859459910385475796
52091234389503251059583028451450194530982322771984982079878455467496436276461696626183648666203671521
57849813839868252871956385736085204193322057641086585308207834609373542674417458791798165097760674863
0342358794378816611199895956667944648382154771584452235751309630861325983230445664681920972502934490
35786589888840520552887864065938982709410986621527371617524492691267222856267444317906706605132503312
4572067834404637951426017333495920626461812732873779400130153571573662376126852832110391120166194811
5587795504039408651223604194975793571897974081155637346720742742415773774044410912384855845196736483
350488868309931389823442124854956202339198190060354898518044067135803314087241326581558085633552923525
6505562434062217386358710059109166690201106008519210620615217298898708361337945882584198929728135937
8463864081462097321574887645854545290564856934762540928991066919225602464254529100149820094514753886
9058507821574990164238325826612333030842360171331330197043092815169746006112974143499543611415071
789015523161636275821160734521938551112723008960033993710873634008474458582811692917284014715929271251
719738223535539639876459247383693261128037429940138758051276175069360473808825916076549966028549415839
794291304417893741349812851299433175675824407673417695102424433117320805419812250315547648257870165081
865766702309785271213432609608282808472227355161279802179432486389890602925191544808852944924913238525
7532148964128445650583365153439839435370632369086818474891634082930286367197857966041601215228834091
8644950300613637984752788790195341774348444672748004363483276180823399161008740511223516596774451783081
8192167027512514354946996687266873708278491919916820259037114776598260684642971682887151964827056579
3794566304849953425828271002028745751425313487886988808012391825275474862935920257972500281821795109
94629178378511034667160915765076894240576434882324541062551714557685235574081545523266017764320947901
0156452054668753256510525147976320111226602675248332139550899312683263493013685192428403094260238790
5323204787679384881578179912075883998951182420162549244398557529250292683380891296512472328149902698231
02378886144353189879921507200178726947659321610518524050768636489123517516719370737742164358862940
6235947704312005226606256976250968421781211488829880026616044059222329331624176122908743379022287804501
61701357723750619521603426862806290537864968871393385712562416964079324475831369885918272992757829412
92957513048250436660285323714020549644733807382457755582570927007591535821362248787395198064475365091
228338732197378945098948812224316665010697839616672989920964420596817569619231839866191793408474257801
68154615941458938640602296132950038120038389767450208633855782667985016590369990816578277851637829371
4599361943366980652979221522153666288839940268038621838784138954997920072289371169507757061724002344
87289868388889469326838782343569212898587188566870906012738621939873420322420960659606991760
20054536038165896602143871771287553098370990207133084712303917975744838100568328095118932927219123
165494090664021456835987446321626557573979283730870286061293977236838581491993925841574254914633515401
82041412850526116414384738621579485025909591694016701922272152051592446384786736844041019760225485021
058596203745202103401958672160817127019264607046079599287131210748035118825068233530449698182126552095601
708088454194102253519913136835291159722281977965191751091412574906675271987992843990372741064588671601

981183509101205683498176773100954846991006642170375101296402795266807926201346490826570837312730887703498538168830180415910735087802789781444525165407708127488473796550331829893602610515170900920110222071001669479948606498841609925577821033292542220238243161693794552445077116612781957202899490592371787630713791620773280519004359506302783720524286071686319756831994495359654617582379331954922183557140638217262170118990624363016468349879943067324897484029900626756368663934498668702957463295592763582274173904767366468327098725400825658274079373013421250157872246935935360232027880213344754711692547244278823478857202047848109668249573259465069381839328994408562932954845234695473247072095768155000316328681830587359766552445629233370980039202446539708198088097516775900082945233933825387379751663668484819906191718923053029327528205228873899777985775746443306673842846683423819777822394151522436389874249517010656678530267666437085462406145075109823826508232921923116946593605521624370127430023949246000846441913713453739088171054329719962592178595836890022753546934419927056354499446464846356921147395454234209303509561912596294276033231402838156419581239921684357059261155218643672939081140498842995401350304582616856151191924704248067778748838713189871896738261924739749089221696564899815767170428901744966620596800868199091956648398712799600060660098336650850131726705066780738105362533240436516109801108476755494877494236585163719465279328497990577018451049009170153356861363244389487965903436759034920266424655815244439282782271735945948307893872024847342032229005213360684605312929409749759889320134905015546639978049960919100873715889175956776491872706181150301964728913257678481619091938845977305288817391379634191013912282881868957666919158175066401906642575919185763887548293436299211719127105497738537301557783810188444186065783059241045272431966922764394688190230293660368939329143527900678345492052288961178806578150831049180891775926096257112883270864363467218816274725571485025357509356019445337057042972793165183243697073638748560927282169570759353217982829317630403988543895057250179465534119664384061183281712258058093138653669016324155042355394880397710071250704105678774162025859908400717682098941448674622992276289202550858081621731508389755388405349427919056734487660483097079108665922529310175974785382557147194652962908784519565174095947894396337685688784133633407353907274376637220528022445916053457375406618371695805242171880332186022858375322597888350188042317887568940231975197437444613352597457897400554662442324975934404953762682364015057347269539801101002565825131195753891584938212512967996772536212764607639106722691844111059166671823174812066194772280535025793861898710732931143196219538853900757325454926544053440534485876237997048696398212906230465260237154569746398218504024061606472122473862536914542275828158930325786719200381523130703012314500162038355855970846362887285686618295880381514125979242712228006580721753759956097530281681324319088267581121397864589779915670771233413006140507207378747754227133477731877806041160283389280732294629861094389948042687763090413828200824932764378445694666855969130972809296560296837842483190637664897589402297476523373707075902957322967641074447790285422057108331864160628348328403937671338148094153081003836346209867409231416257725926016424131076838536096774393896453881219871847087835760284657585006624630131835637759843942325631956738892178647425115391464830610561761852266148498621173529930039419626797835115432471979210099902359950101850452263362136629541758790411552911630059509298870937205111995320951917611911117856568563145823742527336348744262178943734255594438827210995258344048078153363012318172504761120985862139119503812768576842227060228808579228022789927013545026898289812867084694808586907736873104882413520925337799281517828072247304329560570662229560439761428470415982002181000253700311000000000000

201911364568427852227113049930847401550832055345022071115182925037824584151595423857290920550990931555
270937157304365071391970662707208366050653592578075387996624278296260271908636785842034261794272299278
3872074227039258647899698885201728543732963891417985649549871231317421078111745847834714810113022006
6918817139063322374639144278713501339101361465478235473132163877978594229259092873266039806170514
510518935261384635649007582348632915202586551031103413814049101573516178860757643964618834101756544
148747743969587125931820683929218116808835597127226537119526746391354728890901025277774299066816005
31987062324755696316479419943187728998958907371744791382210691683031443021088327187369372236371159824
85112777376585439522066498763027698123461345369762104793758443980664968551500182428724139629300028
4239040770171753087400621103886981275161331107671042265609534206562796480309195917122225686050866069
749195871952851172018630141572231112559580580300595904517801620915600339271581356193961524505829409
3830675223104876027568331931395337061594056900825467163407688518738062028393764941416952247897644827
1592439389073858703368311064748295916563631940759787977389368568253656567981194055582946890756209748
639705158628061604955380199298790107269885264068694896100332327284203406188454212897943218689707273
8273627785013727011496387088466407860794265555255485664825367262388553019570089909441814119681782
2521474367302375772647906302136270092435193537520244824650445028177473530313055878404289052956316
695575746030446840020130258437258760343964296428536338909385040595996338205201313459775949549591
52824913675826557373086309853431031656991864774935235587698561676402569436794960826595164936185390
6776134086344949610923085281595759473244692997843787506957796523406627034893439922003933142022159640
47530798772483746739203319033113606753874099262658761458292524546296274478253060713708610093256561091
06290159370323457846510942541565848633675971714103682469068213645950225938235868980342052145842162
109771259419886051741846098105218023309134915930553236402118013539938273907609601881270857002161499
42023522853011978515148250926695185656765334104044454887176606291533342393355883040977080382629431
894438507702390408475017104093236868309790449784933892153985002143587007713428715847006930247167663
1228530283911202961584833228761837312702544027509886977752418671972580398456783452867233726268194259
137689223732798696369957194750828057492990801609296385648875774366813059933003010651656716864331600
381784331809476982492426608263925647221085630282122858435912911420360327200182523237946293109254102
512454170051664917496978501765800128632544573787252932671274715162433436012734039785930222159961775
173614866679396765633221951493429837679037493082517016185284933834424250418323030264100557818831854
28936408740320396600889234387100234092685238849673228445668736570423431566989381131170854980556334
110903902940206987883668650096416369170528156585835647774755048831914684306459899663270339801097062
714315487174370481121700620918608416245963219627581891687594715082363689276171748516314584518070535
63797072327895745053875844710075645587473724567162607587558263162416380175894812372734658328426334
984211990679033276995061878866730634490328278376485909186680653984049317138256966595922367334822368
760950460020695736736959343537344892878945414299449242296541992068707171799082027511232288302637409
33243082840760823659962230745072395482913251001462315228385669963646816199306108035011934098512877
3081595450154979098326100700084351632209140971316683905093077706782579384891592132099286598075166477
627404210228706958031691019769066584929411630144904175524152840798558419204594224722408579545248996
6149963129567459931787449783413501974760020485583093556978815369731363214452511408292841812804502491
9295984561729788649836529671737406535027575784653418707842130980573575850987089232183386027668098
6744587673937042506105294559344800003379484411869103438489819927217800456984088526180027740246971556
963453537058177132496654317079548795257766421120685694340740736941652110536170777514495029664201508
5673418561330879369079908598888195417742618803144147748693529301286286876997963497164414212177384001969
097486279960894609364253067910417459357128319040298311315505930386112061927540034742996012976985672
856800757478682568526588805504465028247234062122672309876509524679555116757607551897367108186648733
9135554730387177148259924982090655636246368887442816354759738020092703372797235756205852019488731175
7364152085688799839625539506720457656370866786849616739928990516639547348064688416321612696232140430
0430349787937658955255912612374931875585152203054887715420612832215367134062615310369584201177700
58113108574067217688447392412151890674297679952843460462085104229892959015388617177859625965902453
747996448057375423930933455291517369017939751600199875836990940534602006006115708129727286924415558859
75024274901019975278569583453344943250025780204344408682890750774396173670553837615787863853870009
3573335902594668119751237389838726653687955430018415044807205276494457025799466803499492941684710
47452363136504711526982781055205996265002245440734287139919498025383338505885393649941733636963199
8036953211463174617170203880708663490647863404224584691355904242451402814259720943368039404264695762

20351976052537466919686468405748522732221411263468200732680991283596804337124898651248471333866599958
15570362412843119237138052069855463052239628600169326092476187523212570099594164545401759791304831585
22690092440553118658153197845931405135496797501971591305636407967874274388697431812159633202242453695
09081085401074867453223366948874174475845601897763958449021749345971047703154197947217559031048955155
07130337509226428947436615011461711285404898362878232177554033558151308900860023111908928317197946153
27339639814755795610481654721822820928241262244086617316118295314627011962136619959410879358356432093
32964189356289507521834160949562866054760820239439036938294410706907378421593711008435508099349512525
84805561426027948881173577823141092156309775563343568928090062401470430406809674544142850010531291101440
40719318106005561952937599440981612645437443677397892845582361680657305686818893290555248377388669783
83384821269003385255772963294850362572416179456068060625056743983435840688586527984713205682033268018
58401186122537107299459297219831403982495464301364141209379446478463296730804124031579167146810717211
54657975950843790654639268944164367020267173333214286872793000675256808959472460540077392143662377474
03669370647989280683436306662357354918836306740896905345419625492185959482963529914425068124219578494
39762493626997668432011171783094789764534282159211054419253956738906802587429523470246253627205862445
29916142578749920154783492516043423853492438434103038072737707765701474373435807984511214989021387722
6113074931251848497128991745909500393219062568077239325454548469217673500211427825159153713922475215104
26195751812555899192237756055625186257778715204024235643008015440647378686471774548533756851330395770
3050554298410274520848824563801811743244115088666941720292251387140525933292189039302348495217837323
53344653262693777473250410920550827627013601076268805734928341061501432117912584109328122674911529494
6919441405798335403820079492052726207312385833278588756477905672110416544166147076128836100062438413
0531050140081010798755577315225042463587242081346517078196813266528766870097539010235484005434193
10303058505073284856662189230736100169798730109604578627259697182149777459319121724208520832322309
77437336272600910791708541590654006954047576946452693527389589089465660935532169427129426114018933688
1755821233660928809368683610841295931638976682546341618079733799311434919945061976741403836739591043
3250837896097654632163442968447504718087753879660115946910584698569346315467151967310540189434725327
35101233055566492446253089798552598896344445382941448838257096712360533891982931364903499133212284218
96608736575713694328636333838749656941544771070613803436739395432994554889460443428621170422810297576484
9010846130362580988520444565288925386561835543746690507579482110698111609436227278171194688442390136
43555822522433014093711548401136621388410824798090054297249286908770786394338356975808913448554837531
76777196589593060587154327502949340116362862841933060481711093923998791900884203487294343721496123970
17034303370791698316245766050613624590549193588055245203073129842425880818709607849181635833763214177664
55264846608626744947775131163374609853261567716821601314781904456057708920308018521540881268882244014
854206843331277584809219544438966711311444617893147412936651281987902593291945652736873448363988933
389984361211680689865797567488516544886337690035375291987576930810573551739514437952727038004470490
0728573092526263167309907400068490459975878713209353348147998072978030685925274921540325080620629679
36802909636571119655454749832657557609467247229240892061371056217009793399227932066567094589212083904949
8460475864801144553232781360281453445795438733659918540295506010018789625823206446714509639808919967566
6614659823701412873664680385940326652224087508860528841067197999140854485700729302201722026030478638
0710886172631415313923748994778191781040775452555369360545903637816192863942002266964803975868226345
37581285355079620606463620227634100156253911994632578678836087025243725262830330102104489432622627533
22073667652910910281998887191616768669769861721068950090236440592917572185945847630689212470436502753666
328350650043346183189703050828391535850605251722344229331896294372577616315222687395005813591563799
00907045720072609689883787536982426218631495121398835716735638806300562903254751451996618172677828076484
896272799165637748029910225094724409419801468690188622585100423366590664307650167100236787351898041766
50864760380556088271198478863911696605712575811161433220316239861953996081064489115129903832189244964966
71151819857985027704469785184162807329531875221707373542780783674506060843688776930498303041436466916984
398370082693946060456288164169529166544373908475728196961464119571580663688131248487829600519236538496
8166969144131643782128080377458223191425066537273186601585576310005458819146086341512013406605686188
305399109284222097222727266642007098758231596762949129515634967203809897247230420378327835086046176484
4116048285204200743959607793967665745457415734381437629295610953115848209200019968349222746222332034
92297879772093259373458931853036322002185233292266043263277738699295254400374606548044762949827240424
29146560052904598661490530533034118423277471347528763241774686005103519680259504893354161773876502384
9319180840665213804667466529643577196052289272879058233362671718010478707415097865327441555228155079794

```
0153143256991410913299525027699164121847989035534173418028880785943704747003746871669807291369878105191348231743199718175697324716934112404219323832375815834070503251210242327215999762255953608171639659554592152006263493832779387174509876695534287878977466443638551706502650448577147587895166390526126187671738754045938775924495697246870221984680515191826034314463515337517354398346584036650078008253337619581112539605958941718804721787463604668507755956087664614979975907257254669309501812266059756338320451208463864464994775564741104882581862458118114821602417311351155337639418016198608293258482850297215641249354354937014722183714909313527464516140422988403472587358480471003049701103637861996903966400318902701410218747143973930479667127146967452485821961590443588575047471161083558276186011598886799252327767007791348806270678430582303768443224083728557775850821627326777552157538549313918894433113709718797654930990630437083812107927316041468874104427329403027277436378828482437769488472346291732896432364895042303033952547723285292225182097863294127922776106997648396446154980303910368747636700744207201348680420978346670780850085124892600912708157237896817953897147166531635417927413249374555104906770883053829104660986180133549273647157882317517022952922557467629438071843232876387893878730585071721698784087093608489127582967318450352330091008245008946000168372896985534781567708986066370243799181871271374835942442963643209472572711041804433460130501721782472919884095444291559247296793147664168646799094563770460369988700796128573467050871763579926726419077586482979580150961497179864393231187059023097451683435712523358744257165025130783843967812489541028789968672155583518182197672923726750882719132592890457053921699623157341359810162606343841974111595598712194855707915404099126084315344947293618259714666352094030430179436126307970778095387707949828645366676326353342206894534303062697485729601880846564902097499552566734013380282307886298068187052054125204011999043042891991240154630648960755234800192945487528805570655045425348791789559874402509130074164191809399729468221035701898118676215490449502844625996837678251688727079295334993551398511723804911695566612441880493518211954231454579329549732901154976329770042572529288516760556706957888189166689264962782660684281788855135682201059864864426897310404206420388620214159313343356506079776483728611747854101318193588209632637397818675015670201035161382762235549907816708176258006326519090723097311326126453948061274615763974697290381991575806307458753851733483344986760889646227012140401657955979081551364317926971432781160003950929530530156640538544014468195674168914395005060129899532520624256402560997334056335685117062631293788209655763055783267561616292217039451858995939277954633373520501688984643648882073146139928560157646190608827005218388294490528530518564065043364175350139349857051003400442327755328051663250155536001685586063261810167882288985927773980708234430121876498298821950764879374527324977571646794376825978653807770990315826698932185675410221811370662891505071916924055717515327979854679547716944046087285834020071682558785035025806898027943094618467258018625571769991436669256776624788825636710239050892957598248952227094184674434146644119119762486308867522691737956084643416367635588308512954865371125074490373228827199257271519966022166693895605038279519066233710710296452611253548201808162340593161238338328721545090905442719803200644225323600124989344843636759371914232277851596265768452530758485537358308416515247778499835560996791452905537128993380417358033223313804348260191619102805347598662385411253089560611327005564966892813167551297996763366535540470907939888668953608578101733026605688536895608211807721622589219192431173048925224971255312282117588225282650923582292122520413837008286387959940875331422920255378831925901788817597894307727113160489156785086783738812236288755855266118657336744601226117363328802255620686149584672266053779357525583609340989163829636599800730784500136994588212051974271662951889363661788724519387988391498507463670116146235590918089146487824767698323775978709634559315457068005526207062946431076384817183641242844883222416345306417776492806003167897001913734414599529081813008273571120244541781460612372671440187538225274535151155242450079725368799187115671028453031873256531128191873581420504230774629722540369635233574830656204856108640934273833183346343273512715926423939004912797302888378463744236044644196581584384053191082903232393915006370525377701065942921907663931808108787655759690707840731732397323550634413585567460602928122819448262636896185581367216041354633187907256169308154099694376002147684823666895958338645854339139734195773954458947379964993965017931801875821543881858049244015386570718769787890605894934397057243960807662330777502154247770823779044126904120706061717514583290656812401908880652059214459722368760226173024555463740562402748081393937467009941252221532734488417063681524435825611869662613638340292866449700660379967501793763167639180694376784338621449069623561820105640614012377895098833660708473114371688891394268479485387646650984117195433702892158483507258601976515246041543560676274641178103795805511058527509424472965571294588954450204682256720106209862067718216674868855967781333673048941388830396566121918933058714047784551328767628
```

0301432209927052902106177139212773759052612468035783233612231672451314301832827898769529904655069884
8503343483983359922741648106833596317970504048017163575811696110898875265050394554508189045782033398
8805527336174058906607625678876023448996558182195071306798279843747014713056703678040027908882626096
8751798433062161683649757733948961813441882416886502289967808162797562569227498087402088768810359436
9299203957265613050681288387614411959186240022364452480039479994244058253172682465135209489596658726
9366348840150992837598464647534240305615590651054491691242186078781176800389930760906904835067279512
1403003404594882920845335672716329501200711214683765449414070692596214332856787457446838018539304546
3136985598732642070733736209526825330022465956542110218831635555322102232583541986992666491353192963
1878234901491587001477499219108946501280172618658424119597747844638088817923589672263649615822753533
9969841302939493205679621375776989665420961561839357548510610187366314026732506199565814384095884354
5927104275247448553201536290028879027363715117097611575104474448570023258581485607889851283350955612
2124413532387816233318165611929257620999185168792428624230801705860076558534645099212221386291932006
2916671040534441325309940503148420160032899923191032840172480360326411739587737364393158054756344611
6674421959053416946656368004997460891763261639369972680056711919008111646000296009990629766649508010
8508005158670385819718130552317324630137352876930349789853304607612670691519805210418792169386199911
1368284102584387483086310227556652408281412368889519506244729375224036690315921818643234026919293237
6887151770807674952389899214924574176299185804334862960608936311062581001413806306012314943627930687
3326876877147445496118241966037173011732267211548894144715807643463644764575907033408588793793885511
7519467423534045394122524064570712144659046656734809242615388417502636499676403979640395064536033058
4671065916086949364276706384187525076369689615603173119292686338238674463351811330836037439130354372
7907141030795282271658841103444348578821809080282082955442812200380195266021863595246105666063149576
1270251582303335224947078904661050788518613260070870531252100924188140333110980343254472187535643394
3220404659540269285044885566461425105265479584721663057299456357882715607728021482175044787001124793
6570705673098921138721930592905808649783986319434663257922428340207520796201076674604694071705609531
5133339937607492713011765060222240784478082149383963988120054778938005665780359904311487101637277035
2144947284480659802102462962863743293337500534210984024858609771594603462650708927576848032018361905
4988522892809537682131515043558251720372860169596095876425139501382209840496122262422817340434028089
3997262257739330366102986819922133775791637356034537807550181325596156935501032832994238499747515243
3810011519501213178050138796465628491543324891943371832694701926816767596061591878897636526032086512
6585226424524119959018981878845082876944376766318492238487992413740076722940680731052803953540236603
5209984205504305921203882755655930480839116597306245017722525278790798854685084258555173838383851994
3442889159122536441186696441712424001358879607219106134942308330297896634430882711074670005236299743
2610231802714226622618750572543969077381474263522155244832400804375696699071029472640517803015161 89
1026800926358769818418013034526647105519950731606432675504048745317728164797498493168163518881132546
1499640318314012084999975450565440656511483583871743810710444468199573634628689300271371764306960414
7832273275678903050809576914347830867035401616202818491144123200843999282131818484332288134255124888
6686544852708420342849808831385549010037940264847623636375364651105500810294066099152481479263173 08
7744064207095391999167556393167624758908322427072948295443281512295290316475060981071517094932166168
1302220084899907335192848409001433236988693791759977923872805644858636350047169644366020445259704868
2215144197815912322747578772163986557527609840989893737750893740406584561154045344889826356794496 28
6422470716132658749959558344008024494005256475311275829458245552383993886160216709540903950962298 43
6975360579443816778281566351717190156086785010229710669378772149098891576964380072597300797493811034
2945358118921302910485720499673562211673336635007643267240588295448615756961432047896330591725250029
6559546804876536262815497576768027787559023678734816250424531570274183539064997237143086239536428 33
3798525880936563580878758721141673582000237858579171465416121120263427255738422155180158746305 77677
9395293678912532977700508226250083719741646511468473736572522677890874579968237345277327360629946429
2416996715508622928060803167877150190204161663020493507588377690661867461629647016770563441830896762
6618874403297051778814524340121226794104121776172182888508159721642083849399933398295669940349593790
7282033978008796049701378070294014570618322752960348530283719222610059395644991241509277875461366823
1461289469986726588247118985447614664135974724040011672646403383299904010502635271218579913817518382
1542253047992153889028461653792947236379633347931270846642722737043541076853791213903433119245065345
2076733343809168129039267109298757177148028222733070858500225913929052589740024375710216995532657613
3351851876638602761997003980529305274289337090169102367520745177016964047237538638287654319043029 03

57981930446828632045430189142160750516996685123364451883139431581404652068503559767528406209686486400
14632988026383254956272132582757344853558300022255133185962288649772494481966641528190407028797109050
56777558383647075089292801299214655089846527007269657168897401324328795719821723119028109990922494210
69115194270447735875202660217787299739380432917832163467212887284336979031693485924557721759863321 69
22910131299649345656945683126728480958429250935515613535868203373672201361285171957991790678887948 97
87415579507858280400519879514379310240973513754244522910665873007865462514188208807307192689839135 0
49253775437442026570165148549039037849153357835239195091842294100795817946261304621688184412174680 62
20722871046251493876491783338925853594154391358005859024298540855725044894291031130668410610525215 2
94364058942822561951509029885349670118520896464332041879321533366847500909379474586244050094419795 25
93058084705730441714228077856570371279475809345629087704798834697169323551696059155129039465464919 46
97695658010447721221152917788542420630144935990036470488168696037395664956844680082797406485 93
97628886154206344959520477876479602222481404518711220576212828951209642426243976910777918759895 10916
96748849690140417814624882189920472153978970100441591637463548493777672404896305617608574901 9066
41992085649882441665925913614149797211105709200484363562191529053159495207728572853550227717869 11343
17095074741774046112597710544066392888757183933236000244502603875999517421359497976494040001 4409398
68093193286423323138073107260523470222699550297533641333336376838307699122239114777055859977 84287425
6964525973045897989161844009118754738104698043805595170062963029433750112437691659205902272953 01512 5432
139405443377891627819140621551682088473634534197999887951611726102841063233698534566227140898 2502069
1286704441169025820479657650680608338935450498621143873825659964344978803232721758292694516 59864312673
58751095484558784631407597172019624337085219967792883082041708362821886710429402426005844004 37735875
33107041888142219209246071491335029636905846644883203194741017346112878673517942209414546604 18534030
15518155623214316574733266610798980310968170082688732101936459561785851734505472858980078728 7211541
72567402441979028843225315410192140135091238671110323213731459405115614706721289593263819675 80376907
23130321615824730407013885893346366335976771547070197732495488145171495615889159727040316443 49512185
97470414671715097311329473848085021070730048952123748421540389981859513224901441857291935709 43752415
9215545692963115014493847039489307624355383423543950785791770587588732868726361377231317957 63188119
174939973456829559955961684714478441518985430774145594300916272777064006784526222188606338 1067248472
69024402642674133907219353005842440622594642539483685654784505343490529674305897486495648 29293525069
68728255730738865347975697379637394163125122113572366124201402646831987523491375325919651 5806193872
66619391605104935926527132169220962246396992453394941681487697594502275693160173729782522 59321139227
97264469907870797211292701007289316441413289755405112986071300454244972199825923017335 59399196662588
62848902801610297714172814721799607430468636839435837620966370592178003581516991294767 31548326243472
25298003800959587555541536352485292336603666133452157849202685061519492902146178122560 32422331042
28486352089687974218454008734941728320117627378226479639784677713658735111930207072256 00375074 9407
8103946338951998454416631432297316080844049828135430303833631635314540529914831642560 12510820856569
0016030297291658467891832210586994891004782725780617286586644935759237706601999726 06595525
54332733642503894798336601431993073084809345161508804807646366675290866716936206249 28739814887990436
53338716369116172369702731265374284080609734869729325527885419930190416842823213958 57966024873754065
4392608495318634134694686789235833606388044557618564870113259642775582026319256809 97158944893454073
54516693238449214991185549338282445770766882305254697961282240415996689237159295 09392372 1195478945
0740806774448900380624434575224611555723894226838593055127754976545431808349023 87291984674869 3162608
871792151248292476158935141491415890423510507353496796948749186334434304793625 20365105567215 6988823952
03498052301531223852125132616644947370461248186099014395654637271017556216112 21104722479265060881879
21878564564770201918708174098274426388517851782319529341904819315715640400178 2600804746415 45364258579
6882213147120219506870737039312153332239429647101433881763991811507421555422 60482199024500 8205203155
15880310767656881219857503845120447360279692388489439850407766939191917803 8513117904637 2645787280056
64995015957625302767625302767873032069466976207937109531408787466090719090 0547787150227 57386 1562
28403119997936014817401814072685593464247081865137267612797342776412408940 7024122505791283 3204 48767
5083824823354900622431962572928264805660096775092853257303888341824250441 01944383749082928 9077044151
8151343279012631862709344102805833319718393808451112478775779052879961424 80968537580976667 6370 1569484
348743174754899146388916335043383627398851102955909972689955904715112917 9455591269835942 306738 5743
04869898985594432619896425343492171171761949868813811537360119252837634 81221877710943925 9322 05737095
62698164645264593052541308176804768491799670945909756270994574641668731 2998517771315588 6207655433151

02630236084922353201840024644264949822009388561981417423529421101204488878651762047723100723557737 11
75696454026773786987829323848846586854824307251322459971819517637820651677017349639072911973231521 10
45083889636900343634546497713884180568029841405323097836877878873323574584371677859623193118212996 5442
64227460331165621899580738570914074817090777072060125825537255988182554000170967909097413385517915 05
03462413627962943375279803921216124494228573480554092996174221867552670663871540197164959258049284
57272339435872738491298062505229908230414417964201863239335975640856264721149087102756842328471054 42
04769273722795869343255162372870613062489483176830059503162735392722211555960371912609270563209001 688
44642239974599076283608863451560114679086719522744225341537356304436580765820929448168157562440758 35
42094450414818369400724787199371608074714370480527241227205762001482655673842585276152204225756167 756
63448908355159040347559705527811498513025087412165561605854272923028993316547354990791561217866471 78
13433928249941590501409236320169840868059967723646311800323091723144906596018394433573246799472136 36
67143093322687259227699597866342198486047640383312151598246334815753891362137470506267609493915654 3
44496650307157560190525614934341239865008633497687725820142616035876421886575309174051824174917841 21
53032223830041880663938545588917876200687881404876692760597626388508418767172390688215137534469072 20
52796875938629657498654417762942518703009114961352844389205145007155110873094664959499070899793052 34
01295734938668817859272442308152159066064996075502723760812723870585121372745528886177354454495938 51
58956877519518026877985648252026624094448618828672705420747504353679984584680211816124511917916408 38
82209778864182756810585076775657286484828360370249328715819806043555879980375757476331720000544959 84
98725166885657063033528760680930815901814105937213786078810315129253175041105096097516542573103085 5
17485489928079279216508267024775246374998378504723411487224038878779685621658918415735659396870303 19
35075029813828952996830357304306071207546629980584795107732290419143068162870295090071881413421458 28
41561163276458979779431852446703335722015183008067730098443542815498555943657389719903262861007167 4691
15090265946427923755624937423512174450803121349987410210504026254115631141230640337384023024844739 3
61327771431778326487227872000031324379911584541073200832547176553357788419738811198783081161282533 43
50013791097326458045675356269284834551025317569761378314436852274785430693706314325509640762249472 09
69727621016179816307458647731362102916913190193505391736338772095930772880211384952253085233564200 91
47582113215081416345593732766381646209964150418142792614784856112250969744180739940121864957617087 74
29853908394199011885587733637311313017101357779033475602044395262607677976568538504151780028622026 017
39831535789490454442716570559649205222318835447428311193469603711941218609396474369685321630084113 09
21221376123619315550911877534644560429373792151668962024254716803778182743659379670986207356409342 99433
42717920802887522112543317901141491160047963896033187220471455192593058948693350049922335765207063 93
36657861080859200577595735770605634693457603884910805066955160938106943662128758827331613228648314 31
47176721157046192356146500373405387217627411136601782358558451731002982077899936468177685705771969
04429326564149288895061617432739543482331666399791748498402747835405359120022260943990531207076601 9
66727432146673132505991961537491912061092648781953777906142535189223466139609531960625261784257158 69
92437826609171746497163472047738961314867194248490291989419167583088892339731174155541726809475 3
31027377979970981756504505473602276782106975404505926143883778151617925379010606402291673802696257 3
43430464530042110425276623030552072475739306792726393713188722880126958554904248663228307022774015 55
28034220557317260915929275132872044337723638154660224262722795524640479069128534664745936703901536
66448251186234027804025378088666113535664410691376972388236540537057203264851330711800188621777680 59
79532180654367532102225042800043994061851812889536140737323950663115170700057138631530213293685538 01
84898696963028510893012021795064707248775032099948367568717247002905581456984051446746945071887173 76
36802873473556196853175307566120156930570344309876149723068952866444156407483458808986525661664379 72
02895868442203921819431715127564111776147563714059368640001035880263891259692381706227637167628748 06
28381602275941051146269228880912943302776649594724973844730933763274600371084359078599766718005586 87
02873018322966729256651195926100594158100365089290626039997891076469310195227174464519944361699915 55
64156412151087143820808868075229785081480228624135318439205666391152460890481318445192314929106328
15402792248937822825154576821762459611763956688646174239537158657446266439961554789051637325218257 83
33253564458989290595192605865979867134482744782626667898419196273605935202214966815704365569041670 82
57527445881175281160956148185722436954647505083028440371501707792355571329348761178390813029105991 835
22622374686711575705937749093797579381952473316322662359828765699804734334402616879654751304293461 624
26613460747325269570311488146969164293369071948154548179082929107206942973187597197310154261993356 46
15328361822870151559033107061465304217006688253379701323449506071416835268609881312272205409030946 64

606618585799991415397814484774156408225890354064490646351061543371940040138616035071455973601427862
4514865734796217978467570218989951333364438192919053008577399504523493495718968461271137688957597933
2349533208953814539846770285124109139999624094286153561549520156418899621259300512644209686597252899
4184350366818804807529105972336008365482357019199865093250048765737882951629237418327132367686584
946400059670950677834536100367442594918858195595926902512393110725951212115633824158960673748007 1832
468778413078096938241482915189560427550175420651744208813401454360707135560267634995975759600410 3616
096121377362182022356398010145592493601568971489793336585499186349 304103495007905509710373294892197
6405886995320189664933508204310048852305942984868017855565645389715296386871398239389278862831305388
9870441633874853236553065382351329115406941185580040575610780435791910515438952930178607056885781 7217
67329443347903909096450020152545297422360933487377485709060186480705169912575593325182030441 20573389
1169249497937444418172101800486952748158248607577122172413829852529767035268850421330346370320590111
2769270842312247374039903446761895700102591785896614704561188690554318000135741145453848091623845601
93982145769801540367447309334224141647275552190877396917417373506414595184607851240181377454588376298
5179066094251799695036587235132911540694118558004057561078043579191051543895293017860705688578117217
4213915509530721197089841522154253164791930461604981176009940434131909911892155136512265115501 81311
073519406748964186094028486928305502211992434386630966122998376165898127473066900407133131 5325819303
2028149674557028927119805023083424907246105491095799550789366026346984656628188055490104387898 95744
09314152964143377690260506436409832682176336287098826272397430230035063851675289226483750950886 13721
98333534606984890685568590244467888633643960437818236493160750697952536617770448078628521046820 93268
2668289722071591098008197780192649525383047246307958939173700369329663580220506598028533870299 6809
22286754271291338699940266357736086375404720211499273339955963867139414159506355508316227131799287
614329892459586632102280507272017663282902813951362463925987940841197742421478497488792853481 3292261
7580429695360568496416333588361246477604717663033985377267173732323245319297973342376460670072590069
784778225901022471861849551087004140152276349224305850649791746999412247016670031010443276253099301
52842068424685952359105309696810584311855103760808536810333170953490813488313117223593774387414 62183
9265017156090327940281899356126944963967143320782904731916666780851825519771728800627735453991592789
0340107862889636611570807579263712515753212564345879767582279860562178539046344387826022476983164473
091167731376986543944139748134480038182981037549508853983542914632275329122606239178293199621398691
881771118424419627178992305735044724577538311943385179322128576603522126877901140477658976778 4356
96351365329151493096380391047545011699675878027999755389558550059045533297935637026407703348112055
96791096088046544581991175696733538170402029774208446714055476253800166579619972620080781668202
8859158624857236155994016255477707914111600676407826080771078947347289911567613068507322496315912 31
6341975884627647288192023676271637519476695332542049089161024916483733496591727080014711527101 290890
2961211040472462046562280963283626076088897284648491945054852414755813392377362692127662809010 7039603
29946262725094714117691214291335397513015143177467168584029059686222170801110366607146302062064 22073
9673674027544451115318680357371197061263214355234685554438245653255194962230924422262716181 07063 5327
1218486711038748633241670789046885223329211115019790098723766740155479167534474685891162812 6686823
222994356076826978331735176394113756818731185310939147331613471464295748025886612098433336264478 9232
7799217189381104902575089833295752311385116384118101924499132930087784725362736588016792732391 195667
7377346029231167147252754387732395409644747449308810335690168994473265062939368124074685916892 54650
92110914231643396643496535539990522603047114871175019510860362143788729840744982527033242516917 19534
3239338053753415428344399420057275968191874428188588466626602813391591226508703295562974312 10060
8476462382406120209740885851097134502445345626967484521749379519983660135959959980442105533930 57963
54121565926037395454813070900268161647358075309070057946985951218576692820433593133365802 10439358016
107908279426644878023528015749847777187536663886871469284922335979702018592163752637064707239232 8071
17749755236536241706263154632700590260304024739804533530204093913049739713079171815148863238 5631
4091871517272596320603977518189877379429833548962149298830651687972617323342951860291979123542 091466
176185808129635850975540518126245478553871423498728245076280218555416439373557287341317707953318264
106958023181267827292621724790478673313230260287901476485433580999324437234918849958599486258 3067600
012204733634466686800302177442830895673212065731090929852126853082935335203316260961238719270474910316
9411516483884747456771243433550746091268446143275337106037702381158730688623886939413236304060 5042
899652004510603748676961364917251172141710453972369837657482509286253199176103796050507004745 2751987
06924383079720813365107458086253398704529503657739479437519432553660014210556464148224360616467 70791

716585117656108592356346094854976447796211655113187009699029140731514839039089918159185783326502779539578418251970561524675181074563304570829594428891506667159297604128033547415510049394933991135740036810821452010037166333769521213375323593290644515065233379074750428578159695275696181784704238178420315992417112157281753138255289908317222708031933401849974624661506864137178679359480593272851964335736880274143158690076520872345466373639831869120209656207541348874115504351794570520219208628621570465012955951312793744072467620419226656744533344729681714873544938733848016654282642378338488317565438333617440873218792199714309719390756152899799193481684566486989431576014380286263353331361857237931672366063675494380052529671399740350994071219337375857120455594960284445640461306033622626321629341224576151165419387916848132809626445725944569546221250879118935398322196378999498705755174877188610510452587091200155027181112140083303345999772865870452341916673040685570047172861117263358849682710717450035389033631066658091122161122795352059735631542387862792211740027929927660272309100878896448671977510644852854236760680678328702716021491220890738359867916779079846546847654432886332754592689976471361182191936371970943091897609589330741950915357899815945626817403109118621361123870326632874592512380172218592375964203971780119733013545486303115628764539733010535519936890917165821184472025394047093178330601239641672709312163693791933239184259773052761479229302123013163652956137623330528445637744966783857241630555328610532755207843894044247233087001494007564853949389708563666247235115549684263707422419853407218843317118086247851099998176232258058120204907270236751535996038558466728397347325959612710449694896926287007408723556135501883486098273344942119279511596389142170133713625405959158400657637103362185943540907214950797192642474168788661350962013130319398165644318423191036714120512556863328098552077093239955742204583728924383094811084233008764153663084724168976375194193998480863927693179016437278029776888061624908419337641036450961260406512736947334321364751668674541875423533324904525140012619910255049422060890086534891218519778520803538297935164736163639485284975628497148856270364254376152530348567914281833841546765630362935943271568888511396453417550113555234226609517738178180389386443090830539927386531988392370825144349766957951254066405582132495347608244642379595204674037169104022865060164401188212816887278392342736929260620640964091959614590431451723416161791510706177671741511297009743626357169179809791310760755440072743261368226933056135861727526856031286808815689169916046013679356000288784865017387036118613661682337006376249017187035483916530088806575237376799068155478888938646233804336788144738626369751444635331513645033652509877954130939941467601122285012782734557515956198448726728886216911391278644418265010715934333181605528809093137576021954443266891814048761296983574036801175518913300572269947591922872439694710724497700473296751338485372898919851448791269339956272762863015717827057355238450193665288694250301571288649098993055897745148064974007108137602067660610028335398320724359564720594945121684402530561416150472376796871252693156319309816082329795042589816674800878152648677364144935695842879538795111120900413882435069998882091565554032892502280514169678792992662686222467052549066749536250132697003182450110473519298152709116828763161525453361323122680452228896149709173971135532554401236086188154541470853204672299469390714881886033268282617228269647851698409755132809109049299420589020099758680270118297143811306166501656069405094174470841365931729460368323148867837834015846665262779381103471856527342901126469689951352204381389938091814048761296983574036801175518913300572269947591922872439694710724497700473296751338485372898919851448791269339956272762863015717827057355238450193665288694250301571288649098993055897745148064974007108137602067660610028335398320724359564720594945121684402530561416150472376796871252693156319309816082329795042589816674800878152648677364144935695842879538795111120900413882435069998882091565554032892502280514169678792992662686222467052549066749536250132697003182450110473519298152709116828763161525453361323122680452228896149709173971135532554401236086188154541470853204672299469390714881886033268282617228269647851698409755132809109049299420589020099758680270118297143811306166501656069405094174470841365931729460368323148867837834015846665262779381103471856527342901126469689951352204381389938091814048761296983574036801175518913300572269947591922872439694710724497700473296751338485372898919851448791269339956272732991179552447748792024698247835770179058176843376667776890217764906219369958965467659969428721801098713692136744622009747830400927181905137635612325486127214522261680518029325681831093141396659245310344236884339706735287266383000454195146442303262301907189759856124702358650054207598252489819907503165380324950260169372305831481731475243043594249891487918906280263409112272673533448537779853276889704716167261585288351406038225708851992921713307057853674939374555940096761537521778280116269037726528989620344126159880106321682532064438164061291711721200955674738916722296235557461243901559905448832262644162568712687048500344921141575761431548788382262449382571907205282243565403066846433949527866

1978261966212889029317080915069335476093630695038779648380650097087712584207442114997169855615899897
4787651375057853627245365217806628977750732715703498547747167890295666395835111199772543088210830083
8719703001636037548232031811034519634199719570801626375425606969661834362972690706622306143131863618
1161133168418495161296479946354081551662886453122010561796238101443846201413252468510264137934116621
6666044355543396726083900293342498560592304772543016048596898781615324252348894799274995680405750878
5961584656399688277050582480803752624440992284265581071965313962147422234115350770031361866522902424
2427339752232201197300895968910498540544742769756380596262269087884764367655193756819519963044228090
2471965977981411229976113099668948406453043061615428405289846055561051527743167094547976542569994432
5615151270411776840247262990518468739384403174909227786713746504877565400352618233613582209691595165
3100302994702612137983269955154794300452825040411617899229947911176412173992693774165820202835024261
1557953577101928695026460543592411800668078233417498334223525119403957869035786809979575355664634818
4109235356638053216250587339612730165179209152696307741603539343614876508656958944166875931028197227
0842130060698903276812481364340882914506935350078426900283389692890036766306519621256911370825149526
4130730020572342600614347947841846620763374247401965234906393029662233773082064022870408809540394489
2602375593027578381867247119555903626438180369441026989560970224026851892905705634115763456634533530
9178364491270655146521452745160957092696019819351482504230830933240208569382325737324655619783805079
8236783914896441321211903253837193051261214351205434672138024917208445724067560783891183614420617219
6093241887871539065311934562423143050595975813896800145932726803690031531485898178421841408627035413
2340571406372423344162305201146005372433545440858047849152738356053700832984194419408785772894289429
8905564111848901279881742427130941732502246499897761849958444824319633387713606417005057588112062601
8903546125859345154561817568409731473384201495189375815899601208752575627603329500301183188095642910
8679299364914087426322667213868491522412990329146293202682373490956625790320642804533851675572566335
9643282983690679715448949144144284457366131214716525772928322838722519122781850333184575375231181388
9104687301120253329343303322817674447909206656325018838874991783124527795687803251857087877108213218
1754229913702999043640824319822001818143016950158676647723184551735160193539741180681625549863343692
9742793638368312286209015008476329602715420554092342197748755577327275135843792997336755041353900...
6207546017704783200920900004370304772062396931123619969230694519212280751280626100933960085511993...
3625766456058454748929845660151643776323020476293348833136400473571567444997734717821981...
3926294336614853325635257380075373424585696273226443292539121854835008471872615376119359921175544946
8751722095340217149673323008543031277343008442170392235658052374699781195238474449333837385774851142
7462252203934675721232785066105269132797730634628673726224195846716672022151680829100052670223641...
2652274077600461979496685044241492903303752615324755653009315314557741560785488843720415714060087651
2807613311400021517609289824898629450626479863972781208733447929847854531512329334051406847255746928
4826315035477092571914420142218588780257279128331177982212336807793116875865477713399946239543986001
7821714044511587793376458252175919910881923830051663310282837236134127214072246237953912933883641879
3155329932894879874861538613915230746891741006626186077226791348713632214751656850844199178069486...
5460193408937081923214192638277533759194570326450236304347568717345295839955367097394731137451394332
8197791122226939725459124938379823126607096382225967019008381453286290461060658685632097801508542233
4848110590617385229862052871896049500732570440977958310354329585507262140340985962778
6017216895598750303288281768040946852093886403363652364944285765338109795334202587523066009947377791
7483409964056208373304316767108759298266668435467009599704858953748415115221450224994544152838657802
9285301765856291013881441726693837902070500341910121386791346354652287481407153380209019192351467212
6838275100017394805179223575910310629411782671583818637819546488431229736302075907294961331226423551
0849102649984744188701812740398720306793583123154828787803868672076345498495199113445099124424731...
2272527668320660304865738512636931946652992516290262646589416341396091509721872364027550026970
8838683249414212571204886964565829636169864536859883788390280207060702963996208929169242011756462921
2717841443866094448413507132753827418051247564700845614196078604954485925581307161527176818710961...
1702864624451063869927990313298023938322923078600246111212562537492992069623605549739779337090550915
0615995807462647693070614654733657295388010846593077370926439327096173358979875511332985117353358...
9820375607171396495121026056824215353943220637818546333681668379183925431029629978625558313815084...
3460414642850633182078026674085750429654935395449486518527564708814351323195973497899171415169373256
8833893316283389645184887032263989305568945183919124308293251565402367538500430945522752298621936349
9930799560689684466187459894748823413664085185532193673114375894635657021422230371741481201272628291

0573318578392273347952606800413122404444690695700343265791095617342284655138302877081709280043703275264455762009029489870172647182289327617882346799595389668011402866870526336706006304261299460849499563827559906026477765219702537583064118146128754387609857828996342210595022534150439826096187609835216523165433169772144125177003803902159813797489132029292775543871170339116322480752465724972962312476509351794356748381143152864133302908912377714661246904486455116492679934634155621188228175642302405169489544428168314140490438057886059010737006718298499365040749470278557386272032710842602732695690064120155580946913710129842552905449576450645756003740314945879082105473559113639906727806481459191706433870697147736652247844338630255698388102589879309519713128407087189671796967493940026571940572215929586883457866981031818359493810271931161525153017409040319451723832245963305267862642100074573633679726461435297149888460552919078229572134569264638347921759405780513036734887954494733446456067966769127826799049420036288069900026035221616525261892233439185539443465974183103315453291025913036064622666879779455734904546748823275317375995937232273103710445211331153382893042477397241957274401165418484315564894048921358055708557625755849551348891913856437916383424089396022097880195875047614146573843443198087351575166749682003791537961029734944321094760732700463634366125907117926038296577650489833996820052846423420685449469930387124964664248581160442000466693398574168555172983698292635849104471793384468325043384471758725269936686233757079858637995117647437877422102959326217388171799221125649607665490503647530112846059719986422397278433919677740389582319175573259941937900854928259806607678949854843335533052044297814686422621546390705667804793891317765192204993576166388219632235722413875804881872875547783430553930334141642291519181440724910183373607258613130585839379636913731605046386537876161997656835278960391654122119712316370464384350875058804657553196720080481063208311821537956138009835355595260936370006453170806442028883772669082680094247506157736530695369994647344426417990880723658569162389963651757807623731861366280300067759525454698303593502093103401066548823870650963096671525803190270180565107741796599641778895066406027884717068077927555703510222371473067950065096075380534263982026154071272137856032274328861680241733894597905050321379748466149030953017402300954957526179588698360970314291408480458384201770593330872789882921065398608549784177022860019943172312560727966935093784616738081453470811329376345219647441631933117869064998248237271620561502444394472323379106960839688560326743659447613243668623910583435263725870625527272353496543402224782973215846473847079849851417285386752727923065840917432060501099102238929818938645721604168949234020855940480597988871990753899448362457591817958726478548243687178428051181657010359994896167564581441177435999415574156405419809407770607818178732780883923516652729811729470451824894886940253978497040401257850170852522948003264485539823395410250493410544461435613045371236961682202427087546803225777722467645386906917358463290996597892708572413606852947228418998881119769492577756734731492045418824993538607544853832734931602494458301840052011005971211224881899260140903390584301410505980718844415476335609338925955827033536383918920724411566241363467993755416738908930918680803126378923091291660755009890808404308771738687684930623853335091504106000383060163948853687921061238941057439403460624016378154842521771675451639760025505026243961152599429430869304962978375399710612950380438036966005892265852930563788669584667848757260023529183930718547261012014353181230082628245390756526384816628430671241409153532301737357772231705454533185733039863611629092807965140007625802958683252113035625213499854006783295798100262663767805172062475401635370252162187355287204019963596188736069347306728409608128864989228165452185240832827912818493863635272203008598275445998989995835115743687878881270485571738148574030780362924048594206443341579016938395968153358527750878157439719243227798831700650436096761159954373203941299551771974092493728193869104247191680745755805413172816833655379652759510402582937606069379484763050243686693087498612911515579029908914751147143616550977810891689159389043228611630980896160154365423970713173398762556138393349278906057471453816915692648820151026214721832503409165624542935311732839683741755506978877246043985562610853373740287709972880476114915778576510475290891138178065469222072171325415946797805595740544953255877928432324750482025729610721193054272034454311901843265159983295119242549956886629245120615544354851877843376022845731855255330203857806799642333473943287

3255079768143174935290365355235708336227295402976036224596787022467961087290065369158110329772411271
9687184631712015310872022829121678513683286828899841006308305969973295124018343792808075866877889849
6043772754027589522929366853922675139928237161596447373298270175090837568027446669159111149779944667
1135691088924379199309424721308073081984262593144347965790856700825628858836114463307069019106360685
9951853870417106238568043241122994069976976521718948993497188045038643217598286433134023273173503448
7552793783486413104199496595552577066904567184355021562018967279397342682162568608592248811316664734 1
4298910138757127570483051459436630992466106272011244098723999710420756543915068631020135759846014 67
3026511990342986506396760006966879582828924339782590587485678262692633034683723321240661577602951565
3722610682290383661336834150499989593428093201865424703607359076560816215997593438201724618076958 17
8347221271503991239373308059816434946231367174959999463042117638181478301910213344735692656258805710
1644687984556617203758742814909984339304652393122000356442486502802002210387238150855436080610859538
1742785324654979231101510181266741384662946267400340729092430677564917857934277516529509846000986282
1986519350148631413113382340818641810195988722959358603437224236723960151462789656548533533174 0074
1984242601360667355298407544025731777149540219275487625663632097951348389232398473093428272993909 54
9226286852582803762571040814041407092153377982471219334843754450072385259360567659576220
402194519247929124130758546485918128455595125339485377327439546532520168622505372850013045372400046
4744479074597825102944479047597268994937537469280893311554355051420516123636834410073498429470708 65
3487282616826119499544459888850359607914367111963913920911200954133351288550899249393285379476656164
1592545275885347906803485930421014317785771172451137418462432155337224056541214942322346734100328640
9237227571473170380930508546661141052866534929215704384371938758254891849894465897489211239804355925 36
4919086589066917390808867500913233054266548207715736402520816248301635987303608659808378615767364117
736273466976156665392134829345642399129280785997978815204292215190914169078549736187251693099991322
700067499723351557657951436647470237487661496440492613486082997609783626049278237388949792246852097
7599508049882697239249576598722306469511876779916054956726996908515258226529522738588543930217342747
5574391874411376633994128594831234384848812794576013671006676165974395896325545306708168429451211409
1221200910866698998915001020556928448523722554213107166139903448188291751833720841752386967 6
8280591023151992531281445377221647436825950860888636443467208040799574561042901019656088083930982 87
1606160491216360458690862228973756455741357430715910893672423316644773328296824188314921716494972514
0119493690567095296152704329196175641010185140596083954221011253004320324477290450956867286869283797
8994453347325407832005428354880451308874136319369558162874607904656945900407442884187381232567699 6
7164692679986955886052080637298383211238624681202881704348155814062949883062634593344998884038652605
7374223073857866400023774153128859095451257535393369406964442939407522182847100107766580951275670201
4775408298259436553900777906180300371040301910921852932847824116551890922870290124014160045214931709 3
5775360816342085655232014384538895834220684171388239953227063536724261133021723608875319692482011
7906522275084815465360634684320835251951083321606843173343658468059130157408787912295987658228300 20
4252835566932045019819881714582611715984000117232324846223680233784983905719584208339418833025000410
0260037883422114836730547496096779242970449099879904405434971089626589676691300284990960386063046
2400933379790920357551625166640057112187717239030015039609540518458169993864304449804010399161286593
4744955827606683482489093373386292669896469705317415608922966624289143819273723567260303050110341597
0150390759411599156179251165622892441767557720263927108978526059947131533570045904830124535860224570
7716058212332185295875822030519372890200177343206194287342147523788608300702997965315568103011287992
5893918338774064700675202703688772405840664369190902874387633882097145801017495101346458402812780113 1
6813989780650900740767464224096389898045332620765149082876752283904134502684618616814579533717 59
4622683030636661453659920280300843252851498178817712725738675355028513383679230567432436869620272756
9049569472142242467988436041192263169155678827648422239119627403671459897414454318006168862933763562
3975254816109201806289442065086508865174388445174402936151708910665305181913440835241738539089529473 3
1269090022881476173592405472755741008722118602480706552734785464670810033252880494878218846476645138
7194846470027398366396786910872249060894452544993013613598230210096649662658249790741793302604479616
4678961217630473547094410590547677874367269811114648195946765395331326021604518805685201238185383599 3
2509055867304816958939312668888710724516375807869185298046443759849390149864088672912156151469350544
6800392775377162800284661441670887283461332010624784693514171037189813592836544404738893533643422529935 5
6306714864357582266150708472242121395749058812364726080795665391821078069759191962729961376825052719
0168013501825936503904314892374222182997294359105047665101984354967715836390356050902744094547376200

0866255189537989739869552494420945283689372916225586445881857232200509734021124202427013381380975063
8707368786224134616266076141865890367575672804946851392949246947497670448286278503799394283278791220
3329713975438436447227794195248300533133083265941268165481431836724185190716453711839456188537671861
1463451009876355610396882403469322743163868563893669207826287866646316230586562320803446703322414896
5844290862011917977518360789811784708762629615319400347815464050634565985845395933678392047177816196
1151978159915334832397561116212221045289683087138354599880658578013548593749040426395602017295786811 54
9407988789989527859449531291247582481713710885909691407061933036180030332389191321684024237117855941
4793817512261537355292820484620119087835578241076795898728264838018836306051625874581323807170212707
0311601599319566532110559086844637230112193935288299328435969560659719893014841894192469651419504731
0036209138468408714367868832381248718738058221779691668726770528694917232969297574129371576503104861
4982649963942542513355227892655817659328122813519904998628338917695098649870938852865224641624149800
9133604809416167206933424250017253335902412245206966274283860679157097461019323432744228427903009219
7167819796597905954912721055386447240760083100058808181907244787034365745427947504666021168615328207
9367042283157677410978706565288995809215120790091062488938646517486033663048808838585836547541959063
5290169607959866719791951547267530998847762188847851076605567922374551580096127346370372954709966414
4895440357070050975971249707914989405750420165030739210083757394132816578085719802851130424479613451
4504277763666054870573901649799638800674927633566999017421424708604276336870153889542554486051966156 0
1145707431101267836061897633765408595584416739699898991714686664840902419001493111173462062958230778
7057948676463855675872109951364683099777160945465572016812285393776737409942830395155749453169206582
0371452045827753578337982711617535547547595988090288825001513269306218375355881522800804994462199226
3139514759007671504441201028640422323467572146522255433374645407695544296373365183294408114816553112
3178885685344936256510923382232508875197006402178562624050439203931155127424198127866118045752037903
1352263821500210772130502406624183002862247765591114130894749764170442763287774736669514152962797298
4736229019636223154363191418417996968962803592775061551398755785367266353781400817153183187933147980
3166307354183824985477461434758212203503303949134626725084373973133134613497187865221548479521132806
5589745100203248729722392568927537490270425986685614904053752845046626144264259912295769949845956160
8866129342741521686045369508582771055380684247063968607781328993106285511287995439433670099191520888
6144555645744227925330947510327863999882860866447782469776693646959682093306304832523027286162884000
40210758575059523354917451756305580431674912911708620738469600487897826391090562573950948749249507 9
0110901916146590802074849626393582792650583647667083830119688555050518615798097218529807820275467 90
7773528459488554864209571848095773550264183796622010506620176724101647596182317711444198810096102 7947
7760819625246820854579903759391755772555043901644270909099403336820210181891880943144687251194488 39
7607261064289476377050838168092847287045308547161072786663101220366887582906462496532944215040426089
6079559602848376811583310656398188715410222918563637554680861467680060626175912547513262658963341 670
6265117268170549301094065863026692204229894830237641423671421946364900320018910295105537319742039399
3038094287078665529147692881385645878149664779808233148266402166554846713406240185971412175424409787
1277182877853413438373825809953775455566863909370304700917201995246206245339198585921
3450742399985171822495428895132343503182334483788319295953348790109921171899225365294293653336258259
5390946552963581374497293697465118753385374870942177080817457017422516040904388461215741244528502 02
3798390369991138969673477880497042088639312843891579868614995357206376948214920930628125131228080 49
9666262553224283839917352025667452522990840994632586468341113042084589880324142878241486082621057749
4753300377060215168255421685888255205289137193877249869080202539336290407304750500997044070946935819
6990053347830446358119649104831638160680743397475188737745048539320801189209217620032541285900191
8507800878010809612186997215672787878360378342850502233591047386100379003368195821534795312420332192
379118697973810932850103678278060812745282883103918315704414803727152691119589138273502062661678813
6738930058943498719262754217586783775861692911564196954978050050271744148214225317716645648976255943
7582676430949126055292857753556527311492956087911082159619401757024920446262077946805377695544163790
2384442862070600133588293722905895615351099244879237739700126380906210362971005400902332620052688
5512923889365400766396639839257245082449689892624599243943845087655789090281898568343345109961963840
9116437670596054841952533056878723052067916205797398008695875004656119615450401629767770096547018528
4334977644679496028037224229231209258155238064515073175824039784897165956107629449587379456851984590
0936913497322702428283293584347620097295732039739082440490265245937396257295449117729608598859109 7
9831916191923715574314777305613275865030186712879223339994092210626790946258508116928279669238172379

855127629935238608831771468972857155910479691135290085367737548995512471868903957446600048479540144 7
802882232082425670184511475458385946399776041212321829451207207790817682331516184554219598749744558 9
015119927062360518963407353934875640953156034070693595825760816676955874920982404806665275624551923 2
200325145294175539646827190235219576379637897594175052158675420192853655966620145045633855278357125 6
712051282275196092953909245784188554012712828606220318403041625230457334991986203368394185491917443 1
281417027468348053312363654640800952279867998081678614387670229917642042193577030302889240633233371 1
883166689234725665314715426197369248818567744423683067628580803557766708524939432726759182713058028 2
047449868310923884518393656929128269141358022850879202263815575944588577839176304153824777450273030 47
967685689590574290775328240318887419532847250575823699894185736097379347926210756480012760660056 8
485628952962703650334072849924570768216606004308519214499399461592740186790670249250918084254975382
500057364115276562788206831683745613611646610594529609876478761781065732569944130069604044127574848 0
590815431525219147593078101414099180436770663956672634435356246619356407513565467271679094118794884 80
868092309383287382250042851308615564678323134489119765303305903494885070904364085511708968159681885 3
555968367767082875926959001222681306340790521083141753428387501316529218766333572643143710101634779
665888361886098834998328981164007025965502047611990688459306411880137433528093795109830522058657551 90
255331324739355184215726088231016831876409724179664282281705539235733182359972105591467580990960157 12
545362591465735834510808497696708071726694371830812496641392852932420581483522627246937585850957168 7
034502237757778359815858095456674554627297943079756932050856264050328025567228669775444284086209
981389793070925326425964038534996176546386083723031489481001371999013197012628010849690868329080599
525074937754235999978374465778371237535757585267771063814356136789079473247924261815992801915542 36
358409224109269519498675837014008069125945944555925467047320392800470111600155954170202879503592920 1
770396860113456304449233977435455716545727149517353452904622158776693210397256540533868239130580966
002121232483113870490726163498817775250633988037352830441378850403352914195734486262214803329495448 8
890854507924805426837419406691885551675721662211092089973185266767829352661329042761712071734330023 4
525891953735574750926874315296342844964077178567399584381489421046285181038496434273551924438540 11
056003640918609065983002337368459149958401449901293693725893952889834811556455810610307945458047566
465048935767852752111834079945987442056755069846168292974816927434031957448112126921989132435503198 37
054664442496090636337448465587143693424008426830694664726867821605430760555555157113015049963694214 12
014528605671549281450356360885793542044988312557195995587078583365330551210839284998411268479057129 7
220246550138708205244749272341919503603039460347677085153472543388076913543021032331182795994105254 3
731637179189960813384440367350920911106317337657401600018697304252984202448882570331725146924974753
931982350352252261761609438480970535124570875131868926737005074427741207097904073463126205529000639 18
929099033196435333533778277303380095645535200237281189341422221316384124662187626562926213165374744 095
230545401691905912102983325487384431499690081791766624445625710550014060366800133249580906410283487 7
164193564714304095505763808573852209871616795910485222080708395443816652953500877460681136027308248
856626139286677283717030323783233467164064186510599162481676700634565531467640744926589990402492943 38
203476654827303415226579042351013109297637683048481363297639763227467327299098939274496385721441290 84
870644807046616306718326269730189550522617503670443765606386089754748091995263944035360465439982377 3
561902465026931295891402221059887340881632225625868617445380796076249870628475992132726168844888493 91128
745320572483591623606741616344638801312351161263321415735073587373890898646033191010479049510880532 76
243634380195320531364134355459364704198439973273192818302576008576753025832169671464683435880400391 42
037072426708260176163268977152675622198130603529584446464040396885600648176610903305607906503 87689684467316948543438918368935379334168987646104050602410935985553266431299975939026367969692513 76
936926159102285117216541488921572622359773667701463984585521047074763889722549022179557834738536300 8
193343789887481568027697579994951212544157027137649027757150787779949109561489574262273987940332212 8257
532457845619552714971773748923110717580044852610875339714409334236416700073947476053387663802542 9586
355293691303474968689906133362459084735380529991695237406505765703934901547390655652892580819446962 84
012213924551119876038074056609941052311643460192568472205328510725875883788378843407461779760153173 0
495005829381247505250302170023610957367090784022351162225972381301444079847918133032105443244311027 1
979100591373809348377586361399771943672006559245699381470276191846661211804296171866528310695776609 2
437716159831512459361728010390163655204661937092525125363139655920117782142768019428047686534858158

74703119926770979133504911072051652432462312538315744875129541560355275026367961454610934441685357810
02731389969995468673435181034432552595169300233005059257279978159048202235890526047920348250750421710
73455466425312528525947844068421333789797245599838145290249134127724342450971795118644615062815283920
86502372129264368408132689423169315188818953852681800476310307780389924412518719858272225449899967720
87514587916023970766660441333059400177251968288276181389025401424114540322188075221493138730394389430
83863488048858725731296054402267540119444344271239556902379071500713861064198583888795688054863351720
80844464472481327728595610521301309101510626429327631624722285985257102637159462999401322655051880420
46573989800377544218462997919330400605176161879860265557607689149567962403302092035338230064198575120
12805490876812649998662968202160083935774256923900145096765518968300114985803950066246338616802255060
68707591274831975300014554060154119807841134948294656806017074182194482694202914591931179729442535230
51393310112186883167800232663769195193898930495655963440499826362332222131737958633752729015980267620
62438208490894364283124967962162500163529966893044625845804167249071090414279544743227764055860644970
99377481597906129222029306198335261800426117966549675805482901810689473722161202657162630436353302820
12933724508394353437096785432438058312889505278663565172112880369852845487149507985624665557793170507
90289986859714364593077973507014015227544766934198239263898929453534319021801693875850287786879702040
61682319735199280766975865127606468238391696014867111500960384529510516526462562727086373994715025771
81072308962043626411552557152127904583542959059101205185086051998335829520276445242512357735153614513
29122133578341196718707585660635000297664587218996568468098354225556797699786152958316206574320
73721099684406194608527521397200774460280782659809969835866480974386510365642556250842354565301512197
11111673283365507058324791727356514276941098498637066618802528434536119774234900016041091359806582532
51047772348737585884666978580299992241973665041155119628160047321576507005162898453996379194144719612
75368496384841840783539219495160760752984767084138274604403017707579996667675686125361051400391716817
25678050138978371865837968976150172098480602272151208763671116355293561961304020923796418528693604726
51396687556304008753585686831314128686092282555124226959567993035024901136677093640349903748458741980
91089018941705198578124784403557586713931970855498938099970654105791690209598897498734442726137243518
06561649931638973511197033109397804092809479734337726502122497143409823782365196589288627922327799039
41820506856983767116623772151441202276633794917374373422831797279401193745390495797144617183602567322
14021864206728596154389652340248945379811955964968707073021860780513193094678528484442021284324939307
15413564423079362725265852084922694843993948530535272706823358639486081705774075169738521202106289594
17716079254130699134608143824368693523126590734303668095385956084501673932299065428288540978637787221
82590727434195646611659593408713448120299579604000576414684563841920840258326855212379549624891158626
00094009876541358587061925113653719481486085710770837602097237465955311440373394942344861525237226662
53172009816226940002112275918342552982814689971968761014385191901289880742428052835555332523257154858
75614477520119468011155094405429657310193621595918758219482207475615307833480300854994330873982134027
07000311285887927967396273660920712697115138195377155464106337455584919631691755255599189224048979283
12472354538417960784595898647060095284180866467711094641598431500095794702207595874936960934892351311
55308795226753192289517519024600569926959011242800843780897580223727211128142771580921578619031438232
82220158431197363869779722276849589864706009528418086646771109464159383165480575201779780107490052009
86222538671343789879848810445030271773860682106718234856661032815984091857149084770741257737212152962
36289512481939493449231002475529468768814228268826338199210705003148229690781270685023609679524561563
17625378443908683885454717662505486886145388589401906509184101588520883369612987731672726919752638642
76095136945838426185218338957051864132632025229313913483808212879643388136298455390423731285738559006
23287197915909121817103349208828736572596006751034516917303484066477312472853649769032255058028512103
14581379138902540219995258467124772296233047947620417483733963129338846259860546609206872743107989200
93600862996258526495569263422443490758898712075405572391778889837474006283110359893639753683141316379
15535084459949985179035719166459579726405356353579622852232320999562529072956968626564166118217436689
36552658420973588048823563270384629171410427463403270444247217919633892333004523529200287573013563082
11172897133169436372206170585815202908497246369055667628921842626517819765195385246430364256203212959
16089585335981540706502452521657088226438896955320033071238601771994269799874271196603530485283581844
14608549134506444317130866665673247447943228054733913760627581890428365686598954648561709859023357189
36471102219155448014160810863286526572025037347796586980289697568759695665517157299781741491255194508
33779497446698060268643181523422924316776326510155053746770920415647646247512496723011428844839539706
06050724653142273218895838355013985022479801

0616382349453661289940409355917809265868236064198494786394927392551467596218556484340287163998424165
6407922432049921519352709427492550972098376405547995076369623700896158531420828549780644065756946107
4012428301167709175408804880626657504874497001006448172817033718571687699052050431269126758942354642
426321926816126071352559377998468768487666374637370848309130230318759751925243991782626402799661636
7094369112530086284350297886671483877355700954085095109254267237087162850872049910014666606934352543
9681324227750520412084311778362086425913741403701893905849130885307670183377759779815450406004508428
1692653954942424173439682579794296332233121321810780129321979360275038885263104587225787904993301724
9371699290336354529074965146405609127548752881574874850466564788157133643242701577120650882647257091
1545293554151064551035094710720178800179246413597213842994871045977552798427450697742656414883406830
0809230546462603894832672239604062464659457420252210842881282668896752789802434682655789626562663797
4536891092934689209248410970235713162753302389020769867314325842760948183881245857587340140970686718
9561831312229470021978448241581544509819889152942428906934660562607956566439433934279983588235711675
75116329255621499470951855223312458109131586557735917534070369414848815038632640986605244160899442142
3724467318576854052616459608035904616045947747232056287891088699234201957552645554915186288103709855
6289818492084536281760741755782466638790726046775548345174434814890498805637650325035535926271374514
9886240548295624736933344962330902739113708105840712372655403944406661654666981279401611743803144254
58361131387001800510642225047742126749539970759097152710314165536334773200549635491466719499256643771
1135196471224844213383448299147401916929709782733048209657023657441662164041385975771994011075505481
383567509753677020862148022476905759312845796120520106065272212595997445589563929274161013544777148660
2722228078789190310493304860642348889412653650919804680472667929079707702290502025767138443684581563
4829002226658722453892420758678126480742598431037231448350739349163285687087989353089830355384383950
0797078015089931667721215355785047682448066812060081609252490055065606882005394397529404297779977739
095481218233882815997862843934489379186155563845847945929894905438450376976647333225685242402160195
904472401639944492589154522209473819728573560814162944201208296293421367519056921053373592377816006
5668763641463665790435737824436651668510493519577657907919335490054245374836258264105168437864450295
918358983964405685936888263686371050276878388351277770566599560340015723582371471547219056714260143368
7966468436491439499957272215058042899218100989507993449442141990214468925539286769175103982467582463
734380523973508320868118734460641876299320312170407048304612916426319820862850786951729018997001434
28968262441780781916264417195644405475766685632424567458755185913604359995857459882503553846674132820
4525370108050902351148404819530588806416040823552351294128140484544881037085694262600027163923010086
03721424248429126495719569873219052427563394900162246454625934745670300567283750846724982527983673495
617708983651654782788942618610518327692085036036217800339152493371484445501415791162505050689091071382
7580224465050986098688327677971183257939187162167678835622419886753868393157565897768652016394528273
8867440651787566006894982173557480744432557759069273637848180513357096270189715205890970811986200522
76793499133040582845672035856647241058894553087572236461838439640259601125485287660886628483076360128
7006667228026700036149423026125416550458296169931638437058296750705322926910916119674936127372916312
0586368847905250952735154380622219327300289599240793907443573366909352949258689440107342217802437077
1528161242531908822717327154338237401474543621843332568029599407713014983323704510969965453202745335
070217703707061138191051635880307447708198082653195105524034346189080408772855887102619129910922959
50882251819205851887441986348625188245665407803295363481026348430308372418113655627830193910161835
1740322384509791468752162387714423922323245576364107974700125539832471226019053048866664933322848632
905380868351709643504402568667911688444342169402517726916720054236587952458648729193194196398370591
08346595565457374554274722525363872049196484680456121634675557800183911458050720291040917746197882965
0486123555287269755200042382038183207642553640963208933924544967598152309215189473050197853510015299
5735305428112836484236594743595960959569801620275302395941995334462820826497923607942188680411060241
5874150857519458061568880834301854125378195459697414236778741870667215842752231935277070188776280303
2337402862660420730505237852035421042577244255914042700874907643524826938681071637669307303727237417
175424585224773578270295996649854102311510138743237047992994485501682558654515388125485742
5414604291201232285561488953271717240122627994410826869139972987438253568158111262378326261042642491
9146730313932407899673786143290414208441146741535167426897337032190690287477060200884221903212516552
8911717385676477731136615373439175196270795492161709378005434045787378969857940658402609226706663974
7064746191485114465244380741520552128686202006717232636847179672315491533594924534289288748593176642
6993620967340749774505307056843141013326328777592130576231470087437384507626030587577497873242071406

6513617994956945601081928431937367328417989351895954235197022893472976977104975358649956731850469509
8739662801395315247336067459574653673422509659501098766962373414060683935034819838271821186844060176
1576560254701139537357267834526945596079177094497172723469473343678038722377574791686895552041362521
8014282548673776952837038414279340044001305689783556227986071360705966044685334333224708199605172761
5220108066710652379501907097493741821633529938651891700778898834763609236188052690600640828079713434
9789427595928872026107851555411239737304609759231788346880725383510590800218660449028979118967259367
5521396472906857973507367571821697403667069896134578745061097120350147956537516416615312164732544167
8927750724594362819238256233629810375653289132823923222725067941709469113185669962267630084749329492
3772482025168055066631619903457800502962160950971278313497534974849708075014692861779720539208773557
3
5426344654046917507878362978303712221263449953760485706825446567329277539722860367819247326075875375
0363960555755152470447904689279007417440809916215353523795515931682503917084115733894721483377058789
3695415348273160717030232024929094546655312055250126322531741427372940893582323130414035967091049257
1838352019535775111030301893738736672956883475900280466901502804156922062794107897682809626966101213
7488119556311967779804681249306478374053162747606283458684716459438243677534276082695576532344276053
3997129870805208989856144206593472221575351357313531509643256863269976011816768872330904784738782610
8
8300271876502608262925233194476099404167002640065555971369918146697911356778065745105729647119638264
3806989602338811353072149858536878068258711789268796293894691439363012094163553230277734299630152626
3435775125041351983411873620583345473118538074572684337206905220856255010506100937942856140474184559
2117933927270859725115288005694028053614114492402092462087948741836224854453701673479359020150069908
9471119500954771609606451569340980957608723061167985930544742494585927637542655079098507826227452413
4280564196195794701618141018859396702928840881750713269491264514792458713883472209570125453762871154
6135844710131132320149549094464014760003023763285717139536547149001355586963306925811264047920053172
8092117912870096788138937329594906879162309178222864353334053393679160242893274844466315594574856113
3204517830646491662241813246295767509185902988333230655145023629404347405492556117642216093884711734
1895740719985035273669869338669851702573938066020302791062808535254935316619458529388540134761981829
7
9019270269975539762709721332077521428883136382794037795481043639846216952494822984432296896920853535
5530853174095397100274487325285275736247945801278044550361060645585780357362625255636064773490568636
8324600588264572996728670647068819718804899591820953876967241261058123133718832815387305324063517168
0488373186348319448785524534021310596054326978736278990273623581526866772864841376321754066899897348
8261186018002936002236261588495903893818383478150216473108913836953738086831643699087980859301283735
2876220600536227587287679465791680576358143240925305502388654829492572512760977104308414241327149223
0145550249153801165157010725991966088910334458778020184201986872557983485892794115791654898418079655
9816529244002860000928330899598461251541347364124753370565807249607337289686395655103449758580001718
8013929340815934657740749168731401990382842771226233342464506588756739838593500769513118556316845738386
5551229294080306842203625672459181138606350480155226167063564964286732345965669379924358729329116688
4983936420697970390191593194559703612592627063708371713607972229244838973659949263221859430952934955
1705400945927480739388140267085915289450576363870232417063046230932441509575691850480891957
1739191653100024157142935668690970675538485026108040743064574263428325221102071034503745383407217197
2728609307970908786402740375603419620326095180233319446604704393480054063586910294183143819807662636
9229201519626745478890548730085334220081597403289253567824780457234485556388429936517859381542871477
3470540776250407980710868325712720965952470280931298490597903061967508059944421798850698316109638043
1857573493208970279214433931342829009838902927609098917675340055530206556574851358206981718943173
6521873727270338665243420592696839958587716587536293049174582102753301267023622273305213709274757549
2754032248665363232928428878807181194344775443943157463373774219051446263840148384523206013263650278
8
4514717047905831805834894085694942499441151548386334237720406996019335803137534497602844449951540
1381560664132329431053540356633494959500900534136214959747529802823984619728367006210584613977815824
6
7657982601784729657646395894187742496331695884228391191590565640228193496801758163840139429208142088
2045469029946376520599881978317544801271199655622131732442716080219316644607184506702451604612011797
6382723921134833943879896290584017968636094325530065062888973239251361632702390752395982653489399468
0658059487686462751410940649931153419873217299143125910097771886869455712444493582861381259764637551
1342798457376202343562568983122530420590214907099967416032154670758816502965863991553151342905513333
16532483088503470195490556744841010323187658995675839458823826883108141862835473194755252747117575402
5434804661774803878685983787156949134278530829472887354120431902320905952954328610661326976076266926

611352114662527698412773408524191382685828095507583757875182944195381669964759338058109744197040887683325257374385618259110897501964314793725720708094058960553970982858004455630785998611082978451980398982309416250986802635162822807560816507048348964416836183659463297691973326645044332702652964407332608359487129720563680362999226922055550219361313094392925616898258938095311543481328951849165428725462863519781030237300349037991793076886122045326513181013816898791956784667866144331058014381259912791415887667170290675999071222928172745278544341917637751864885446905521418294774540755373435060855634642039617667095752874549490120466035196463687357729297428234750549678564459273606276089892469782418907136691030092667819130350559195116956931478319754079624033842110466446524045818686393260964635303347112921315436957144220672372270190321612831636606835353591402798852609531474419767057640109075306047213865706766549972656139956259081850853030455592840761412465221809965435307163185074886478933137159804019105801024254171356618961206011069761203387121769536274181470240462879594479656892916665615162911773694618494617768316635945285117164140087961096558671942116381654578935594374741659601910402650969653760891088490540908807662422362445253132852178568721171075072872580243702749583566462456351397720596477873476721370969778743722222285444150516258147590016001034987342164287379415209172808743828700687452996758506346236156568380656846858665912883929939874919280450975993576219153034533964024162816375645733798596901282927418796276250380640305799822393935095892199527851029164380478329362091927804077150418730068917857381782537931265322695642984805750570433859369934339436246923259372433043457667147711662426205667161937773757701820493615978206551775960445574271401595850624208614320221027947003826440974985311949339722525601724380678398080630198033038714010823737020217802399919659484742081004162463139998938726781296983713965334369780639446286002415282447875363941292793533836077076083007500854688366848968344738138000194809994631127939729472548092617557123230722879914729496278293181217 19995311913933682914217080039171083920399625324657124267114807621873238886630273263260510227485587685824829627378563037502052629488316960894901291631372634885804981924875455352488732623943936750450165476893402068821458565566171050775119437380589413425696041313947581919700668230263242340902440540958834108768898805836001904880856199491033238845013313960419454688283590614806279170274250569836303868190408076986074650884423677617054622608454397222194035542026520331044552690070468827745821456966366744699942884147341148070234795179430778361527574001167542381049782265169967932701790991325359692526413102120317675960263679551086925989193605152931611639993790605221612826257344736334202630750572642522552550069625030372133805324484446537714971705777121738555021403641091178389721417976354731764860997323702087657166232862736406666652522444844237047431260270194052932539203881245611678392260714780011957184750455611615250384482208269918461549725900194547955474267031195338846336750475341192430517289055830639606042730093178903971343931584059161008586393822138202271708192475778210015039163848866601709081390913595334061901462690424095680250624010707056477766184073659189855981201591486191247910453282084900376962357904520492814748144846581726874291601112567800981177269122200970237851486611243714459198466857630947375120942433520464650532973912516708301825542971302260674660980052603919627557983895090692431900473756401836074549348591017794475577162863150554887610286729181867586476644786596278229403999320990533549691384757430422035802668200501835248565151703420994311072603747508216434958854143204573557419801182940065163845907783130947300992993954182718180599773199552253765623522616879880482847203149589062569689442427840761719747852121117086752944603670533553570333619526994064352330819095743708046560784123500619341564951004973317383620400422734437891578965349586115919181358972444959560707032232278280157914855808866326700920423420313927164689630137056221855039903462294679176978329701524127573584580130897595258639855021546367705090070332907975583265256975309521994233523426743245562878343878039320990147869174131728112546445904277797369126687707533793810529597921956009451964724521456644781129408884259739950228931567320489003595653148818181335248844869698498006125364742775500050420404621034425555880866214452379324645861308670915261439688781633677349695124090077391672641409412421645618536420858380521948089887377463428514000439792350670242467066706973079235532297655668457047629025632259783196183339724915469525525143514347973072505089539350344143010372769330828701075552612213237948915324248501547845700977456836739570646245323167834271605995132533503844764618045298891008925761423645689209371216233577791901016985274650743313394038605588354356051252991146475251049740083923818800409524665697821906750077451273404137469485989032430384217737576080925883639494285249166123974213430366399684399 4573153145921895618542973694163929127458660214919325606290835478894538705890991028778642634490924 2578153782277154455507550828207848836009388575040713380643144643510663577254200107215128214873801278325521947772666696435196399513880976645324332723557768426415313809782298046321315377462523540429060358017056192744688484477598491780 86

745549832965851953461620010812542242676672446591985003737246700094531418328513802244338672642501595
7592114147496245184912047206467560940593359887917979059001367488538645068696206561499083496822578568
113455648395727288903768564734858702787470924437841701540033745698274823246922177670738059985075586
668713787186830968096765583702166111407722078522106402682698887307289605168437835203674022500330129
220879972807335520332084982518794763661742905528375912856416497413728193651143325413215965447975088
96637844058341384482455733859312236750877066637661433837033604334586746501571999949357
26314983135394760109974600000537658144945733267458610633049321502973839393544273709938641506048173
381076058253094394198785753236572332304794359244366345485760174078300407648946768258640951038
437439926955341506926095520996684963621960975419902566892730018303988411552635487584101918628565195
894991754367013775262962123556260890759814472451063453168437420392696341251225494255324466039223854
49218025474882876575364066956656854499484914915280503825352855646720044252289431463574597056054132
113477618345914350068040975165789396711785485165229206778357107525513953145282312224607743264857802
4710696712582490549532520029801187432980385609820847498781051624257385362174946834560794401386804272
597372177995976134849268369259049544728554834855547672855092626966333515439531919926200922901179165
0354053205335415573239628587788501001210509448364369355456565298702510427337311270368643270185393046
22986390184984977908527046638100741060641993467683487921490182379262315357767600452539830948834099531
29731567381103578141738095060268988103260660443176435502995330743512178353637422630730894118909833411
63860551522024968443939425305314272823845197724460166040399583547279137257994876905630996389204706
3760505652182054213457997735548935383680333652820760812514894364249001742501483694891032497863941911
6005402306216185569908610246731548646098802028107347373777704918834398152960696061717154770915580453
91472137944886307027510629100795373999513948617449030229789218152339558821015764964020954591409001
0611658741111198073433510338190204012029929324031443298771647219418758925676343459232199092290155723
3728886071369406177616450934567152280499827475092783026359931981639916855626044956799351948320384470
3965009899853801126586133337947755213032889161978793373276844741328316215382135050223298247951985584
253064406273275670404854215795371733438301365792271697658152542272531902691721585363479065501433932
21184545774337318654527185594102106229488252043473087838122229831687235778373373434515599482329233926
9578929447998242009493926642707394323274713200971603475707441072850307963039075727028380502095916175
50700050166277924293181245052386723996858519317795093804067556513325575821494315351795949879017593
954018699586162066971538933325756001809207795785012715852501324370267983815321683151058802737458939
82251650796556806740552933580416458692852316528925062441034507254751746696957664759912065568479831
883862444164695419191341166383135950942698407897055494658072945983183865462177521063114551043363353
61775730503740634292089127502236209418309382037942049430183256488151462174731493312272496651940332
66946548174825341725229124749705114616160542818054035419897234572559079014198518298148145990271
5143266071026229634919481259342145337898724583097604861888669585313039072917339207722568960983652091
79311482179747506446559761340387575922402239603473570849183378112149892582953233544478531833361017
37217127075561619143805327266024494866487503480752518392245923957178302347141902235035038079411272
748728950400230804740722455600124762073159921250457886191338649903391268514724330910608559436605624
48676475330103363659857748963154641346528176374126158110078173670124954479654116012249009394603499
0999402392749438135040080294489916879393667610867550354692386570894147039461647784558939370011895036
696843007815886653291356205568916972515781275439554162151952105300131336221871938145719744354678463
7988318741333305512922100157473047822274735436233806082009833664536855868925633157469202431516831234
06729773141705079853683002114653362735781206338697324868904859587581672286764887437654650449048380
6518022973211179540566543794461504202156304606080621749083449296578888295379303504726086216823201
984933606588503695960666607613634125466771426576696938253333382968667780185547637528741375614974085
1683629386444454372825282127596573341880697804086375624321359335338276914364031872205194448826998
8678043790301726519533324288476168006516298029907502324788344342406567828812886076963740649602563
15656767505203705308391361662839216418450982844674305122844423983346413015302739217504201192661457
5082813903767336357639254460529716017653757738989469139770671718421230296681287204971364532552899
64012330523226054081611387889253194878617232763152748024076997010841422239979291101028956919632942
3327708353525389035143185882863193742926542221292762852600422543539998100595536678399989229129180
38777361400928153081940786066474842382259576352851784924147444843684342520668215659669197697288057
670362156355124994436122668933058039480454627750978873567271612375825713839139410872431955195160533
651555589235551826979714750066893173531053191521229835917302302675839274164931422439389744310087449

196222448073715852494755275828413716873388568451397193571743513766510489523902901706309271226468406692783483998862859594239679364564173558718401990187525463460676207062627896232203480537183635179469108776385199110783793669022647896142652819899498944095836469265144316565821247179207892033351405936678284940177989652979815054754358031775856225586906101230234010930612955353576358432997463079288408167023366788970128145958847428199428749866143771185970104607861183122285674946913190044825643028262007248787525394397901252220351799067010886444441733329427083564277762959484381499099891262534888002247011268538660594376622270365082672212322235157255037650385409525310175758687338353119969360241660039650928072743807137547048484456886848919687212310981990987307479532054140103959736196967523052442164056190052674275993979137268686227853543051791257504717526573041879418223934849391969295394499883801967266250365175749403311969597941172125762371383165114796260557917844287818812255383408963982704978907257430443034411791081090599505417122077373797477503036812582030995844119486799985794011117139332423952627036199270637724170339781113332827156827734201479054726165434019322671442180196105337050263337427319045518794871348524986266822221111131891445576221042289834739034998125977811080901318642567088910367429803049136365142132498203398923826233114577540076316382512686668485031584111141155886426790721346040922170507459824720702455243140352011956531392458331009142536349587897907439371365970955227255566660070242028391007459462466313074544675113059937775120411928055649729151234915053327810229864306084053764409891744431076998717276003815163442860652183069210091799279417093193629424774580681735533597017598932860314853201568769791565212418967004030959127757081613030186949571407998244363332875434192214073359695226283038565959845208956554977810132745394340864386526954140660420525015130837808664572997492569037617093002816162250318700303273714486051251640723900700882382390681147399580435559035124122182329827436677789038538586239181478158835325813789319130516472390159051500732925827462042891439266753495215494292409278561294142586937295146931427054442270009324161093344478176261888491644169256081359077579447382060015924040120633749989542529116309524485314231824745868679213510269662875170521060640784704974666154518814151035167349815783088805062902524727849132883585857196877031633009755390490488456644897468882484225042274107690691547824158619851842195790945913926945539349707417082601299136137293319908996124476112702770438892717012348861763196368502467208264968706889841975265151171846839743308317260487854033032942786443609114897628797413020336753926893185945801618329179440100958324980587506458366412769529285982657703330062345826549555323166532305637373512195284921489639294238105955982270927599730329947375056874498728129347020606624477615834666170491626975717975872429291141879750748782171153341997452680557322560031417046342203189757820773023738624697850416509797584452716458522043551397592875295089546522806626969434499014880200418118642039774220404072066955446032990925495943520252796488734580054345846849453975353915837929113057037617736633757952397710873933795733211854879043328149584212003879536106338008991522104521065020038005180834770931615105441297268508996642282464489776423231947675602438097694631001688776057256798936928008650248744676082454957501328381000122974730565399913766112726067858345129580303840502530416631173982221137922074743939663003407049670643288219873398773338028597936098215354659100724317097157060971059686886906647906795150810115197051363575163611207596373863757385849998378645305790304439301295041097437837274732158210722267675703966141860877443969096247776142986821572551820738210629142917928985775770023307492988858235399319356394252680619486420670833245104681706704055421341876516419219776188680295892187243673912979296702170026047079540759988069653829470682629475007992091780545010872131836710302140341239939886741014047291317644420908011091980352443211333653582852718284326236725037002411625948225597448808317606737015485628911866544655045630521377904505732818512019796654094302700297443524612224141128329643794032198447927665610927170763559401220553590244673070737840681089160484022631613165378822613062347649322815232409251839601714955706253533930191906745971899007765435853945373057085876733937752255618875908626655726071481360268430480946337810894870533343769315286522818485015039938003366387897338934112883447773394212319997625754387619478290608410492317086979272668565017717845357601444068717168690095280680343189335630427097277870765088633337297031001593202232479700410181494676461336968844790945361149017446298974932300375802535319176955052416250065528426114337307622654420826622135945467533036526184218165664437572096300136885151459679472036012139399463261145414446827526486615867566816023239217047451257013486590616430860058855687920847833606246304195864617470830336605330053420620483686318888324266816035175214007464026900758761089479463515084495961700050182770896682246323747552939134464826876500962762224250729272188378746958539680869873126892374812698135012870259485298709385372271129950556301593716628582188659160520740380572603304513197216792914771867563052935572768823909292266219730580487311401175308133892170118618011733725143663530875657342408

9417048072193359887479736426861987941028542129529410436548061666466560953506812679860837723426852206
1587747745045440859724123572813629395024877213229181447460352824090500401017736626986417021816703518
1897467051204279605436662794521749414856486403423375959054906013609693091684106291636794689232189120
9127401951706180387212284307087960773113725441306054172989950548378827727046663864191137798869740632
7767999608103275565628709067701614858118516715255720673109259432660248555938871841243042256167746151
0838972883424225891485050847290517606189977758300706565252084740820884217337391076739811488077333220
1658891141005158542238840636728656725089712885038452940316291883714453787866130540009170501111547185
4363558310332072119813348586311534236019202937183043526169084401955004814506589376871523847124185820
7056441353048742015561451208666809688655364730701451711555838996458356780034909490393274451441177991 6
3796310338254450612672966298207689283473834827563761713839607939250893828751938890837424828395334225
6640130058887766579178359774040670750177108719391145784682542500121104011567818312956625725510917646
0365967780579961308628200115012516792484947604849884203733393954346866859023545752309540738153041167
5919196403395008232212122031948582192443475529337530173935181814669205300676835498326089762673603011
7845855487527030732200353241229102967066481671955925453221347824974025002702740859981683762141118183
3876083487925806981381516610041464208075202053745495801305135529753878317955606609534527502899952843 0
5259958631497799031259959268523867599757644136022576047065119871932352626491081301935915996762477542
0054683329136080933231845309103266426957027363686158684198635589869662161263694234626987065295160460
3659088977630943695328921497180259712631079198634234336833798542781592437536105952313676705884251472
6866925962256233388254449153389514807803531600962362723626293210938381213442925961689776071296109650
3285812638536532840691870726909508719908487588059768061543438498330588210737329950538595446528725800917
3848216395247618669194284409202032528586359734273652108417792406533769948709190400653628400265799102
7808588169428112419867803212676611908212068769439025689831855029507358136283325881329349787561199657 70
0647703233546013593301573718699852759527884015523130446664670070445701673747302944778425837971339579
8102341927430114166335631047200206230346672004347363620918607406379387410837798372659102262266628166
8368174614508881058679402091696237070267227078556702466966212525923248906556541142321653001230668315
8137095011751649747441770771047867114423120308259649702806523097265225955326092760320464181104012825
7029715290996630179728749671607818738643464043656031260091308019905591344969824305981044812142232919
8832330517487617761060380242248693360782698348799053187190936565718811379882854700036618037764641 6
8080056211065665713544573803557216270206981706659611630266928133512859723422734740435504503018436615
7059758602591897297176293782075851521436630058441375435301528736386393755920249401991229614783120533
9020402152495716237517713920740481216320695616878374406751427661181935704092622544281257244674793566
9902240001616279356997737936222932889951096671881472547244744233245083281611358850626177847522637
74166893067961890698173854262071168345202086617225540215131520301426153635292417624887240193843 28470
3145335685532116346032411910969004980661636537048300440101181291865610897469806955769191358515559383
3791589067981878573696873349166531702934832744826234967893671344072677226840840390785044733709169016
1948341749284476857660558238949976266570726095917281026112370882204240489674179817610597121652434189
8769732540518357639068967524643049459814039960198338618628217058607337201469935697285000023185154135 7
6999413002897984546387206092599165675004257455772138558216066238188432608559086488423057729024583 7
5483327194659221890608225577193031024428450882380411824587454595405991187938986652434677760671624411
1861001010409073491303067136969073215943484812974545321465061611170158707923782677675224366356639191 9
6042731264289021440187347592854707425670344996917675233598478138865565705899983318601234615036475938
1711586038569704782049935904441279760418898209130348330214953067826196903024060825099184094962411 71
4750193668254719668444733985153038785005110697902006497353545593857570788333676889246110793462714441
9802729030596670946542694668000365572482505385372657003464552984375485605766436548468959198703255509
0859093742098048624130992674323728636117618097163687365885252847254292899868407065674091310523331169139
0175906117700840555041040973012675087716768600432640471731730874894457953682806516682884188097638768
7751677225401507003933693798837135823136755015852875240337553986870947839756157946308525991462120723
8560952292201024194226436550096437281621236592956492120341931085580480579252070560970733110036108725
7336551244363974172687751532206542256393391424987919220299243040151353261830425163982175998799403271
7706306529601965946603316609193225213978517420275604532822558009110705570160851977806571014316301211
8809125580649860349804914467610772074550263450695615815328307044196446240419789530416738083329238
4654513573513331684033152563896589471100879881182953236468004613394537670149121770428219428285050662
2188464305088040978357011532665491625524952638509627579674947577160349234735962801761556267439062330 3

470044538119409528697550938580679702661145684844846308991494315152488178024416974099890603908439441 8
502530473572469373056161853793406882946014254022114233731397240857449458623776193225518556891103646 3
768407051600360614064357081184779777040599564120010460141300089078808937759529574504765503905318359 9
146854554075702541944535881728230217184087340156065947706698217296638653913212771651925166629122416 00
260824045564181828242071555930414128479212381485951448399656715057646027136104073454888170722718008 3
296814012039662237524091762593050119645743753899286461890218741075016941551730748796555794341221340 1
861945711114145143476791105087385843795432281462392510321525178061022009850611715115229246844062336
709270892264532410781786234930020364861529930701368469705066369587621303871918496736262861287731353 0
772131662228046580869011784522319949365322724431774790440805210124268282577605768520721103235336355173 3
222846585385579072592997658497868869601193452927464227525558504878241741596277439272693508630037391972
324328096423320985806746974551157236703879559324457584531226002395645186342413700109861502651950965 0
127633382819673659764355940000898370507219226837562534245796421520814153814035283423274574528218865 3
998454027744122254522942265414540142662883885257064639199041273858637472377197018166150155593065525804
024490929824495350132778095325642634262634327989951686692296919787694623076382134287918534016638265 8
613898105566506740106342082853947818002045364225367901634799177968142697019624314183701793323920820 6
963911456865343575178937024664269595259600654326091604270903277334124879376220897854569431847419434 5
106203974665551472056170610194612226866815969048383942904309928316773544144923722192576392177792342 2
377339714848818191016998201827826113181284536326398438621565509677653198854178314354882187582179300435053
451170737372867280188233545785306959967788067419764300979384132540544812380831903058654085153227854 2
313542837423537587216882363220550706527064952513569636752121046320641843223935637795209989654547200 1
696835107430770193988419787070947694987486420085547074572670601271274095669266543143833776902401314 2
789997756778724179571357223060632305238563514763132557264467597733776986284175094033938121692142673 5
638646235434020669430635075139402744288976582370030756493010657345186982126947578850411919613610460 9
343164407415337439577243588525650231378094754319577305451878520720986386116227304334532958757447485512
733832797221918503504787189788424928710689618210899417158677612083801516188865145400128585971646301 3
644965160551493822792834449083767432819892114291043106111086869216552792079820164932937458134215076 5
511878489276382548793511783235161120855178410837621175281500785185477818607679428641484323323319109
726685003293412891404585820443155761077878576257742383848993157237395983811060781246778635856899652
736884524094252870459643592076057934668432414532559635364874882117912081839703478464175492914224861 9
881698358368191292319024071712957000768746334505854605361897413652911487487826670127663175041210072
031578108892437664036773091175530904092817119613621053923413873344225937678768526257135122434134824 49
243786331885276827537431049305344554221435713693138564029040006569936253681779918785587184473077058 25
297435057486641270846544260384727445331835298489368389889780828901862307448404708449085289490303934 34
295546385527408457590152688160003747725228117811611574237139819507406741753431841463505443424383574 51
924861158088003960668343940190898737819244040021983228452076515479211369255074787416583660242218543
194643077515218613558151988754046355176140959054373857005250135809390485967216202160698441615690787 7
168406812741437661409111813963108559915166140795445153249599213668254996327123735625684147454143 12
312060488195654935082835797994376777571835665900690204179933699017282410496363123712442494160142851 6
092807790636038335841419927091542276842581774992219930572580325951237712012994424840701227968679444
734180679258265793495891476788837891544093263165670060889473109325610561988030860548652087745645452 4
748535315405011168124284749579043721594155394367029071257783536897542289496401161850244731314083463
035435806477212711435043136254843866092436155003196510855005090715833704150255688910245840048189333 52
025212449845716767925568221802326396231570964589079686994941373432621181495473897856248821720105582
789845333331365484894508887856751904044595926531952158119244898113529022134550383565982719369493176 5
657818676004700773169181614550341947667743801815903330993025665956680601025092653011515016622468616
389719309021432141704002191457726858074876520972216809806340340974308690729882230975195198310029866126
704890817788276877579611783302232901846001907160874768423152240507743607793306997142082661884079404 6
925806327424725045167242546714723309962603957286538590552888005911222577468976219282389040655521371
974945223212768338451551457684661127766882078834758858600648658735763165275569994119108346614456186 7
068127494633928642736615115589122408685970789270075033942198758365359734300410695592267610355330599 3
105917631279351162971347709506501253293619401366506307941570050211188043581494533791320858732548430463
659635754543702368957640754770441557149317686793022777531198821848373019117862893069401115893599512
748299120530681295519298249382721477955410129443773451756425117165262161669991858356468746493579240 9

772455133280236293667570990769785251422664119119358543450137409607937600508655283422806572647802204
206130795169963953324761662289154668469409691451525256341542697672709654289236696840319253509490673840
5090202972431809123127018255195138346002937882176207767661470179105698709532770660030841618105920658
7486560051483952184824992543254856132508585197436417072553803196406928071563221322173345451876615575
526390318339396568420029460701119129003740322343976701062591895494110816617775621921623121137090313913
71995109300060103007153333452901597889358798595884784180052537960087983471109265787548954190753861063
6220189959789540593437873947063842367677502183369288728660527834544522024058057113629600759746651977
711112630969469503423846510531433667095110760686290587829078822087133464200364390366332980988500851393
37814963166188577108066312558044324207698647512261658234061848144083152527539605062700669652630
1902593848744034126063605374377192472521516522939406695524534224812999183265659442375912055988200042844
642042060707422287275907409713982157237963214549916729673080286486448406832219033684902698992971020099
20411865780702517859183575055270334768877563858560440075209750997614740057219289818296672620120311458260
8158553826922251013825611025429443067962436400899781005440067802134276707642549925936710102284674866
225941175294051667558186269417505939971153767539966509833016157369709227005573096959556492723851825
757875534178887526397885559646244749232640748216285423836335949374899526806016754142197883509023733
10576875032819182590521253318493183061007022026785805278515630243241955533938565711330622522462341479
71144793257899373626522022284579923034601177104157440941246819729500291141398676115503525998194807352
3915452858111822298181564944792756355332239360316632960149436871828093274304915238932256255191114
73007531481288072064268422269811361855201544798087601072876094502692494540614879525977809863600696699
1013778141234900206869198389262415376722551768749515204014883031204113020278064596458948058713779967
6887349391870379821439167206459086965189722569169850999031020309665736330187721579107878142644572561
7411584185877699635635291576611517161175561912146377213801136522636271278503521453330658424042719346
57080561805642862699883193272737424658080806372534586774314536471343299136108458711456701768060241563
987452403858336379398356339349444595030714427694213921659335151610028068260018621710802758990916345464
246022698586717454231097729730601950455750212178667292686336651258071050500174721336920726980719277903
0633012919033429182201301171302068612463736153617639480556112267903595998345820736803032150650836640
16691546402608726050606519849075749621296331192046086424701059966275204067990852111887393952622217173
183945674358436625025940756778745520688317702101822639289293319946189556214339398753777418233490776
385599354008719353215578106343492646921668019730695877934717222544807911811119639263927648001235177253
5719274788305783971136690606455254331919032228919360954971843249100904540662729238502740563383854859
0188268149435843645843980262611116171765842816097957652506787611795116323992635170032619534150929750
055340488664096583519165979491935086348821926159814484369243754556511220419480552682307533247697408878
472778743545235927210880405262583536868919931984460989716084467426367359168005528478634062691271721775
31717560616771714634755616198078843903113584777164260510474576636143832085499367219745739979786652
277503539806189140888838590932137437527203362302578779561047292856386085130915778464960008736333920309
104897781691990454837211576932614722101693733956770865691376111086915327835405568948605071082295424809
0918055808095852840660735287814803865380214664675714389647580860434512955393551309586932110862993111
0605839939424965760106574952402644946365524442407305990365284908966648004046794551760568902763171719179
87687277257489033656717785638232165305692129115050326412815732707501138355197893094089107488034262109
0882741413711940912309437167869613630672476571046234861504060068547046457718789166038214014347509730
536910311084407969550456237753811982755215952136501877563397073543958012047196601951288251054503317395
0516216341090518192260525546312264355322592957475728820016262708082360424445903581361990459604491675
403755372720618198899951477716149327607979993540532317930374352758499542681718721374730002593335613694
192111122916389162184695669562033564970596509332371688551878420333041807503066505560625174160505262
3316640919252238258870955218902881295750521711655979171308253404608433079774654768816669196814476897
32848391792177676427103215174524479507358807632890541673160318112962024170038177565961182654152518
09679876026163430172783703279617329250813470476548576560590107276735213854598276892987332975832994461

853656599192720272312841968966632593594772667223500113719502646730844926286095985260520522409521822359
2003147069826977926672250423655929192492054350354442390940880576201050465309269773134941085727997638
0113049297939865584198988765833201593343961046875079635201784729873173044084272669658406096180546546
5631930250501495988401925085596318809233280147303879129195795825101292043776533474108918075807247127
72402761666296862622222311660704487522921471507146159607733512382721669154552729130788761367040033447
71052077005942899727117736591924299132120809706489631255884391194426424834555020727461522672099255644
5652834674987994906603417413684761557731344073469800379804214122671320372465321073217357376049193202627
5564676654903913028689978015277912027247510532924595527398642066245295718008680915733555397019651293
1004832314704135049429359651182657240498204431397567031470537098506131461559915459679080382063327112
7053976438946130683352466915676444805847905318562649578393545683629709750886407255782366929906509812
70643207674253904368857138109407458556596741811348102672978012876597705816267284575615932812660064575
3286983567541694373351718691544942980395328095625332967176474278419217105496385341428321986214861895
6791483040045423024372446249427695881885130478794805150924022184727268743260289620485985683180743752
14862909921339139921806953807443634711062302102023990801664283121979039310889890286812774491398778816
0123696349353837904833788616497398864240856403886001217356037112563403882844139371526035502553565
0068469137995125330157088169317619809576609611328145362486381437470813709973392793407138710218356 0
44379465559576321085972836061135173862779961866282810658000636060566965160500275463200064283833999047
0686106215897801359187080238837576895579111713270218719138612446092285096466217880091235664671425284 5
8131688323343861702634545310635732686171417326825210992719583249078323219289804851229823033793785926
9722693195895033354120667763684519202799011294353129008525896049061348176846124481858634526732494412
39503370242895528566745576377654831303915449072417446990333489630103269608512640536887822212146 21521
9423825090781889402643536751568910448568329212836025815748009984583205487653084082425613508 93597229
603185883885803835818566441215386708767601248310104630844744238118014479093367474688447856414817 44590
9124539810323008859206371015635875651643095961273964015861769781314087950737142883177604366898 19926
4134750069719405651950455051677423979401968998625279008523744580732886706974177396357254506098542 34
56789852042507322860709420263948418286692566061665444072045775683939323122654078124783166688018 03018
422504520053551868684850558453085253849754261205794305353308074775082649608852944557278500344139 5589
3793350511840302252987061629150596422599960688526875233653179869146369699444278665789125256842604 147
98110865685571739072133078933108525903333113718587728703492629027156732916866271667749081993182 58316
6832828500157570780161193169221931514754937755159804654092839991094937420103717085608605881785 449005
70410413604043513764246899781526806092554011234653295043491805374773561666710469298305678203 1640820
6393173922124840551203478063211131681337322406321645541558823784609194273808850283831236226 65 4974430
0558099898299574258432353764298631465630555285360477090757473282636770439630979234562979493 44566413
960851464371303932136774212470904452721540687921542630642597260229018994655298115142612604900 7638671
4173023572727683904155972345066669386646058829201247114417883178223482153389187605836327618 181943322
76955531125819048475174629056200134689964407161982320543471846110205113155509302268251074919 90149608
17856260510885903658474515037638459151340003295163991062192400557280081035217619796168322139 816924083
576395562116571261121929050871632552585849616638254193591482181876195923292056995506376458 18268557
52211527870111820992984471575367620774978540541330313633542319541780574776561666710449638852 79841490076464
697648540094796358954975461448137636970591635699836811985725054793069335320757076678014844 47014247162
4190816682249007420711186488154772891718653596776539579933503342728214605416964960098470 697958559264
304287036366471307131478332313196411794199132224206460998988307626089041749990478467610 7604241784215
06285175573529996478625529542836742987066457943375801014070421161861448432976574426342 8528728477 85563
08309631435278783041945019702946575777732816746858087459316093972533158992805794346314 08735 86086177
88263349227746151184911655130681846713677348823341085316403947939208876863363394613 82358344794081569
6109142938773471389342377361910964605642444747799820760496602713561689541064448321 3659808293890 97296
1891211834291490616389638610693752089534688398334446718982124347807238740745769755 45074368467 13502
485881839966556819634452881194183172636825050611864900394125520574571203603557802 51419043526 71837 21
9213848299058032246958424323158984432510396544353505353543229216747040778614848597 6255744615351188003
143056995492784716745449726791283933251838197222328360705522781292813010656941262 9487306342 68837338
18174217060864754827639424239140275321804295190341163517046980742335155605785756 24509992532017 87496
366407347703898558730650760638709977318431281098978988208543559550943253 90237189521 68202334442 4557257
5307879263398550901645594237339662522335164875058955694217297244895998825089 232112034795894154654603

037878617591571661398869326873749684730549653293782147564810579380828530053244708050656929422340010959348294614539078890661264021501307353300331920745637263770770999399922886212243248802062634850888530360107234368901360642758142528398785949179979611219637975765192452186709608809213711977500087815930430729344883930957574159241375285977797291893453850508038311986774590025186579172370808574164297153807884060713068680361982419715774763895072534684045691927595319372237022290155800656076047385473599044779967487499697694271376686955331951253377640985870966838632639261649456086841403745684207194059507017430354691821509004664939985517413893851975731215682616228622318810967297476060130283311937161140874727067625585677751199566674861519649129701933180849941096181392964927893609021253544332737506426062429941203273625582441749834509473094534366159072841631936830757197980682315357371555718161221567879364250138871170232755557793022667858031999308108305763076523320507400139390958079016377176292592837648747901772741256781905555621805048767469911408399779193765423206233747173247033697633579258915152603156140333212728491944184371506965520875424505989567879613033116462839963464604220901061057794581151

Made in the USA
Lexington, KY
25 November 2012